POWER QUALITY ENHANCEMENT USING CUSTOM POWER DEVICES

THE KLUWER INTERNATIONAL SERIES IN ENGINEERING AND COMPUTER SCIENCE

Power Electronics and Power Systems
Series Editor
M. A. Pai

Other books in the series:

COMPUTATIONAL METHODS FOR LARGE SPARSE POWER SYSTEMS ANALYSIS: *An Object Oriented Approach*, S. A. Soman, S. A. Khaparde and Shubha Pandit, ISBN 0-7923-7591-2

VALUATION, HEDGING AND SPECULATION IN COMPETITIVE ELECTRICITY MARKETS: *A Fundamental Approach,* Petter L. Skantze and Marija D. Ilic, ISBN 0-7923-7528-9

OPERATION OF RESTRUCTURED POWER SYSTEMS, Kankar Bhattacharya, Math H. J. Bollen and Jaap E. Daalder, ISBN 0-7923-7397-9

TRANSIENT STABILITY OF POWER SYSTEMS: *A Unified Approach to Assessment and Control,* Mania Pavella, Damien Ernst and Daniel Ruiz-Vega, ISBN 0-7923-7963-2

MAINTENANCE SCHEDULING IN RESTRUCTURED POWER SYSTEMS, M. Shahidehpour and M. Marwali, ISBN 0-7923-7872-5

POWER SYSTEM OSCILLATIONS, Graham Rogers, ISBN 0-7923-7712-5

STATE ESTIMATION IN ELECTRIC POWER SYSTEMS: *A Generalized Approach,* A. Monticelli, ISBN 0-7923-8519-5

COMPUTATIONAL AUCTION MECHANISMS FOR RESTRUCTURED POWER INDUSTRY OPERATIONS, Gerald B. Sheblé, ISBN 0-7923-8475-X

ANALYSIS OF SUBSYNCHRONOUS RESONANCE IN POWER SYSTEMS, K. R. Padiyar, ISBN 0-7923-8319-2

POWER SYSTEMS RESTRUCTURING: *Engineering and Economics,* Marija Ilic, Francisco Galiana and Lester Fink, ISBN 0-7923-8163-7

CRYOGENIC OPERATION OF SILICON POWER DEVICES, Ranbir Singh and B. Jayant Baliga, ISBN 0-7923-8157-2

VOLTAGE STABILITY OF ELECTRIC POWER SYSTEMS, Thierry Van Cutsem and Costas Vournas, ISBN 0-7923-8139-4

AUTOMATIC LEARNING TECHNIQUES IN POWER SYSTEMS, Louis A. Wehenkel, ISBN 0-7923-8068-1

ENERGY FUNCTION ANALYSIS FOR POWER SYSTEM STABILITY, M. A. Pai, ISBN 0-7923-9035-0

ELECTROMAGNETIC MODELLING OF POWER ELECTRONIC CONVERTERS, J. A. Ferreira, ISBN 0-7923-9034-2

MODERN POWER SYSTEMS CONTROL AND OPERATION, A. S. Debs, ISBN 0-89838-265-3

RELIABILITY ASSESSMENT OF LARGE ELECTRIC POWER SYSTEMS, R. Billington, R. N. Allan, ISBN 0-89838-266-1

SPOT PRICING OF ELECTRICITY, F. C. Schweppe, M. C. Caramanis, R. D. Tabors, R. E. Bohn, ISBN 0-89838-260-2

INDUSTRIAL ENERGY MANAGEMENT: *Principles and Applications,* Giovanni Petrecca, ISBN 0-7923-9305-8

THE FIELD ORIENTATION PRINCIPLE IN CONTROL OF INDUCTION MOTORS, Andrzej M. Trzynadlowski, ISBN 0-7923-9420-8

FINITE ELEMENT ANALYSIS OF ELECTRICAL MACHINES, S. J. Salon, ISBN 0-7923 9594-8

POWER QUALITY ENHANCEMENT USING CUSTOM POWER DEVICES

by

Arindam Ghosh
Indian Institute of Technology

Gerard Ledwich
Queensland University of Technology

KLUWER ACADEMIC PUBLISHERS
Boston / Dordrecht / London

Distributors for North, Central and South America:
Kluwer Academic Publishers
101 Philip Drive
Assinippi Park
Norwell, Massachusetts 02061 USA
Telephone (781) 871-6600
Fax (781) 681-9045
E-Mail: kluwer@wkap.com

Distributors for all other countries:
Kluwer Academic Publishers Group
Post Office Box 322
3300 AH Dordrecht, THE NETHERLANDS
Telephone 31 786 576 000
Fax 31 786 576 474
E-Mail: services@wkap.nl

 Electronic Services < http://www.wkap.nl>

Library of Congress Cataloging-in-Publication Data

Ghosh, Arindam.
　Power quality enhancement using custom power devices / by Arindam Ghosh, Gerard Ledwich.
　　p. cm -- (Kluwer international series in engineering and computer science ; SECS
　701. Power electronics and power systems)
　Includes bibliographical references and index.
　ISBN 1-4020-7180-9 (alk. paper)
　　1. Electric power systems--Quality control. 2. Electric power systems--Equipment and supplies. I. Ledwich, Gerard. II. Title. III. Kluwer international series in engineering and computer science ; SECS 701. IV. Kluwer international series in engineering and computer science. Power electronics & power systems.

TK1010 .G46 2002
621.31'7--dc21

2002073058

Copyright © 2002 by Kluwer Academic Publishers

All rights reserved. No part of this work may be reproduced, stored in a retrieval system, or transmitted in any form or by any means, electronic, mechanical, photocopying, microfilming, recording, or otherwise, without written permission from the Publisher, with the exception of any material supplied specifically for the purpose of being entered and executed on a computer system, for exclusive use by the purchaser of the work.

Permission for books published in Europe: permissions@wkap.nl
Permissions for books published in the United States of America: permissions@wkap.com

Printed on acid-free paper.

Printed in the United States of America

Dedicated to my father Bholanath and the memory of my loving mother Arati.

Arindam Ghosh

I dedicate this book in memory of my father Harry and give thanks to God for His continued blessings.

Gerard Ledwich

Contents

Preface   xv

Acknowledgements   xix

1 Introduction   1
  1.1 ELECTRIC POWER QUALITY   3
      1.1.1 Impacts of Power Quality Problems on End Users   4
      1.1.2 Power Quality Standards   6
      1.1.3 Power Quality Monitoring   7
  1.2 POWER ELECTRONIC APPLICATIONS IN POWER TRANSMISSION SYSTEMS   8
      1.2.1 HVDC Transmission   8
      1.2.2 HVDC Light   9
      1.2.3 Static Var Compensator (SVC)   10
      1.2.4 Thyristor Controlled Series Compensator (TCSC)   12
      1.2.5 Static Compensator (STATCOM)   14
      1.2.6 Static Synchronous Series Compensator (SSSC)   16
      1.2.7 Unified Power Flow Controller (UPFC)   16
      1.2.8 Other FACTS Devices   17
  1.3 POWER ELECTRONIC APPLICATIONS IN POWER DISTRIBUTION SYSTEMS   18
  1.4 DISTRIBUTED GENERATION   22
  1.5 REFERENCES   23

2 Characterization of Electric Power Quality — 27
2.1 POWER QUALITY TERMS AND DEFINITIONS — 29
  2.1.1 Transients — 29
  2.1.2 Short Duration Voltage Variations — 33
  2.1.3 Long Duration Voltage variations — 35
  2.1.4 Voltage Imbalance — 36
  2.1.5 Waveform Distortion — 36
  2.1.6 Voltage Fluctuations — 39
  2.1.7 Power Frequency Variations — 39
  2.1.8 Power Acceptability Curves — 39
2.2 POWER QUALITY PROBLEMS — 40
  2.2.1 Poor Load Power Factor — 41
  2.2.2 Loads Containing Harmonics — 42
  2.2.3 Notching in Load Voltage — 45
  2.2.4 DC Offset in Loads — 45
  2.2.5 Unbalanced Loads — 46
  2.2.6 Disturbance in Supply Voltage — 52
2.3 CONCLUSIONS — 53
2.4 REFERENCES — 54

3 Analysis and Conventional Mitigation Methods — 55
3.1 ANALYSIS OF POWER OUTAGES — 55
3.2 ANALYSIS OF UNBALANCE — 60
  3.2.1 Symmetrical Components of Phasor Quantities — 60
  3.2.2 Instantaneous Symmetrical Components — 64
  3.2.3 Instantaneous Real and Reactive Powers — 67
3.3 ANALYSIS OF DISTORTION — 72
  3.3.1 On-line Extraction of Fundamental Sequence Components from Measured Samples — 76
  3.3.2 Harmonic Indices — 84
3.4 ANALYSIS OF VOLTAGE SAG — 86
  3.4.1 Detroit Edison Sag Score — 88
  3.4.2 Voltage Sag Energy — 88
  3.4.3 Voltage Sag Lost Energy Index (VSLEI) — 88
3.5 ANALYSIS OF VOLTAGE FLICKER — 90
3.6 REDUCED DURATION AND CUSTOMER IMPACT OF OUTAGES — 92
3.7 CLASSICAL LOAD BALANCING PROBLEM — 93
  3.7.1 Open-Loop Balancing — 94
  3.7.2 Closed-Loop balancing — 98
  3.7.3 Current Balancing — 102
3.8 HARMONIC REDUCTION — 104
3.9 VOLTAGE SAG OR DIP REDUCTION — 108

Contents

3.10 CONCLUSIONS	110
3.11 REFERENCES	111

4 Custom Power Devices: An Introduction 113
  4.1 UTILITY-CUSTOMER INTERFACE 114
  4.2 INTRODUCTION TO CUSTOM POWER DEVICES 116
    4.2.1 Network Reconfiguring Devices 117
    4.2.2 Load Compensation using DSTATCOM 121
    4.2.3 Voltage Regulation using DSTATCOM 126
    4.2.4 Protecting Sensitive Loads using DVR 127
    4.2.5 Unified Power Quality Conditioner (UPQC) 130
  4.3 CUSTOM POWER PARK 131
  4.4 STATUS OF APPLICATION OF CP DEVICES 134
  4.5 CONCLUSIONS 136
  4.6 REFERENCES 136

5 Structure and Control of Power Converters 137
  5.1 INVERTER TOPOLOGY 138
    5.1.1 Single-Phase H-Bridge Inverter 138
    5.1.2 Three-Phase Inverter 143
  5.2 HARD-SWITCHED VERSUS SOFT-SWITCHED 146
  5.3 HIGH VOLTAGE INVERTERS 153
  5.4 COMBINING INVERTERS FOR INCREASED POWER AND VOLTAGE 154
    5.4.1 Multi-Step Inverter 155
    5.4.2 Multilevel Inverter 162
    5.4.3 Chain Converter 167
  5.5 OPEN-LOOP VOLTAGE CONTROL 169
    5.5.1 Sinusoidal PWM for H-Bridge Inverter 169
    5.5.2 Sinusoidal PWM for three-phase Inverter 174
    5.5.3 SPWM in Multilevel Inverter 175
    5.5.4 Space Vector Modulation 178
    5.5.5 Other Modulation Techniques 180
  5.6 CLOSED-LOOP SWITCHING CONTROL 182
    5.6.1 Closed-Loop Modulation 182
    5.6.2 Stability of Switching Control 183
    5.6.3 Sampled Error Control 185
    5.6.4 Hysteresis Control 187
  5.7 SECOND AND HIGHER ORDER SYSTEMS 188
    5.7.1 Sliding Mode Controller 192
    5.7.2 Linear Quadratic Regulator (LQR) 193
    5.7.3 Tracking Controller Convergence 195

5.7.4 Condition for Tracking Reference Convergence	198
5.7.5 Deadbeat Controller	200
5.7.6 Pole Shift Controller	202
5.7.7 Sequential Linear Quadratic Regulator (SLQR)	203
5.8 CONCLUSIONS	210
5.9 REFERENCES	212

6 Solid State Limiting, Breaking and Transferring Devices — 215

6.1 SOLID STATE CURRENT LIMITER	216
6.1.1 Current Limiter Topology	216
6.1.2 Current Limiter Operating Principle	217
6.2 SOLID STATE BREAKER (SSB)	220
6.3 ISSUES IN LIMITING AND SWITCHING OPERATIONS	223
6.4 SOLID STATE TRANSFER SWITCH (SSTS)	225
6.5 SAG/SWELL DETECTION ALGORITHMS	232
6.5.1 Algorithm Based on Symmetrical Components	232
6.5.2 Algorithm Based on Two-Axis Transformation	233
6.5.3 Algorithm Based on Instantaneous Symmetrical Components	234
6.6 CONCLUSIONS	238
6.7 REFERENCES	239

7 Load Compensation using DSTATCOM — 241

7.1 COMPENSATING SINGLE-PHASE LOADS	242
7.2 IDEAL THREE-PHASE SHUNT COMPENSATOR STRUCTURE	245
7.3 GENERATING REFERENCE CURRENTS USING INSTANTANEOUS PQ THEORY	249
7.4 GENERATING REFERENCE CURRENTS USING INSTANTANEOUS SYMMETRICAL COMPONENTS	259
7.4.1 Compensating Star Connected Loads	260
7.4.2 Compensating Delta Connected Loads	265
7.5 GENERAL ALGORITHM FOR GENERATING REFERENCE CURRENTS	268
7.5.1 Various Compensation Schemes and Their Characteristics Based on the General Algorithm	269
7.5.2 Discussion of Results	270
7.6 GENERATING REFERENCE CURRENTS WHEN THE SOURCE IS UNBALANCED	276
7.6.1 Compensating to Equal Resistance	278
7.6.2 Compensating to Equal Source Currents	280
7.6.3 Compensating to Equal Average Power	282
7.7 CONCLUSIONS	285

Contents

7.8 REFERENCES	285

8 Realization and Control of DSTATCOM ... 287
   8.1 DSTATCOM STRUCTURE ... 288
   8.2 CONTROL OF DSTATCOM CONNECTED TO A STIFF SOURCE ... 291
   8.3 DSTATCOM CONNECTED TO WEAK SUPPLY POINT ... 296
      8.3.1 DSTATCOM Structure for Weak Supply Point Connection ... 299
      8.3.2 Switching Control of DSTATCOM ... 302
      8.3.3 DC Capacitor Control ... 308
   8.4 DSTATCOM CURRENT CONTROL THROUGH PHASORS ... 310
      8.4.1 Case-1: When Both Load and Source are Unbalanced ... 311
      8.4.2 Case-2: When Both Load and Source are Unbalanced and Load Contains Harmonics ... 313
      8.4.3 Case-3: Both Load and Source are Unbalanced and Distorted ... 314
      8.4.4 DC Capacitor Control ... 319
   8.5 DSTATCOM IN VOLTAGE CONTROL MODE ... 321
      8.5.1 State Feedback Control of DSTATCOM in Voltage Control Mode ... 322
      8.5.2 Output Feedback Control of DSTATCOM in Voltage Control Mode ... 327
   8.6 CONCLUSIONS ... 330
   8.7 REFERENCES ... 330

9 Series Compensation of Power Distribution System ... 333
   9.1 RECTIFIER SUPPORTED DVR ... 335
   9.2 DC CAPACITOR SUPPORTED DVR ... 340
      9.2.1 Fundamental Frequency Series Compensator Characteristics ... 341
      9.2.2 Transient Operation of Series Compensator when the Supply is Balanced ... 346
      9.2.3 Transient Operation when the Supply is Unbalanced or Distorted ... 348
      9.2.4 Series Compensator Rating ... 350
      9.2.5 An Alternate Strategy Based on Instantaneous Symmetrical Components ... 355
   9.3 DVR STRUCTURE ... 359
      9.3.1 Output Feedback Control of DVR ... 360
      9.3.2 State Feedback Control of DVR ... 365
   9.4 VOLTAGE RESTORATION ... 370
   9.5 SERIES ACTIVE FILTER ... 372

9.6 CONCLUSIONS	376
9.7 REFERENCES	376

10 Unified Power Quality Conditioner — 379
10.1 UPQC CONFIGURATIONS — 380
10.2 RIGHT-SHUNT UPQC CHARACTERISTICS — 381
10.3 LEFT-SHUNT UPQC CHARACTERISTICS — 388
10.4 STRUCTURE AND CONTROL OF RIGHT-SHUNT UPQC — 391
    10.4.1 Right-shunt UPQC Structure — 391
    10.4.2 Right-Shunt UPQC Control — 392
    10.4.3 Harmonic Elimination using Right-Shunt UPQC — 398
10.5 STRUCTURE AND CONTROL OF LEFT-SHUNT UPQC — 401
    10.5.1 Left-Shunt UPQC Structure — 401
    10.5.2 Left-Shunt UPQC Control — 402
10.6 CONCLUSIONS — 405
10.7 REFERENCES — 406

11 Distributed Generation and Grid Interconnection — 407
11.1 DISTRIBUTED GENERATION – CONNECTION REQUIREMENTS AND IMPACTS ON THE NETWORK — 407
    11.1.1 Standards for Grid Connection — 408
    11.1.2 Key Requirements in Standards — 408
    11.1.3 Grid Friendly Inverters — 409
    11.1.4 Angle Stability for Inverters — 410
    11.1.5 Issues for Distributed Generation — 410
11.2 INTERACTION AND OPTIMAL LOCATION OF DG — 411
    11.2.1 EigenAnalysis and Voltage Interaction — 411
    11.2.2 Simulation Results of EigenAnalysis and Voltage Interaction — 415
11.3 POWER QUALITY IN DG — 417
    11.3.1 Mitigation of Voltage Dip during Motor Start — 417
    11.3.2 Harmonic Effects with DG — 419
    11.3.3 Voltage Flicker and Voltage Fluctuation — 421
11.4 ISLANDING ISSUES — 422
    11.4.1 Anti-Islanding Protection — 422
    11.4.2 Vector Shift — 423
    11.4.3 Dedicated Islanding Operation — 423
    11.4.4 Rate of Change of Frequency (ROCOF) — 424
11.5 DISTRIBUTION LINE COMPENSATION — 425
    11.5.1 Line Voltage Sensitivity — 425
    11.5.2 Case-1: Heavy Load — 426
    11.5.3 Case-2: Light Load — 435

11.6 REAL GENERATION	435	
11.7 PROTECTION ISSUES FOR DISTRIBUTED GENERATION	435	
11.8 TECHNOLOGIES FOR DISRIBUTED GENERATION	437	
11.9 POWER QUALITY IMPACT FROM DIFFERENT DG TYPES	437	
11.10 CONCLUSIONS	441	
11.11 REFERENCES	441	

12 Future Directions and Opportunities for Power Quality Enhancement — 443

- 12.1 POWER QUALITY SENSITIVITY — 443
  - 12.1.1 Costs of Power Quality — 444
  - 12.1.2 Mitigation of Power Quality Impacts from Sags — 446
- 12.2 UTILITY BASED VERSUS CUSTOMER BASED CORRECTION — 447
  - 12.2.1 Dips and Outages — 448
  - 12.2.2 Harmonic, Flicker and Voltage Spikes — 449
- 12.3 POWER QUALITY CONTRIBUTION TO THE NETWORK FROM CUSTOMER OWNED EQUIPMENT — 450
  - 12.3.1 Issues — 450
  - 12.3.2 Addressing the Barriers to Customer Owned Grid Friendly Inverters — 451
- 12.4 INTERCONNECTION STANDARDS — 451
- 12.5 POWER QUALITY PERFORMANCE REQUIREMENTS AND VALIDATION — 452
  - 12.5.1 Commercial Customers — 452
  - 12.5.2 Regulator Requirements — 452
  - 12.5.3 An Example — 453
- 12.6 SHAPE OF ENERGY DELIVERY — 454
- 12.7 ROLE OF COMPENSATORS IN FUTURE ENERGY DELIVERY — 455
- 12.8 CONCLUSIONS — 456
- 12.9 REFERENCES — 456

Index — 457

Preface

Reliability and quality are the two most important facets of any power delivery system. A power distribution system is reliable if all its customers get interruption-free power for 24 hours a day and 365 days a year. The term power quality is often referred to as maintaining near sinusoidal voltage at the stipulated frequency of 50 or 60 Hz at the customer inlet points. It could be argued that maintaining voltage levels and frequency are the responsibility of generation. However, it will be shown in this book that there is no guarantee that the customers get quality power, even if the generation quality levels are met.

The aim of the book is two-fold – to introduce the power quality problems and to discuss the solutions of some of these problems using power electronic controllers. To achieve these aims, we discuss the power quality problems and their impacts on the end users at the beginning of the book. In the remainder of the book we present the custom power solutions to some of the power quality problems. We define those devices that provide power electronic solutions to the power quality problems as custom power devices.

The power quality problems in power distributions systems are not new, but customer awareness of these problems has increased. Similarly there are sets of conventional solutions to the power quality problems which have existed for a long time. However these conventional solutions use passive elements and do not always respond correctly as the nature of the power system conditions change. Custom power offers flexible solutions to many power quality problems.

In recent times, the issues involved with power quality issues and custom power solutions have generated a tremendous amount of interest amongst power system engineers. This is reflected by a large number of publications in IEEE Transactions on Power Delivery and Industry Applications and other journals like Proceedings of IEE, Electric Power System Research etc. Also power quality and customer power are regularly discussed in IEEE and CIGRE conferences. From this point of view, we hope that this book will be able to provide an insight into these two very important aspects. It is however to be remembered that every book represents the viewpoint of the authors and cannot be treated as the final word on the subject. We shall therefore be delighted if this book generates increased research and development in custom power devices and their application.

A large number of numerical examples are presented in the book. Many softwares are commercially available for simulating power electronic circuits. We have found that Manitoba HVDC Research Center's EMTDC/PSCAD is a very useful tool for simulating power systems and related power electronic circuits. Also for system level simulations using mathematical models, MATLAB, a product of Math Works Inc., is most suitable. The advantage of using MATLAB is that complex control algorithms can easily be incorporated in the models. All the simulation results that are presented in this book have been prepared using either of these two packages.

The book is organized in twelve chapters. In Chapter 1 we introduce the concepts of power quality and custom power solutions. Some of the flexible ac transmission systems (FACTS) devices are also discussed in this chapter as they can be considered as precursors to the custom power devices. We also introduce the concepts of distributed generation and grid interconnection.

In Chapter 2 we discuss power quality terms and their definitions. We also discuss the impacts of poor power quality on the end users.

Chapter 3 presents the analysis and indices of the power quality problems. In this chapter we present some of the important concepts that are used extensively in the later chapters. Also the conventional mitigation methods of some of the power quality problems are presented in this chapter.

We introduce the custom power devices in Chapter 4. These devices are categorized into two broad classes – network reconfiguring devices and compensating devices. The network reconfiguring devices include SSCL, SSB and SSTS, while the load compensating devices include DSTATCOM, DVR and UPQC. We also discuss the concept of custom power park in this chapter.

Chapter 5 deals with the structure of power electronic converters and their controls. Since most of the custom power compensating devices are realized by power electronic converters, this chapter elucidates their topology, operating principles and control to make the book self-contained.

Preface xvii

Chapter 6 discusses the topology and operating principles of the network reconfiguring devices and illustrates how these devices can be used to protect distribution systems from abnormal operations.

Chapter 7 discusses the theory of shunt compensation. It illustrates how an ideal shunt compensator can be used for load balancing, power factor correction and active filtering. Most emphasis is given to discussing the theory behind instantaneous correction of disturbances, as these developments facilitate the generation of compensator reference currents based on the measurements on instantaneous currents and voltages.

Chapter 8 deals with practical shunt compensator structures and their applications. It illustrates how a DSTACOM can be used in a distribution system for load compensation when the supply voltage is stiff or non-stiff. It also discusses how a DSTATCOM can be controlled to regulate the voltage of a power distribution bus.

In Chapter 9 we discuss the principles of series compensation. Here we illustrate how a series device can regulate the voltage at a load terminal against sag/swell or distortion in the supply side. We also illustrate how a series device, in conjunction with shunt passive devices, can be used as active filter.

In Chapter 10 we discuss the unified power quality conditioner. Two different structures of this device are discussed in this chapter along with their merits or demerits.

Chapter 11 discusses the distributed generation and grid interconnection issues. It presents a range of issues from standards to grid friendly inverters to islanding.

The book concludes in Chapter 12 where some future directions and opportunities in power quality enhancements are provided.

Arindam Ghosh
Gerard Ledwich

July 2002

Acknowledgements

I thank my friend and colleague Prof. Avinash Joshi for his help in writing the book. He has made many constructive suggestions, corrected many mistakes and has kindly allowed me to use some of his class note materials. Without his help this book would have been incomplete. I also thank my graduate student Amit Jindal who has painstakingly corrected the entire manuscript, verified many derivations and given me great support during the printing. I also thank two of my former graduate students Dr. K. K. Mahapatra and Dr. Mahesh Mishra who have increased my curiosity in this area and for the discussions that I had with them to clarify many doubts.

I thank my wife Supriya for carefully proofreading the entire manuscript. I also thank her and my son Aviroop for giving me mental support and tolerating my long hours of absence from home during the preparation of the manuscript. I also thank my father Bholanath and uncle Biswanath for the encouragement I have received from them, not only for writing the book but throughout my life.

Finally I wish to thank Prof. M. A. Pai for motivating us to write the book. I also thank Mr. Alex Greene and Ms. Melissa Sullivan of Kluwer Academic Publishers for their helpful hints during the preparation of the manuscript.

AG

This book has arisen from a long friendship with Arin and an abiding excitement for both of us in the way active control can enhance power system operation. I am thankful for the material developed in conjunction with Dr Mohammad Kashem and Ph.D. student Khalid Masoud. I am thankful for the continued support of my wife Catherine in this distracted time of manuscript development. My thanks to the sponsors for my Chair in Power Engineering in Powerlink Energex and Ergon. These engineers have helped develop my understanding in power quality issues and distributed generation.

GL

Chapter 1

Introduction

Modern day power systems are complicated networks with hundreds of generating stations and load centers being interconnected through power transmission lines. An electric power system has three separate components – power generation, power transmission and power distribution. Electric power is generated by synchronous alternators that are usually driven either by steam or hydro turbines. Almost all power generation takes place at generating stations that may contain more than one such alternator-turbine combinations. Depending upon the type of fuel used for the generation of electric power, the generating stations are categorized as thermal, hydro, nuclear etc. Many of these generating stations are remotely located. Hence the electric power generated at any such station has to be transmitted over a long distance to load centers that are usually cities or towns. Moreover, the modern power system is interconnected, i.e., various generating stations are connected together through transmission lines and switching stations. Electric power is generated at a frequency of either 50 Hz or 60 Hz. In an interconnected ac power system, the rated generation frequency of all units must be the same. For example, in the United States and Canada the generation frequency is 60 Hz, while in countries like United Kingdom, Australia, India the frequency is 50 Hz. In Japan both 50 Hz and 60 Hz systems operate and these systems are interconnected by HVDC links. An HVDC converter station converts power at 50 Hz ac to dc power at transmission voltage. An identical converter station converts the dc to power at 60 Hz ac. In this book we shall consider only 50 Hz ac systems in all the examples that are presented.

The basic structure of a power system is shown in Figure 1.1. It contains a generating plant, a transmission system, a subtransmission system and a distribution system. These subsystems are interconnected through transformers T_1, T_2 and T_3. Let us consider some typical voltage levels to understand the functioning of the power system. The electric power is generated at a thermal plant with a typical voltage of 22 kV (voltage levels are usually specified line-to-line). This is boosted up to levels like 400 kV through transformer T_1 for power transmission. Transformer T_2 steps this voltage down to 66 kV to supply power through the subtransmission line to industrial loads that require bulk power at a higher voltage. Most of the major industrial customers have their own transformers to step down the 66 kV supply to their desired levels. The motivation for these voltage changes is to minimize transmission line cost for a given power level. Distribution systems are designed to operate for much lower power levels and are supplied with medium level voltages.

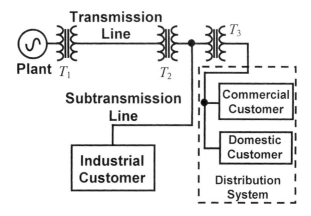

Figure 1.1. A typical power system

The power distribution network starts with transformer T_3, which steps down the voltage from 66 kV to 11 kV. The distribution system contains loads that are either commercial type (like office buildings, huge apartment complexes, hotels etc) or residential (domestic) type. Usually the commercial customers are supplied power at a voltage level of 11 kV whereas the domestic consumers get power supply at 400-440 V. Note that the above figures are given for line-to-line voltages. Since domestic customers get single-phase supplies, they usually receive 230-250 V at their inlet points. While a domestic customer with a low power consumption gets a single-phase supply, both industrial and commercial consumers get three-phase supplies not only because their consumption is high but also because many of them use three-phase motors. For example, the use of induction motor is very common amongst industrial customers who run pumps, compressors, rolling mills etc.

It is to be noted that the voltage levels quoted above are not standard and vary from one country to another. Let us consider for example the Indian

1. Introduction 3

power system in which the generation level varies between 11 kV to 25 kV. The domestic customers get supply at 415 V (line-to-line) or 240 V (line-to-neutral). The distribution side voltages are either 11 kV or 33 kV. In some places the distribution voltage is 22 kV or even 6.6 kV. The subtransmission voltages are either 66 kV or 110 kV or 132 kV and the transmission voltages are 220 kV or 400 kV. A new 800 kV line has also been installed recently.

It can therefore be seen that there are various stages between the point of power generation to the stage when electric power is delivered to the end users. The correct operation of all components of a power system is absolutely critical for a reliable power delivery. There are many issues involved here such as the maintenance of power apparatus and system, the stability of the system operation, the operation of power distribution system, faults etc. Some of these problems are power transmission related – a subject matter that is not treated here, as this book deals exclusively with problems related to power distribution systems and their solutions.

In this chapter we shall introduce the concept of power quality and discuss its impact on the end users. We shall also discuss the use of power electronics in power system in which we shall present an overview of high voltage dc (HVDC) transmission and flexible ac transmission systems (FACTS) as they are the major areas of use. We shall then introduce the concept of custom power as well as power electronic systems for general distribution quality enhancement.

1.1 Electric Power Quality

Even a few years back, the main concern of consumers of electricity was the reliability of supply. Here we define reliability as the continuity of electric supply. Even though the power generation in most advanced countries is fairly reliable, the distribution is not always so. The transmission systems compound the problem further as they are exposed to the vagaries of Mother Nature. It is however not only reliability that the consumers want these days, quality too is very important to them. For example, a consumer that is connected to the same bus that supplies a large motor load may have to face a severe dip in his supply voltage every time the motor load is switched on. In some extreme cases, he may have to bear with blackouts. This may be quite unacceptable to most customers. There are also very sensitive loads such as hospitals (life support, operation theatre, patient database system), processing plants (semiconductor, food, rayon and fabrics), air traffic control, financial institutions and numerous other data processing and service providers that require clean and uninterrupted power. In several processes such as semiconductor manufacturing or food processing plants, a batch of product can be ruined by a voltage dip of very short duration. Such customers are very wary of such dips since each such interruption cost them a substantial amount of money. Even short dips are sufficient to cause contactors on motor drives to drop out. Stoppage in a portion of a process can destroy the conditions for quality control of the product and require restarting of production. Thus in this changed scenario in which the customers increasingly demand quality power, the term *power quality* (*PQ*) attains increased significance.

Transmission lines are exposed to the forces of nature. Furthermore, each transmission line has its loadability limit that is often determined by either stability considerations or by thermal limits. Even though the power quality problem is a distribution side problem, transmission lines often have an impact on the quality of power supplied. It is however to be noted that while most problems associated with transmission systems arise due to the forces of nature or due to the interconnection of power systems, individual customers are responsible for a more substantial fraction of the problems of power distribution systems.

1.1.1 Impacts of Power Quality Problems on End Users

The causes of power quality problems are generally complex and difficult to detect. Technically speaking, the ideal ac line supply by the utility system should be a pure sinewave of fundamental frequency (50/60 Hz). In addition, the peak of the voltage should be of rated value. Unfortunately the actual ac line supply that we receive everyday departs from the ideal specifications. Table 1.1 lists various power quality problems, their characterization methods and possible causes.

There are many ways in which the lack of quality power affects customers. Impulsive transients do not travel very far from their point of entry. However an impulsive transient can give rise to an oscillatory transient. The oscillatory transient can lead to transient overvoltage and consequent damage to the power line insulators. Impulsive transients are usually suppressed by surge arresters.

Short duration voltage variations have varied effects on consumers. Voltage sags (also knows as dips) can cause loss of production in automated processes since a voltage sag can trip a motor or cause its controller to malfunction. For semiconductor manufacturing industries such a loss can be substantial. A voltage sag can also force a computer system or data processing system to crash. To prevent such a crash, an uninterruptible power supply (UPS) is often used, which, in turn, may generate harmonics. The protective circuit of an adjustable speed drive (ASD) can trip the system during a voltage swell. Also voltage swells can put stress on computers and many home appliances, thereby shortening their lives. A temporary

1. Introduction

interruption lasting a few seconds can cause a loss of production, erasing of computer data etc. The cost of such an interruption during peak hours can be hundreds of thousands of dollars.

Table 1.1. Power quality problems and their causes

Broad Categories	Specific Categories	Methods of Characterization	Typical Causes
Transients	Impulsive	Peak magnitude, rise time and duration	Lightning strike, transformer energization, capacitor switching
	Oscillatory	Peak magnitude, frequency components	Line or capacitor or load switching.
Short duration voltage variation	Sag	Magnitude, duration	Ferroresonant transformers, single line-to-ground faults
	Swell	Magnitude, duration	Ferroresonant transformers, single line-to-ground faults
	Interruption	Duration	Temporary (self-clearing) faults
Long duration voltage variation	Undervoltage	Magnitude, duration	Switching on loads, capacitor deenergization
	Overvoltage	Magnitude, duration	Switching off loads, capacitor energization
	Sustained interruptions	Duration	Faults
Voltage imbalance		Symmetrical components	Single-phase loads, single-phasing condition
Waveform distortion	Harmonics	THD, Harmonic spectrum	Adjustable speed drives and other nonlinear loads
	Notching	THD, Harmonic spectrum	Power electronic converters
	DC offset	Volts, Amps	Geo-magnetic disturbance, half-wave rectification
Voltage flicker		Frequency of occurrence, modulating frequency	Arc furnace, arc lamps

The impact of long duration voltage variations is greater than those of short duration variations. A sustained overvoltage lasting for few hours can cause damage to household appliances without their owner knowing it, until it is too late. The undervoltage has the same effect as that of a voltage sag. In the case of a sag the termination of process is sudden. But normal operation can be resumed after the normal voltage is restored. However in the case of a sustained undervoltage, the process cannot even be started or resumed. A sustained interruption is usually caused by faults. Since the loss to customers due to any sustained interruption can be in the order of millions of dollars, it is necessary for the utility to have a good preventive maintenance schedule and to have agreements or regulations to encourage high supply reliability.

Voltage imbalance can cause temperature rise in motors and can even cause a large motor to trip. Harmonics, dc offset and notching cause waveform distortions. Harmonics can be integer multiples of fundamental frequency, fractions of the fundamental frequency (subharmonics) and at frequencies that are not integer multiples of the fundamental frequency (interharmonics). Unwanted harmonic currents flowing through the distribution network can causes needless losses. Harmonics also can cause malfunction of ripple control or traffic control systems, losses and heating in transformers, electromagnetic interference (EMI) and interference with the communication systems. Ripple control refers to the use of a 300Hz to 2500Hz signal added to distribution lines to control switching of loads such as hot water heaters or street lighting. Interharmonic voltages can upset the operation of fluorescent lamps and television receivers. They can also produce acoustic noise in power equipment. DC offsets can cause saturation in the power transformer magnetic circuits. A notch is a periodic transient that rides on the supply voltage. It can damage capacitive components connected in shunt due to high rate of voltage rise at the notches.

Voltage flickers are caused by arc discharge lamps, arc furnaces, starting of large motors, arc welding machines etc. Voltage flickers are frequent variations in voltage that can cause the light intensity from incandescent lamps to vary. This variation is perceived as disturbing by human observers, particularly in the range of 3 to 15 times per second. The voltage flicker can have adverse effects on human health as the high frequency flickering of light bulbs, fluorescent tubes or television screen can cause strain on the eyes resulting in headaches or migraines. The voltage flicker can also reduce the life span of electronic equipment, lamps etc.

We can therefore conclude that the lack of standard quality power can cause loss of production, damage of equipment or appliances or can even be detrimental to human health. It is therefore imperative that a high standard of power quality is maintained. This book will demonstrate that the power electronic based power conditioning devices can be effectively utilized to improve the quality of power supplied to customers.

1.1.2 Power Quality Standards

Geneva based International Electrotechnical Commission (IEC) and Institute of Electrical and Electronic Engineers (IEEE) have proposed various power quality standards. A review of various standards is given in [1]. Table 1.2 lists some of these standards that are given in [1]. We shall

1. Introduction

discuss the indices for the measurements of the various power quality components in Chapter 3.

Table 1.2. Some power quality standards of IEC and IEEE

Phenomena	Standards
Classification of power quality	IEC 61000-2-5: 1995 [2], IEC 61000-2-1: 1990 [3] IEEE 1159: 1995 [4]
Transients	IEC 61000-2-1: 1990 [3], IEEE c62.41: (1991) [5] IEEE 1159: 1995 [4], IEC 816: 1984 [6]
Voltage sag/swell and interruptions	IEC 61009-2-1: 1990 [3], IEEE 1159: 1995 [4]
Harmonics	IEC 61000-2-1: 1990 [3], IEEE 519: 1992 [7] IEC 61000-4-7: 1991 [8]
Voltage flicker	IEC 61000-4-15: 1997 [9]

1.1.3 Power Quality Monitoring

Power quality variations are classified as either disturbances or steady state variations [4]. Disturbances pertain to abnormalities in the system voltages or currents due to fault or some abnormal operations. Steady state variations refer to rms deviations from the nominal quantities or harmonics. In general these are monitored by disturbance analyzers, voltage recorders, harmonic analyzers etc. However with the advancement in the computer technology, better, faster and more accurate instruments can now be designed for power quality monitoring and analysis.

The input data for any power quality monitoring device is obtained through transducers. These include current transformers, voltage transformers, Hall-effect current and voltage transducers etc. Disturbance analyzers and disturbance monitors are instruments that are specifically designed for power quality measurements [10]. There are two categories of these devices – conventional analyzers and graphics-based analyzers. Conventional analyzers provide information like magnitude and duration of sag/swells, under/overvoltages etc. Graphic-based analyzers are equipped with memory such that the real-time data can be saved. The advantage of this device is that the saved data can be analyzed later to determine the source and cause of the power quality problems. In addition, these analyzers can also graphically present the real-time data.

Harmonic data is analyzed with the help of harmonic or spectrum analyzers, which can graphically display harmonic data. These are usually digital signal processor (DSP) based data analyzers that can sample real-time data and then perform fast Fourier transform (FFT) to determine the amplitudes and phase angles of the harmonic components. These analyzers can simultaneously measure the voltage and currents such that harmonic power can be computed [10]. They can also sample the signals at a very high rate such that harmonics up to about 50th order can be determined. Also note that the magnitudes of the higher order harmonics are typically much smaller than the magnitudes of the lower order harmonics. Therefore for the signal conversion and detection of the higher order harmonics, these analyzers have built-in high-resolution analog to digital converters.

Currently, dedicated power quality measuring instruments are manufactured that can combine both the functions of harmonic and disturbance measurements. These are graphical instruments that can also transmit data over telephone lines [10].

Flicker monitoring is done through IEC flickermeter [1, 9]. These meters measure the instantaneous flickering voltage. This is called the instantaneous flicker level (IFL). The recorded IFL is then stored and statistical operations on these data are performed to determine short term (10 min) flicker severity index and long term flicker severity index.

1.2 Power Electronic Applications in Power Transmission Systems

The application of power electronics to power systems has a long tradition. It started with bulk power transmission through high voltage direct current (HVDC) transmission. Static var compensator (SVC) systems were employed later for reactive compensation of power transmission lines. Subsequently, devices like thyristor controlled series compensator (TCSC), thyristor controlled phase angle regulator (TCPAR), static compensator (STATCOM), static synchronous series compensator (SSSC), unified power flow controller (UPFC) were proposed and installed under the generic name of flexible ac transmission systems (FACTS) controllers [11]. Since most of these devices are the predecessors to the power quality enhancement devices, we briefly review them below.

1.2.1 HVDC Transmission

The schematic diagram of a double-poled HVDC transmission system is shown in Figure 1.2, in which only the rectifier side is shown. This contains two pairs of converters – one for the positive pole and the other for the negative pole. The converters are realized by thyristors. These two converters are supplied from the three-phase ac side through two transformers. Also at the point of coupling of the ac system and the dc system, tuned ac filters are provided such that harmonics generated by the converter are prevented from entering the ac system. On the dc side a smoothing inductor L_s is connected to the output of each converter to smooth

1. Introduction

the ripples in the dc current and dc filters are also provided to cancel harmonics from traveling down the dc transmission line. Figure 1.2 shows only one side of the line – the same configuration is repeated on the other end of the line, except that the converters at the other end inverts the power back to ac. The direction of the power transfer can be reversed as well by changing the operating principle of the converters. All practical HVDC converters are 12-pulse or higher. A 12-pulse converter is realized by connecting two 6-pulse converters through phase shifting transformers. For more details on HVDC transmission refer to [12,13].

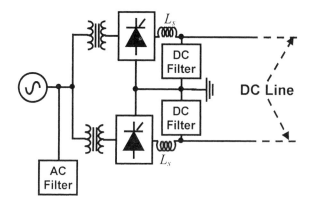

Figure 1.2. A double-poled HVDC transmission system

1.2.2 HVDC Light

HVDC light is the recent addition to HVDC technology that uses Insulated Gate Bipolar Transistors (IGBTs). These IGBTs are used to realize voltage source converters. The converter switches are operated in high-speed pulse width modulation to obtain a better control bandwidth. Also associated with HVDC light are extruded polymer cables suitable for direct current transmission [14]. By changing the PWM pattern of the converter, it is possible to almost instantaneously create any phase angle or amplitude of voltage for connection to the ac system. The use of PWM offers the possibility of controlling both active and reactive power independently. HVDC light uses cables that can be buried under the ground by a plowing tractor. Unlike the overhead lines, the cables are not subjected to storms, snow and ice and there is not right of way problem either. It is therefore claimed that the HVDC light is a technology for the future dc transmission.

In Australia a 180 MVA HVDC light project interconnects the Queensland and New South Wales networks through a cable length of 65 km. Through this interconnection, any capacity shortage in Queensland can be offset by the surplus capacity in New South Wales without the risk of endangering the system stability.

1.2.3 Static Var Compensator (SVC)

There are two main building blocks for SVCs – thyristor switched capacitor and thyristor controlled reactor. In a thyristor switched capacitor (TSC), a capacitor is connected in series with two opposite poled thyristors as shown in Figure 1.3 (a). Current flows through the capacitor when the opposite poled thyristors are gated. The current through the device can be stopped by blocking the thyristors. To achieve controlled reactive power a TSC always comes in a group as shown in Figure 1.3 (b). The effective reactance of the group can be changed by switching a TSC on or off. For example let us assume that four identical TSCs, each having a capacitance value C, are connected in parallel. Then the equivalent reactance when all the TSCs are gated is given by

$$X_{eq} = -j\frac{1}{4\omega C}$$

Similarly when one TSC is switched off, the equivalent reactance drops to be equal to $-j1/(3\omega C)$. Thus the effective reactance of the device is given by $-j1/(n\omega C)$, $n = 0, \ldots, 4$.

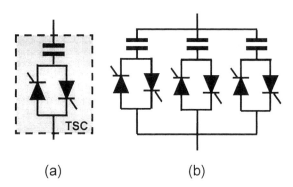

(a)　　　　　　(b)

Figure 1.3. (a) Schematic diagram of a TSC and (b) multiple TSC connection

One of the main issues while using a TSC is switching transients. Since a TSC blocks current through it when the thyristors are blocked and allows it to pass when the thyristors are gated, it is obvious that severe switching

1. Introduction

transients will occur if a TSC is switched off while the current through it is not zero. Similarly, the device must be switched on at a particular instant of the voltage cycle. The transient free switching can be obtained when the voltage across a capacitor is in either its positive peak or negative peak such that the current through the capacitor is zero [15].

In a thyristor controlled reactor (TCR), a reactor is connected in series with two opposite poled thyristors. One of these thyristors conducts in each positive half cycle of the supply frequency, while the other conducts in the corresponding negative half cycle. The schematic diagram of a TCR connected to an ac voltage source is shown in Figure 1.4. The gating signal to each thyristor is delayed by an angle α (often called the firing or conduction angle) from the zero crossing of the source voltage. This is shown in Figure 1.5 in which typical voltage-current waveforms in the steady state are also shown. The conduction angle must be in the range $90° \leq \alpha \leq 180°$. For a conduction angle of $\alpha = 90°$, the current waveform will be continuous and for an angle of $\alpha = 180°$, the current will be zero.

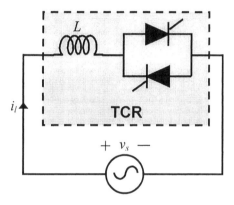

Figure 1.4. Schematic diagram of a TCR

A practical SVC circuit often contains both TSC modules and TCR as shown in Figure 1.6. In addition to them the SVC also contains tuned filters to suppress harmonic current from flowing into the ac system. Additionally there are firing and control circuits which are not shown in this figure. The SVC is connected in shunt to an ac line through a step down transformer. Through reactive power injection, the SVC can regulate the voltage of the ac bus. For line power and voltage modulation, the SVC is placed in the middle of a transmission line. There are several advantages of such a placement and these are listed in [16]. SVCs placed close to loads can be very effective in providing voltage support, thereby avoiding voltage instability.

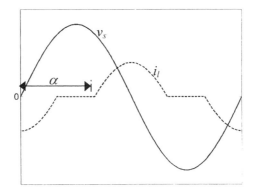

Figure 1.5. Steady state voltage-current waveforms of TCR

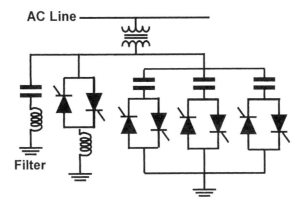

Figure 1.6. SVC connected to an ac network

1.2.4 Thyristor Controlled Series Compensator (TCSC)

The schematic diagram of a TCSC compensated single machine, infinite bus (SMIB) power system is shown in Figure 1.7. The TCSC here contains an ac capacitor that is connected in parallel with a TCR. Because of this topology, this configuration is sometimes called a fixed capacitor-thyristor controlled reactor (FC-TCR). It is to be noted that since this is a series compensation device, its placement is not that crucial and it can be placed anywhere along the line.

Let us denote the voltage across the fixed capacitor as v_C and the current through the TCR as I_P. Then the voltage-current characteristic of the device is shown in Figure 1.8. As shown in this figure, the firing is delayed by an angle α from the zero crossing of the capacitor voltage. The equivalent

1. Introduction

reactance of the parallel combination of a TCR with a fundamental reactance of $X_L(\alpha)$ and a capacitance with a reactance of X_C is given by

$$X_{eq}(\alpha) = j\frac{X_C X_L(\alpha)}{X_C - X_L(\alpha)}$$

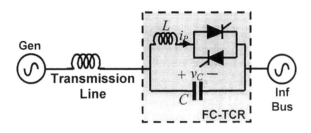

Figure 1.7. TCSC compensated SMIB system

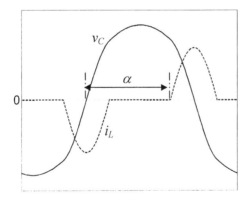

Figure 1.8. Voltage-current characteristics of a TCSC

It can be seen from the above equation that by varying $X_L(\alpha)$, the reactance $X_{eq}(\alpha)$ can be made inductive or capacitive. Then as in the case of TCR discussed in Section 1.2.3, $X_{eq}(\alpha)$ will vary as α changes from 90° to 180°. The operation of the TCSC will be capacitive when α is closer to 180°. Again when the value of α is just above 90° the operation of the TCSC will be inductive. In between, depending on the values of L and C chosen, the value of X_{eq} will be excessively large as the TCR and capacitor will go through a fundamental frequency resonance.

There are many ways of computing the fundamental frequency reactance of $X_{eq}(\alpha)$ [17,18]. A typical plot of the fundamental frequency reactance of the TCSC as α varies is shown in Figure 1.9. For this X_C is chosen as 0.5 per unit and X_L is chosen as 0.1667 per unit. In this figure the zone between the inductive and the capacitive regions is the resonance zone and the TCSC is never operated in this zone.

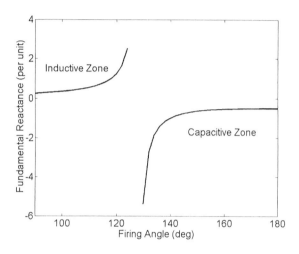

Figure 1.9. Variation in the fundamental reactance of a TCSC with α

1.2.5 Static Compensator (STATCOM)

This is a shunt device that does not require passive elements like inductors and capacitors. The schematic diagram of a SMIB power system that is compensated by a shunt compensator is shown in Figure 1.10. The STATCOM is built around a voltage source inverter, which is supplied by a dc capacitor. The inverter consists of GTO switches which are turned on and off through a gate drive circuit.

The output of the voltage source inverter is connected to that ac system through a coupling transformer. The inverter produces a quasi sinewave voltage V_0 at the fundamental frequency. Let us assume that the losses in the inverter and the coupling transformer are negligible. The inverter is then gated such that the output voltage of the inverter V_0 is in phase with the local bus voltage V_m. In this situation two ac voltages that are in phase are connected together through a reactor, which is the leakage reactance of the coupling transformer. Therefore the current I_q is a purely reactive. If the

1. Introduction

magnitude of the voltage V_m is more than that of the voltage V_0, the reactive current I_q flows from the bus to the inverter. Then the inverter will consume reactive power. If, on the other hand, the magnitude of V_0 is greater than that of V_m, then the inverter feeds reactive power to the system. Therefore through this arrangement the STATCOM can generate or absorb reactive power. In practice however the losses are not negligible and must be drawn from the ac system. This is accomplished by slightly shifting the phase angle of the voltage V_0 through a feedback mechanism such that the dc capacitor voltage is held constant.

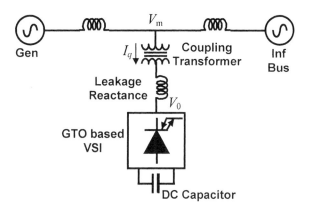

Figure 1.10. A STATCOM connected to an SMIB power system

The structure of the GTO-based VSI must be so chosen that the lower order harmonics are eliminated from the output voltage. The VSI will then resemble a synchronous voltage source. Because the switching frequency of each GTOs must be kept low, overall switch ripple needs to be kept low without use of PWM. This is accomplished by connecting a large number of basic inverter modules. The construction of a 48-step voltage source inverter is discussed in [19]. In this inverter, eight identical elementary 6-step inverters are operated from a common dc bus. Each of these 6-step inverters produces a compatible set of three-phase, quasi-square wave output voltage waveforms. The outputs of these 6-step inverters are added through a magnetic circuit that contains eighteen single-phase three winding transformers and six single-phase two winding transformers. This connection eliminates all low-order harmonics. The lowest order harmonic on the ac side is 47th while that on the dc side is 48th. The line-to-line output voltage of the 48-step inverter is shown in Figure 1.11 along with the fundamental voltage. It can be seen that the output is a stepped approximation of the fundamental sinewave. The construction of a multilevel synchronous voltage source is given in [20].

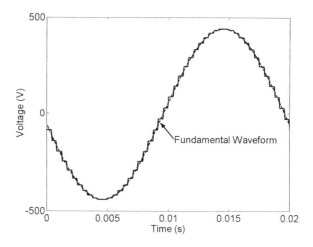

Figure 1.11. The line-to-line output voltage of a 48-step inverter

1.2.6   Static Synchronous Series Compensator (SSSC)

The static synchronous series compensator is a series device in which a synchronous voltage source injects a fundamental frequency voltage in series with the transmission line. The synchronous voltage source is realized by a multilevel or multi-step voltage source inverter as shown in Figure 1.12. Since this is a series device, it can be placed anywhere along the transmission line. Ideally the inverter is operated in quadrature with the line current such that the voltage source either behaves like an inductor or a capacitor. In this mode the inverter does not consume or generate any real power. However, in a practical circuit the inverter losses must be replenished by the ac system and hence a small phase lag is introduced for this purpose. There are two modes of operation of this device – one in which the injected voltage is proportional to the line current and the other in which the injected voltage is independent of the line current. These modes and their respective power-angle curves are given in [21]. The non-capacitor like behavior and the superior operating characteristics makes this device very attractive for power transmission application. The main limitation of application is due to the losses and cost of the converter.

1.2.7   Unified Power Flow Controller (UPFC)

The schematic diagram of a UPFC is shown in Figure 1.13. This contains two voltage source inverters that are connected together through a dc link capacitor. There are three different ways of operating the UPFC – as a shunt

1. Introduction

controller, as a series controller and also as phase angle regulator. The main quoted use is for control of power flow when there are alternate paths with different ratings. The operating characteristics of UPFC are given in [22,23].

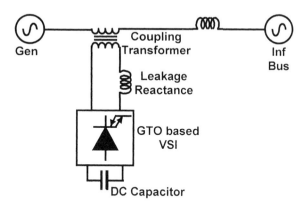

Figure 1.12. An SSSC compensated power system

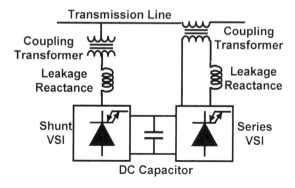

Figure 1.13. Schematic diagram of a UPFC

1.2.8 Other FACTS Devices

In addition to the ones mentioned above, there are other power electronic controllers that are members of the FACTS family. One such device is the thyristor controlled braking resistor (TCBR) [24]. A power transmission line has a largely reactive impedance. Thus the power transmitted from a remote generator over the line reduces drastically during a fault and this causes the acceleration in the generator rotor angle. A TCBR is connected at the generator terminals. It is switched on once a fault is detected thereby allowing an amount of real power to be dissipated in the resistor during the fault. It therefore restricts the machine acceleration. Even though this device does not enhance the power transfer, it increases the system stability limits. Therefore a generator can operate at a higher steady state load angle which enhances the power transfer capability.

A thyristor controlled phase angle regulator (TCPAR) injects voltage in series with a transmission line. As opposed to an SSSC which injects voltage in quadrature with the line current, the TCPAR injects voltage in quadrature with the line voltage [25]. Therefore by adjusting the magnitude of the injected voltage, the phase angle between the sending end and receiving end voltages can be adjusted. Like a TCBR, the TCPAR also does not increase the transmittable power through a transmission line. However it increases the stability limits of the power transfer allowing the system to operate at a higher power angle provided the thermal limit is not reached.

An interline power flow controller (IPFC) is a FACTS controller proposed for providing flexible power flow control in a multi-line power system. In the interline power flow control scheme, two more parallel lines are compensated by SSSC. These SSSCs are connected to a common dc link. Thus the SSSCs can provide series compensation to the line to which they are connected. In addition, they can also transfer real power between the compensated lines. This capability makes it possible to equalize both real and reactive power between the lines, to transfer power from an overloaded line to an underloaded line and to damp out system oscillations resulting from a disturbance [26].

1.3 Power Electronic Applications in Power Distribution Systems

The flexible ac transmission technology allows a greater control of power flow. Since these devices provide very fast power swing damping, the power transmission lines can be securely loaded up to their thermal limits. In a similar way power electronic devices can be applied to the power distribution systems to increase the reliability and the quality of power supplied to the customers. The technology of the application of power electronics to power distribution system for the benefit of a customer or group of customers is called *Custom Power* (CP) since through this technology the utilities can supply value-added power to these specific customers [27]. Other applications of power electronics are to improve the power quality to general customers in a region.

Custom power provides an integrated solution to the present problems that are faced by the utilities and power distributors. Through this technology the reliability of the power delivered can be improved in terms of

1. Introduction

reduced interruptions and reduced voltage variations. The proper use of this technology will benefit all the industrial, commercial and domestic customers. In this book we shall discuss this technology and its implication to the customers.

The custom power devices are basically of two types – network reconfiguring type and compensating type. The network reconfiguring equipment can be GTO based or thyristor based. They are usually used for fast current limiting and current breaking during faults. They can also prompt a fast load transfer to an alternate feeder to protect a load from voltage sag/swell or fault in the supplying feeder. The following devices are members of the family of network reconfiguring devices:

1. Solid State Current Limiter (SSCL): This is a GTO based device that inserts a fault current limiting inductor in series with the faulted circuit as soon the fault is detected. The inductor is removed from the circuit once the fault is cleared.
2. Solid State Circuit Breaker (SSCB): This device can interrupt a fault current very rapidly and can also perform auto-reclosing function. This device, based on a combination of GTO and thyristor switches, is much faster than its mechanical counterpart and is therefore an ideal device for custom power application.
3. Solid State Transfer Switch (SSTS): This is usually a thyristor based device that is used to protect sensitive loads from sag/swell. It can perform a sub-cycle transfer of the sensitive load from a supplying feeder to an alternate feeder when a voltage sag/swell is detected in the supplying feeder. An SSTS can also be connected as a bus coupler between two incoming feeders.

The compensating devices are used for active filtering, load balancing, power factor correction and voltage regulation. The active filters, which eliminate the harmonic currents, can be connected in both shunt and series. However, the shunt filters are more popular than the series filters because of greater ease of protection. Some of these devices are used as load compensators, i.e., in this mode they correct the unbalance and distortions in the load currents such that compensated load draws a balanced sinusoidal current from the ac system. Some others are operated to provide balanced, harmonic free voltage to the customers. The family of compensating devices has the following members:

1. Distribution STATCOM (DSTATCOM): This is a shunt connected device that has the same structure as that of a STATCOM shown in Figure 1.10. This can perform load compensation, i.e., power factor

correction, harmonic filtering, load balancing etc. when connected at the load terminals. It can also perform voltage regulation when connected to a distribution bus. In this mode it can hold the bus voltage constant against any unbalance or distortion in the distribution system. It is however to be noted that there is a substantial difference in the operating characteristics of a STATCOM and a DSTATCOM. The STATCOM is required to inject a set of three balanced quasi-sinusoidal voltages that are phase displaced by 120°. However the DSTATCOM must be able to inject an unbalanced and harmonically distorted current to eliminate unbalance or distortions in the load current or the supply voltage. Therefore its control is significantly different from that of a STATCOM.
2. Dynamic Voltage Restorer (DVR): This is a series connected device that has the same structure as that of an SSSC shown in Figure 1.12. The main purpose of this device is to protect sensitive loads from sag/swell, interruptions in the supply side. This is accomplished by rapid series voltage injection to compensate for the drop/rise in the supply voltage. Since this is a series device, it can also be used as a series active filter. Even though this device has the same structure as that of an SSSC, the operating principles of the two devices differ significantly. While the SSSC injects a balanced voltage in series, the DVR may have to inject unbalanced voltages to maintain the voltage at the load terminal in case of an unbalanced sag in the supply side. Furthermore when there is a distortion in the source voltage, the DVR may also have to inject a distorted voltage to counteract the harmonic voltage.
3. Unified Power Quality Conditioner (UPQC): This has the same structure as that of a UPFC shown in Figure 1.13. This is a very versatile device that can inject current in shunt and voltage in series simultaneously in a dual control mode. Therefore it can perform both the functions of load compensation and voltage control at the same time. As in the case of DSTATCOM or DVR, the UPQC must also inject unbalanced and distorted voltages and currents and hence its operating characteristics are different than that of a UPFC.

One example of how the power electronic based custom power devices can protect a sensitive load is given in [27]. Consider the distribution system given in Figure 1.14. This contains a sensitive load in addition to other regular loads. The loads are supplied by two independent incoming feeders A and B. Normally the SSTS is connected such that the sensitive load is supplied by the feeder A and the other regular loads are supplied by the feeder B. Therefore any fault upstream or downstream in Feeder-B does not affect the sensitive load. For a fault upstream in Feeder-A, the solid state circuit breaker 1 opens and the sensitive load is transferred to Feeder-B in

1. Introduction

less than a cycle by the SSTS. In the same way, the sensitive load can also be transferred to Feeder-B in case of voltage sag/swell in feeder A. Also the voltage of the sensitive load can be regulated by a DSTATCOM. This DSTATCOM can eliminate any fluctuation in the load terminal voltage. In case of a fault at the distribution bus, the SSCB 2 opens to isolate the fault quickly and the DSTATCOM supplies power to the load. However this can only be a temporary arrangement as a DSTATCOM has that much stored energy to ride through during a fault. Once the mechanical breakers clear the fault and SSCB 2 is closed, the sensitive load starts getting its supply from the feeder. The dc capacitor of the DSTATCOM then gets charged by absorbing power from the feeder.

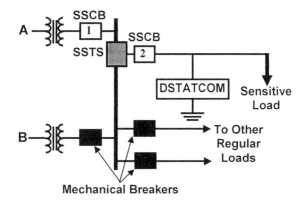

Figure 1.14. A hypothetical distribution system equipped with custom power devices

High quality power to commercial customers can be supplied using the various custom power devices in a custom power park [28]. Such a park gets its supply from two incoming feeders that are coupled by a solid state transfer switch. Every consumer of this park must pay a premium tariff for the electricity consumption. Moreover there will be gradation in the price structure depending upon the service they are provided with. For example the lowest grade consumers will get almost uninterrupted supply. Their supply is guaranteed unless there is a catastrophic failure in which both incoming feeders are lost. The next higher grade customers are supplied power through a diesel-generator set when both incoming feeders are lost. Therefore they get almost uninterrupted power except for the time required to start up the diesel generator. Even higher grade customers, in addition to the services provided to the lower grade customers can have the benefit of a DSTATCOM or DVR or even a UPQC. Their power will be totally uninterrupted, as these compensating devices have to ability to provide the ride through during the start up time of the diesel generator set. Furthermore, the highest grade customer will also get supply voltages that are free from harmonics or unbalance due to the presence of the compensating devices.

1.4 Distributed Generation

Throughout the last century and in the present, the size of generating plants has been increasing. A new trend however is emerging currently, in which significantly smaller sized generating units are being connected at the distribution level. Some of the factors that contribute to this trend are listed below.

- Greenhouse gas issues have become very significant in many countries to consider dispersed energy sources such as solar, wind and wave that operate with smaller sized units.
- Local generating units using gas microturbines are becoming more economical when the transmission and distribution overheads are taken into account. Fuel cells are still more expensive but have been showing great promise for low cost reliable small size generation units.
- Even though solar cells currently require a very large area and substantial investments, they can be used for power in large office buildings during business hours.
- The move to open competitive markets in electricity has increased the uncertainties of supply. A notable example is the Californian market in 2000/2001 where customers saw increased cost of energy and rolling blackouts. In response to this uncertainty of central supply there was a massive increase in demand for back-up generation with the possibility of generation back into the grid when conditions suited.
- In medium sized industrial plants or large buildings with a significant heating load, co-generation is becoming more attractive. The local generation of electricity also provides waste heat that can supply much of the heating needs of the local processes.

The following means are used for distributed generation [29,30]

- Reciprocating Piston Engine Generators: These are the most popular types of distributed generation (DG) units that are used world wide. The size of these units varies between 5 kW to 25 MW. These units run on fossil fuels like petrol (or gasoline) or diesel and run the alternator at low speeds. Even though diesel is the most popular form of fuel for these sets, its use in these units increases the environmental pollution. Therefore either these units are placed in sparsely populated areas or their exhausts

1. Introduction 23

are treated to extract the harmful gases. Preferably both these methods should be used.
- Gas Turbine Generators: Combustible gases rotate the turbines of these generators. Note that the same technology is used for large alternators of capacity 500 MW or more. However for DG smaller alternators in the mini and micro range are used. These turbines produce hot exhaust gases. These gases can be trapped for cogeneration purposes or for non-electric use.
- Fuel Cells: These are like chemically powered batteries that produce dc currents by converting chemical energy of an electrolyte into electricity. A typical fuel cell requires both gaseous fuels and oxidants. The most preferred gaseous fuel is hydrogen because it is highly reactive and does not require much catalytic agents. Electricity is produced when the hydrogen fuel reacts with an oxidant like oxygen. The byproduct in this case is water. Such a fuel cell is ideal from the environmental point of view and it is expected that this will be used by the automobile industry in the near future.
- Renewable Energy: A vast majority of the energy sources that are already available on our earth is not tapped because the technology has not advanced enough to harvest them at a reasonable cost. These energy sources include solar, geothermal, wind wave and tidal. The biggest challenge of the 21st century is to utilize these abundant resources in an environmentally friendly way. The various renewable energy conversion options that already exist are solar thermal power generation or solar pond by trapping solar energy through reflectors, photovoltaics cells to convert solar energy into electricity, windfarms to utilize wind at the seashores to run wind turbines, geothermal power generation through the trapped energy under the ground etc. In addition to these energy can also be trapped from the oceans through wave, tidal or ocean currents. However significant investment in research and development is required to reach the stage when all or some of the above-mentioned processes become viable for the production of inexpensive electricity.

This expansion of distributed generation has the potential to significantly change the nature of the distribution system and the associated power quality issues.

1.5 References

[1] J. Arrillaga, N. R. Watson and S. Chen, *Power Quality Assessment*, John Wiley, New York, 2000.

[2] IEC 61000-2-5: 1995, Electromagnetic Compatibility (EMC), Part 2, Environment, Section 5: Classification of Electromagnetic Environments.

[3] IEC 61000-2-1: 1990, Electromagnetic Compatibility (EMC), Part 2, Environment, Section 1: Description of Environment – Electromagnetic Environment for Low-Frequency Conducted Disturbances and Signalling in Public Power Supply Systems. First Edition, 1990-05.

[4] IEEE 1159: 1995, IEEE Recommended Practices on Monitoring Electric Power Quality.

[5] IEEE c62.41: 1991, IEEE Recommended Practices on Surge Voltages in Low-Voltage AC Power Circuits.

[6] IEC 816: 1984, Guide on Methods of Measurement of Short Duration Transients on Low Voltage Power and Signal Lines.

[7] IEEE 519: 1992, IEEE Recommended Practices and Requirements for Harmonic Control in Electric Power Systems (ANSI).

[8] IEC 61000-4-7: 1991, Electromagnetic Compatibility (EMC), Part 4: Limits, Section 7: General Guide on Harmonics and Inter-harmonics Measurements and Instrumentation for Supply Systems and Equipment Connected Thereto.

[9] IEC 61000-4-15: 1997, Electromagnetic Compatibility (EMC), Part 4: Limits, Section 15: Flickermeter – Functional and Design Specifications.

[10] M. McGranaghan and C. Melhorn, "Interpretation and analysis of power quality measurements," *PQ Network Internet Site*, http://www.pqnet.electrotek.com/pqnet.

[11] N. G. Hingorani, "Flexible ac transmission," *IEEE Spectrum*, Vol. 30, No. 4, pp. 40-45, 1993.

[12] E. W. Kimbark, *Direct Current Transmission*, Vol. 1, Wiley-Interscience, New York, 1971.

[13] N. Mohan, T. M. Undeland and W. P. Robbins, *Power Electronics: Converters, Applications and Design*, John Wiley, New York, 1989.

[14] G. Apslund, "Application to HVDC light to power system enhancement," *IEEE Power Engineering Society Winter Meeting*, Singapore, 2000.

[15] T. J. E. Miller, *Reactive Power Control in Electric Systems*, John Wiley, New York, 1982.

[16] L. Gyugyi, "Power electronics in electric utilities: Static var compensators," *Proc. IEEE*, Vol. 76, No. 4, pp. 483-494, 1988.

[17] N. Christl, R. Hedin, K. Sadek, P. Lutzelberger, P. E. Krause and S. M. McKenna, A. H. Montoya and D. Togerson, "Advanced series compensation (ASC) with thyristor controlled impedance", *CIGRE Int. Conf. Large High Voltage Electric* Systems, Paper No. 14/37/38-05, Paris, 1992.

[18] A. Ghosh, A. Joshi and M. K. Mishra, "State space simulation and accurate determination of fundamental impedance characteristics of a TCSC," *IEEE Power Engineering Society Winter Meeting-2001 (in CD)*, Columbus, Ohio, 2001.

[19] L. Sunil Kumar, *Design, Modeling and Control of a 48-Step Inverter Based S^3C*, M.Tech. Thesis, Indian Institute of Technology Kanpur, 1998.

[20] R. W. Menzies and Y. Zhuang, "Advanced static compensation using a multilevel GTO thyristor inverter," *IEEE Trans. Power Delivery*, Vol. 10, No. 2, pp. 732-738, 1995.

[21] L. Gyugyi, C.D. Schauder and K.K. Sen, "Static synchronous series compensator: a solid-state approach to the series compensation of transmission lines," *IEEE Trans. Power Delivery*, Vol. 12, No. 1, pp. 406-417, 1997.

1. Introduction

[22] L. Gyugyi, "Unified power flow control concept for flexible ac transmission systems," *Proc. IEE*, Vol. 139, Pt. C, pp. 323-331, 1992.

[23] L. Gyugyi, C. D. Schauder, S. L. Williams, T. R. Reitman, D. R. Torgerson and A. Edris, "The unified power flow controller: a new approach to power transmission control," *IEEE Trans. Power Delivery*, Vol. 10, No. 2, pp. 1085-1097, 1995.

[24] Y. Wang, R.R. Mohler, W. A. Mittelstadt and D. J. Maratukulam, "Variable-structure braking-resistor control in a multimachine power system," *IEEE Trans. Power Systems*, Vol. 9, No. 3, pp. 1557-1562, 1994.

[25] S. Nyati, M. Eitzmann, J. Kappenman, D. VanHouse, N. Mohan and A. Edris, "Design issues for a single core transformer thyristor controlled phase-angle regulator," *IEEE Trans. Power Delivery*, Vol. 10, No. 4, pp. 2013-2019, 1995.

[26] L. Gyugyi, K. K. Sen and C. D. Schauder, "The interline power flow controller concept: a new approach to power flow management in transmission systems," *IEEE Trans. Power Delivery*, Vol. 14, No. 3, pp. 1115-1123, 1999.

[27] N. G. Hingorani, "Introducing custom power," *IEEE Spectrum*, Vol. 32, No. 6, pp. 41-48, 1995.

[28] N. G. Hingorani, "Custom Power and Custom Power Park," *Flexible Power HVDC Transmission and Custom Power, CIGRE Australian Panel 14*, Sydney, 1999.

[29] H. L. Lewis and W. C. Scott, *Distributed Power Generation: Planning and Evaluation*, Marcel Dekker, New York, 2000.

[30] A. M. Borbely and J. F. Kreider, Ed., *Distributed Generation: The Power Paradigm for the New Millenium*, CRC Press, Boca Raton, 2001.

Chapter 2

Characterization of Electric Power Quality

The term *electric power quality* broadly refers to maintaining a near sinusoidal power distribution bus voltage at rated magnitude and frequency. In addition, the energy supplied to a customer must be uninterrupted from the reliability point of view. It is to be noted that even though power quality (PQ) is mainly a distribution system problem, power transmission systems may also have an impact on the quality of power. This is because the modern transmission systems have a low resistance to reactance ratio, resulting in low system damping. Usually, a well-designed generating station is not a source of trouble for supplying quality power. The generated system voltages are almost perfectly sinusoidal. Moreover in many cases the utilities operate with a spinning reserve which ensures that the generating capability remains more than the load may demand. In some cases, a temporary shortfall in generation is overcome by reducing the peak of the generated voltage to reduce power consumption.

As mentioned above the PQ problems start with transmission systems. In order to transmit power over a long distance, the generated voltage is stepped up by transformers. However, this high voltage transmission has its own problem due to corona and other losses. These high voltage lines are hung overhead between two tall transmission towers. Near the towers, they are supported by long porcelain insulators that are connected to steel towers. The towers and the lines are exposed to nature. Therefore they are ideal targets for lightning strikes that cause spikes in the transmitted voltage. Moreover, high wind may cause two sagging transmission lines to come near each other causing arcing, momentary transients in voltages or voltage sag/swell. Flashover is a typical problem of dusty and arid regions. Usually the insulators in these regions are covered with dust for most part of the year, especially after a dust storm. A few droplets of water from a light shower or mist can mix with the dust to form a conductive path. This causes a flashover in the conductors resulting in a voltage sag followed by a voltage swell.

Most PQ problems occur in distribution systems. In most metropolitan cities, the distribution feeders run underground in the central business districts. In most other places the feeders run overhead. As a result these lines can easily come in contact with trees. Furthermore, they are likely to be hit by lightning or suffer from interference from birds and smaller animals. Moreover distribution systems feed loads directly. It is at these low voltage connections that the power quality becomes significantly worse. As we shall see later, a single customer can impose its harmonics and the effect of unbalanced loads on other customers. The utility has very little control over the loads. Furthermore, switching on of a large induction motor can cause a large inrush current to flow in that circuit causing a voltage dip in other parts of the system. In addition, some of the loads may have poor power factors causing unnecessary power loss in the distribution feeders. However, as we shall see in the later chapters, modern power electronic based systems provide solutions to some of the problems created by customers.

Based on the above discussions, we can summarize that there are two different categories of causes for the deterioration in power quality. The first category contains natural causes such as

- Faults or lighting strikes on transmission lines or distribution feeders.
- Falling of trees or branches on distribution feeders during stormy conditions.
- Equipment failure.

The second category contains the man made causes that may be due to load or feeder/transmission line operation. Some of these causes are

- Transformer energization, capacitor or feeder switching.
- Power electronic loads such as uninterrupted power supply (UPS), adjustable speed drives (ASD), converters etc.
- Arc furnaces and induction heating systems.
- Switching on or off of large loads.

Before we discuss the PQ problems, let us briefly review the terms and definitions used in the context of power quality. For more details, refer to the book by Dugan et al [1]. Also various aspects of power quality issues are regularly discussed in the Electrotek Concepts' Internet site PQ Network [2-5].

2.1 Power Quality Terms and Definitions

The power quality standards vary between countries. However, it is needless to say that poor quality power affects almost all consumers. It is therefore important to list the terms and definitions that are used with power quality. In particular, we shall consider the following

- Transients.
- Short duration voltage variations.
- Long duration voltage variations.
- Voltage imbalance.
- Waveform distortions.
- Voltage fluctuations.
- Power frequency variations.

2.1.1 Transients

A transient is that part of change in a system variable that disappears during transition from one steady-state operating condition to another. Transients can be classified into two categories – *impulsive transients* and *oscillatory transients*. An impulsive transient is a sudden, non-power frequency change in voltage, current etc that is unipolar in nature. The polarity of such a transient can be either positive or negative. Impulsive transients have a very fast rise time and also a very fast decaying time. These transients are mainly caused by lightning strikes. Impulsive transients usually do not conduct far the point of their entry into the power system. The distance to which an impulsive transient travels along a feeder depends on the particular system configuration. Let us consider the following example.

Example 2.1: Consider the radial power system, the single-line diagram of which is shown in Figure 2.1. In this the source supplies two buses to which loads are connected. A fixed capacitor can be connected to load Bus-1 through the switch. The system frequency is assumed to be 50 Hz and the peak of the supply (or source) voltage is assumed to be 1.0 per unit such that the instantaneous source voltage is given by the expression $1.0 \sin(100\pi t)$ per unit. The loads are assumed to be inductive. The per unit load and feeder impedances are given as

Load at Bus-1 = $2.0 + j1.5$, Load at Bus-2 = $2.55 + j1.25$
Feeder-1 impedance = $0.05 + j0.3$, Feeder-2 impedance = $0.075 + j0.4$

The capacitor has an admittance of $j0.0157$ per unit.

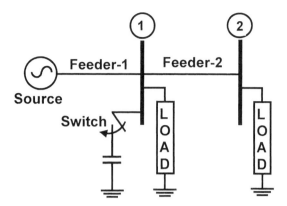

Figure 2.1. Single-line diagram of a radial distribution system

Let us first assume that the capacitor is not connected to Bus-1. We want to investigate the impact of a lightning near the supply point on the system voltages and currents. The voltage impulse is shown in Figure 2.2 (a). The resulting waveforms are shown in Figure 2.2 (b-d). It can be seen that there is a momentary spike in the bus voltages. However, the spike in the Bus-2 voltage has a reduced magnitude compared to the voltage of Bus-1. The impact is felt on the feeder current as well.

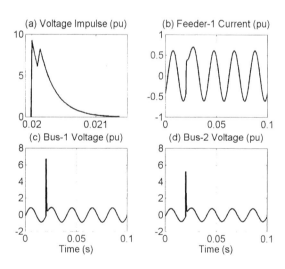

Figure 2.2. System response to a voltage impulse when the capacitor is not connected

Let us now investigate the system response to an impulsive transient when the capacitor of Figure 2.1 is connected to the system. This is shown in Figure 2.3 when the same transient as shown in Figure 2.2 (a) is applied at

2. Characterization of Electric Power Quality

the same location (i.e., near the source). It can be seen that the transients in the feeder currents and the bus voltages have a different nature in this case than shown in Figure 2.2. Comparing the Feeder-1 current of Figure 2.2 (b) with that shown in Figure 2.3 (a), it can be seen that the current in this case has a peak of about 2.0 per unit and also undergoes a prolonged transient. Interestingly, Bus-1 voltage in the previous case (Figure 2.2 c) has a peak of about 7.0 per unit while it has a peak of less than 4.0 per unit in this (Figure 2.3 c), even though the oscillation sustains for about 3 cycles in the present case. Similarly, there is also a sustained oscillation in the bus 2 voltage as shown in Figure 2.3 (d). However, as in the previous case, the overshoot has reduced as we move away from the location of the impulse strike.

ΔΔΔ

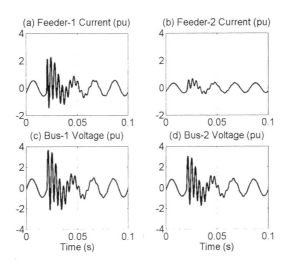

Figure 2.3. System response to a voltage impulse when the capacitor is connected

An oscillatory transient is usually bipolar in nature. It has one or more sinusoidal components that get multiplied by a decaying term. A typical oscillatory transient is shown in Figure 2.4. These transients may have more than one oscillating frequency depending on the modes that get excited. For example, the transient of Figure 2.4 is of the form $e^{-at}(\sin \omega_1 t + \sin \omega_2 t)$. This transient has oscillation frequencies of 750 kHz and 700 kHz and a decaying time constant of 2 ms.

Oscillatory transients are classified in accordance with their frequency. An oscillatory transient with a primary frequency greater than 500 kHz is considered high frequency transients. A transient within the frequency range of 5 kHz to 500 kHz is considered a medium frequency transient and anything below 5 kHz is termed as a low frequency transient. The principal cause of low frequency transient is ferroresonance and transformer energization [1,2].

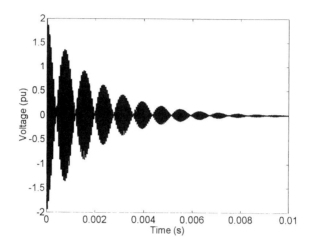

Figure 2.4. A typical oscillatory voltage transient

Typical causes of oscillatory transients are capacitor or transformer energization and converter switching. Sometimes an impulsive transient causes an oscillatory transient. For example, consider the waveforms shown in Figure 2.3. Low frequency oscillation is exhibited in both voltage and current in this case. These oscillations are caused by the presence of the shunt capacitor in Bus-1. Similarly, transient oscillations are exhibited in a system when a capacitor is suddenly energized. For example, consider the distribution system of Figure 2.1. Let us assume that the system is operating in the steady state when the capacitor across Bus-1 is suddenly connected to the bus by closing the switch.

The results are shown in Figure 2.5. It is assumed that the capacitor is not precharged. Hence the transient behavior will depend upon the time on closing the switch. For example, if the switch is closed near the zero crossing of the bus voltage, the transient oscillation in the bus voltage and the capacitor current will be minimal as shown in Figure 2.5 (a) and (b). On the other hand, if the switch is closed near the peak of the bus voltage, there will be very severe oscillations in these quantities as shown in Figure 2.5 (c) and (d). Note that the system frequency chosen for the study is 50 Hz and the peak of the source voltage is 1.0 per unit.

2. Characterization of Electric Power Quality

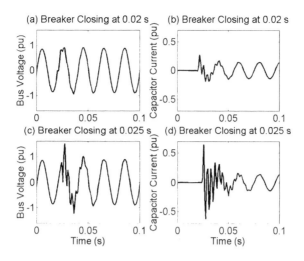

Figure 2.5. Transient caused by capacitor switching

2.1.2 Short Duration Voltage Variations

Any variation in the supply voltage for duration not exceeding one minute is called a short duration voltage variation. Usually such variations are caused by faults, energization of large loads that require large inrush currents and intermittent loose connection in the power wiring. Short duration variations are further classified as *voltage sags*, *voltage swells* and *interruptions*. These are shown in Figure 2.6.

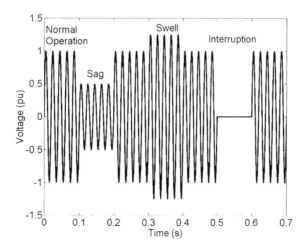

Figure 2.6. Short duration voltage variations

A voltage sag is a fundamental frequency decrease in the supply voltage for a short duration. The duration of voltage sag varies between 5 cycles to a minute. Voltage sags are typically caused by system faults, but can also be caused by energization of heavy loads. Voltage swells are defined as the increase of fundamental frequency voltage for a short duration. Voltage swells are not as common as voltage sags. One possible reason for their occurrence is due to the temporary rise in the voltage of an unfaulted phase during a single-line-to-ground fault. The severity of a swell that will be experienced by a load depends on its proximity to the fault location, system impedance and grounding. Let us consider the following example.

Example 2.2: Consider the distribution network shown in Figure 2.7. It is assumed that the network is supplied by a 3-phase, 33 kV, 50 Hz source with grounded neutral. The source is connected to a Δ-Y (33 kV:11 kV) transformer with ungrounded neutral. Each of the two feeders supplies a three-phase Y-connected balanced RL load. The per phase system parameters in a 1 MVA base are:

Load at Bus-2 = $2 + j2$, Load at Bus-3 = $0.686 + j1.4$
Feeder-1 impedance = $0.1 + j0.4$, Feeder-2 impedance = $0.5 + j2$

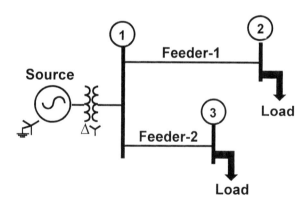

Figure 2.7. A three-bus distribution system

With the system operating in the steady state a single-line-to-ground (1LG) fault has been created at Bus-2. The fault occurs after 2 cycles of steady state operation. The duration of the fault is 2.5 cycles. After this period the system is restored to its pre-fault state. The three-phase voltages of Bus-3 are shown in Figure 2.8. It can be seen that while the voltage of phase-a exhibits a voltage sag, the voltage at the other two phase swell. In

2. Characterization of Electric Power Quality

fact the peak of the voltage in phase-c during the swell is higher than the peak of the voltage in phase-b during this period.

ΔΔΔ

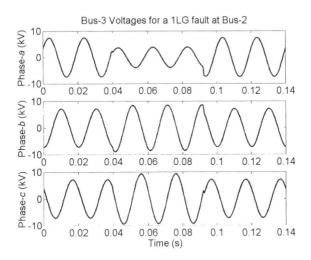

Figure 2.8. Voltage sag/swell due to 1LG fault

An interruption occurs when the supply voltage (or load current) decreases to less than 0.1 per unit for a period of time not exceeding 1 minute. It is caused by system faults, equipment failure and control malfunction. The time duration of such an interruption is dependent upon the operating time of the protective device.

2.1.3 Long Duration Voltage variations

These are defined as the rms variations in the supply voltage at fundamental frequency for periods exceeding 1 minute. These variations are classified into *overvoltages*, *undervoltages* and *sustained interruptions*. An overvoltage (or undervoltage) is a 10% or more increase (or decrease) in rms voltage for more than 1 minute. In a weak (i.e., poor voltage regulated) system the switching off of a large load or the energization of a large capacitor bank may result in an overvoltage. An undervoltage is the result of an event, which is a reverse of the event that causes overvoltage. The term *brownout* is often used to describe sustained periods of undervoltage due to specific utility strategy to reduce power demand. When the supply voltage is zero for a period of time in excess of 1 minute, the long duration voltage variation is called sustained interruption. Human intervention is required during sustained interruptions for repair and restoration.

2.1.4 Voltage Imbalance

This is the condition in which the voltages of the three phases of the supply are not equal in magnitude. Furthermore, they may not even be equally displaced in time. The primary cause of voltage unbalance is the single-phase loads in three-phase circuits. These are however restricted to within 5%. Severe imbalance (greater than 5%) can result during single phasing conditions when the protection circuit opens up one phase of a three-phase supply. For example consider the distribution system of Figure 2.7. Let us consider the case in which the phase-a circuit breaker of Feeder-1 is opened accidentally. The Bus-3 voltages are shown in Figure 2.9. It can be seen that while the phase-a voltage swells significantly, the voltage of phase-b sags. The magnitude of the phase-c voltage remains almost constant.

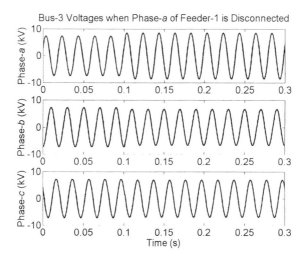

Figure 2.9. Voltage imbalance due to loss of a phase

2.1.5 Waveform Distortion

This is the steady-state deviation in the voltage or current waveform from an ideal sine wave. These distortions are classified as *dc offset*, *harmonics* and *notching*. The major causes of dc offsets in power systems are geomagnetic disturbance and half-wave rectification. The offsets due to geomagnetic disturbances are especially severe in higher latitudes. Poor grounding can also result in dc offsets. The presence of a load drawing dc current results in a dc component of the current in the secondary of a distribution transformer. This current will cause a dc bias in the sinusoidal flux of the transformer core. The increased peak value of the flux may push

2. Characterization of Electric Power Quality

the transformer towards saturation. As a consequence there will be heating in the transformer coil and core due to excessive magnetizing current and core losses. One remedy for this problem is to use a larger area in the magnetic circuit of the transformer core to accommodate dc bias. This is usually done in special rectifier transformers but hardly ever in distribution transformers.

Power electronic loads like UPS, adjustable speed drives etc usually cause harmonics in power system. A measure of harmonic content in a signal is the *total harmonic distortion* (*THD*). The percentage *THD* in a voltage is given by [6]

$$THD = \frac{\sqrt{\sum_{n=2}^{\infty} V_n^2}}{V_1} \tag{2.1}$$

where V_n denotes the magnitude of the nth harmonic voltage and V_1 is the magnitude of the fundamental voltage. A similar expression can also be written for current harmonics. For example, consider the distorted waveform shown in Figure 2.10 (a). It contains a 50 Hz fundamental, plus 3rd, 5th, 7th, 9th and 11th harmonics with their magnitudes being reciprocals of their harmonic numbers. Here the harmonic number implies the order of harmonics, i.e., 3rd harmonic has a harmonic number of 3, 5th harmonic has a harmonic number of 5 etc. The fundamental voltage waveform is also shown in Figure 2.10 (a). The harmonic spectrum of the distorted waveform is shown in Figure 2.10 (b). From this we calculate the THD as

$$THD = \sqrt{\left\{\left(\frac{1}{3}\right)^2 + \left(\frac{1}{5}\right)^2 + \left(\frac{1}{7}\right)^2 + \left(\frac{1}{9}\right)^2 + \left(\frac{1}{11}\right)^2\right\}} \times 100 = 43.83\%$$

Usually for good quality power it is recommended that the THD be less than 3%. We shall however define limits of harmonic distortions in Chapter 3.

Notching is a periodic voltage distortion due to the operation of power electronic converters when current commutates from one phase to other. During this period there is a momentary short circuit between the two phases that distorts voltages. The maximum voltage during notches depends on the system impedance. The frequency components that are associated with notches are usually very high. Assume that a three-phase diode bridge rectifier load is supplied by a source through a feeder. The voltage of one phase at the terminal in which the rectifier is connected is shown in Figure 2.11 along with the feeder (line) current in the same phase. It can be seen that the voltage has two types of disturbances – one due to the current in the same phase while the other is at the end of commutation in some other phase. The notches in the phase voltage coincide with the rising or falling edge of the line current of the same phase in both half cycles. In addition to this, high frequency oscillations occur even when line current in the same phase is constant. These high frequency oscillations are associated with sudden change in output voltage due to commutation in the other phases. The oscillations during these disturbances and their damping are a function of the RC values in the snubber circuit used in the rectifier [7].

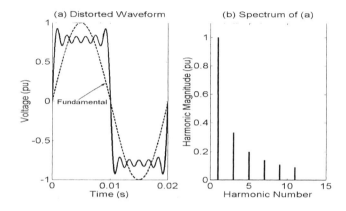

Figure 2.10. (a) Distorted voltage waveform and (b) its harmonic spectrum

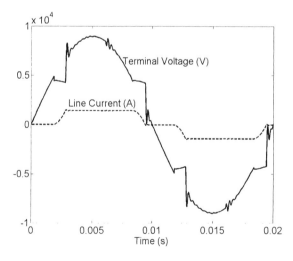

Figure 2.11. Notch in voltage due to rectifier action

2.1.6 Voltage Fluctuations

These are systematic random variations in supply voltages. A very rapid change in the supply voltage is called *voltage flicker*. This is caused by rapid variations in current magnitude of loads such as arc furnaces. In an arc furnace, a large inrush current flows when the arc strikes first. This causes a dip in the voltage of the bus to which the furnace is connected. Therefore other customers that are supplied by the same feeder face regular severe voltage drops unless the supply bus is very stiff. Voltage flicker is discussed in details in the next chapter.

2.1.7 Power Frequency Variations

These variations are usually caused by rapid changes in the load connected to the system. For example, the supply frequency may drop during the operation of large drag lines in a comparatively low inertia system. It is however desirable that the supply frequency does not deviate too much from the nominal frequency of 50 or 60 Hz. The maximum tolerable variation in supply frequency is often limited within ± 0.5 Hz. The frequency is directly related to the rotational speed of the generators supplying them. Thus a sustained operation outside the tolerable frequency range may reduce the life span of turbine blades on the shaft connected to a generator. Furthermore, if the frequency falls below a certain threshold, an under frequency relay may trip to protect the turbine blades.

2.1.8 Power Acceptability Curves

These curves quantify the acceptability of supply power as a function of duration versus magnitude of bus voltage disturbances. One of these curves was originally developed by Computer Business Equipment Manufacturers Association (CBEMA) to set limits to the withstanding capabilities of computers in terms of the magnitude and duration of the voltage disturbance. The CBEMA curve has however become a de facto standard for measuring the performance of all types of equipment and power systems.

In the CBEMA curve shown in Figure 2.12 there are two traces – one for overvoltage and the other for undervoltage. These show the percent bus voltage deviation from the rated voltage against time. The region below the upper trace and above the lower trace is the acceptable range. This region defines the tolerance level. For example an overvoltage of very short duration can be tolerable if it is in the acceptable region. Again a computer may be able to sustain about 5-6% overvoltage for a prolonged period of time. Similarly operating below the lower trace is also not permissible.

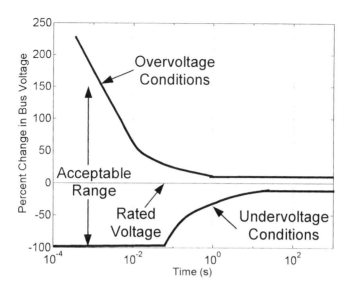

Figure 2.12. The CMEMA curve

The CBEMA curve was originally designed in 1970 for mainframe computers. But the curve has been used to quantify the voltage tolerance limits of adjustable speed drives, fluorescent lighting, microprocessor based controller etc [8]. The Information Technology Industry Council (ITIC) redesigned the CBEMA curve in the later half of the 1990s. The ITIC curve describes the acceptable range in steps rather than smooth curves used in CBEMA. In addition to these two curves the other power acceptability curves are FIPS curve for automatic data processing equipment and IEEE standards [8].

2.2 Power Quality Problems

Of the terms and definitions of PQ that are listed in the previous section, some of the major concerns of both customers and utility are

– Poor load power factor
– Harmonic contents in loads
– Notching in load voltages
– DC offset in load voltages
– Unbalanced loads
– Supply voltage distortion
– Voltage sag/swell
– Voltage flicker

2. Characterization of Electric Power Quality

We shall now discuss their implications separately.

2.2.1 Poor Load Power Factor

Consider a distribution system in which a source is supplying an inductive load through a feeder. The feeder has a resistance of R_s and a reactance of X_s. The feeder current is denoted by I_s and the load voltage is denoted by V_l. The load power factor is lagging and the power factor angle is denoted by θ_l. The system phasor diagram is shown in Figure 2.13 (a). In this diagram the load current is resolved into a real part $I_{sp} = |I_s|\cos\theta_l$ and a reactive part $I_{sq} = |I_s|\sin\theta_l$. Of these two components, the work done depends only on the real power.

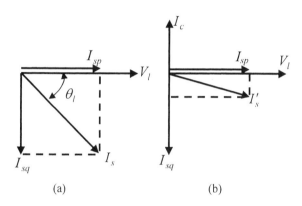

Figure 2.13. (a) Poor power factor and (b) its improvement by a shunt capacitor

Now suppose the load power factor is poor, i.e., the load has a large X/R ratio. Then the power factor angle θ_l will be large. This implies that the reactive component of the current is large and hence the magnitude of the load current $|I_s|$ is also large. This will not only cause a significant drop in the feeder voltage but there will also be a large amount of $|I_s|^2 R_s$ loss. This loss is associated with high heat dissipation in the feeder. Excessive heat may reduce the life span of the feeder.

To correct the large feeder drop, let us assume that as a remedial action we connect a capacitor in parallel with the load. This capacitor draws a current I_c that is in phase opposition to I_{sq}. The resulting current drawn by the capacitor-load combination is denoted by I'_s. This is shown in Figure 2.13 (b). It can be seen that even though the real component of the current remains the same, the magnitude of the current drawn from the source has reduced considerably. This is because the reactive component of the current drawn has reduced considerably and, as a consequence, the power factor angle has decreased. Therefore, to operate the feeder in an optimal fashion, the power factor at the load should be maintained near unity. In an ideal situation, the load power factor should be unity. However this may always not be achievable. With the improvement in the power factor, the line drop decreases resulting in better voltage regulation at the load as well.

2.2.2 Loads Containing Harmonics

It is well known that any nonsinusoidal but periodic signal can be decomposed into a fundamental component (50 or 60 Hz for power systems) and its integer multiples called the harmonic components. The harmonic number usually specifies a harmonic component, which is the ratio of its frequency to the fundamental frequency. For example when the fundamental frequency is 50 Hz, a harmonic with a number of 3 (3rd harmonics) will have a frequency of 150 Hz. The harmonic components that are integer multiples of the 3rd harmonic (e.g., 6th, 9th etc) are called *triplen*. In power systems, the electrical components are symmetrical. Therefore, the current drawn in the positive half cycles is the exact mirror image of the current drawn in the negative half cycles. Such symmetrical waveforms cannot contain any even harmonics. Transformer saturation and rectifier loads are examples of components typically exhibiting these symmetries. There is another form of symmetry in a 3-phase, 3-wire system. Assume that the harmonic current in phases-b and c are identical to that of phase-a but is delayed by $2n\pi/3$ and $4n\pi/3$ respectively where n is the harmonic number. The currents at each triplen frequency are then in phase with each other. Without a neutral they have no return path to flow just like a zero sequence current and thus must individually be zero. The triplen currents may however circulate inside a Δ-connected winding of a transformer. The triplen currents may also be present in a three-phase, four-wire system as the neutral wire provides a path for them to flow. Usually in power system even harmonics are less common. There the harmonics in a three-phase system are of the type ($6q \pm 1$) and $3q$ where $q = 1, 2, 3, \ldots$

Power electronic loads are the major source of harmonic generation in power systems. Consider an example where a new main frame computer system has been installed in a multistoried office building. At the same time, to protect the computer, a very large uninterrupted power supply (UPS) has also been installed. The UPS employs power electronic switches and as a result it can cause interference to the loads that are connected in parallel with the UPS. Assuming that all the loads of the office building are placed on the same bus, the UPS can cause screens of many smaller computers to flicker or roll and can even cause these computers to freeze. It can also cause other electronic circuits to malfunction. For example, it can change the timing

2. Characterization of Electric Power Quality

sequence of the elevator control circuit. In the Indian Institute of Technology Kanpur campus in Northern India the power is supplied by a 33 kV feeder. The incoming voltage is stepped down by a 33 kV/11 kV, 5 MVA transformer and power is then distributed to various facilities through five substations. In one such substation, the UPS of the main computer center is connected and so is the computerized telephone exchange. In an incident in the late 1990s, the ac input voltage to the telephone exchange became triangular with a peak of 600 V due to harmonic contamination by the UPS when the expected nominal fundamental voltage has a peak of 325 V. The power supply of the telephone exchange was damaged due to this. Harmonic contamination can also upset ripple control systems thereby causing street light control system or hot water control system to malfunction.

Let us now consider the impact of a harmonic current on a power distribution system. Consider the three-bus radial distribution system shown in Figure 2.14 in which three separate loads are being supplied by a single source. Load-1 is connected to Bus-2 while the other two loads are connected to Bus-3. Two feeders join the three buses. Now suppose out of these three loads, Load-2 is drawing harmonic current. This will cause a harmonic current to flow through both the feeders. Due to the presence of the feeder impedances this harmonic current will cause a harmonic voltage drop at Buses 2 and 3. Bus-1 is connected to a source and hence its bus voltage will not have any harmonic component. We shall call any such bus a stiff bus. Since both Bus-2 and Bus-3 voltages are distorted, the currents drawn by Load-1 and Load-3 will also get distorted as a consequence even if they are linear loads. This is undesirable and might even be unacceptable. Let us consider the following example.

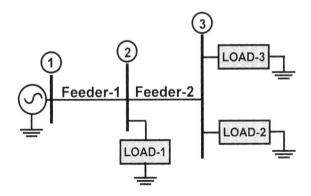

Figure 2.14. Single-line diagram of a power distribution system

Example 2.3: Consider the circuit of Figure 2.14 where the three-phase balance supply voltage has a magnitude of 11 kV (L-L, rms) and the system frequency is 50 Hz. In a base of 11 kV (L-L) and 1 MVA, each of the two feeders has a per unit impedance of $0.1 + j0.1$. Load-1 and Load-3 are grounded Y-connected passive RL loads with per phase impedances of $2.0 + j3.0$ per unit and $3.0 + j3.0$ per unit respectively. Load-2 constitutes a three-phase diode bridge rectifier, the dc side of which is connected to a 151 Ω resistor. This rectifier will cause distortions in the system quantities.

Figure 2.15 depicts the phase-a voltages and currents at various parts of the circuit. Figure 2.15 (a) shows the voltage at Bus-3 while the voltage at Bus-2 is shown in Figure 2.15 (b). The current drawn by the rectifier load is shown in Figure 2.15 (c) and the current through Feeder-1 is shown in Figure 2.15 (d). It can be seen that all these quantities are distorted. Therefore we can conclude that the presence of a nonlinear load can cause distortions in voltages and currents of a distribution network. If the current drawn by the nonlinear loads is higher compared to those drawn by the linear loads, then the distortion in the bus voltages at various parts of the network will be significant. As a result distortion in the linear load currents will also be high making the THD of these quantities unacceptable.

ΔΔΔ

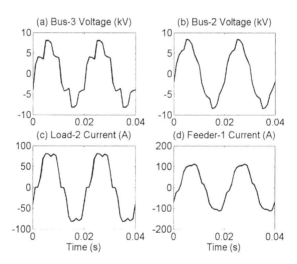

Figure 2.15. Harmonic distortion caused by a rectifier load

Harmonics can affect loads through several mechanisms. For example,

— the presence of harmonics can cause additional losses in induction motors, especially when they are operating close to their rated values.

2. Characterization of Electric Power Quality

Moreover, due to the additional losses that are created by harmonic currents, the overall heating may increase leading to premature failure of the motors.
- the supply voltage is used for timing purposes in many cases. For example, supply voltage cycles are counted to get timing information for digital clocks. Similarly many items of power electronic equipment, such as phase controlled thyristor circuits, use the zero crossing of the supply voltage to generate trigger pulses for the semiconductor devices. A distorted voltage waveform can create false triggering of the timing circuits.

2.2.3 Notching in Load Voltage

With rectifier loads there are commutation periods where the line to line voltage falls to zero. This effect is due to the finite inductance in the supply. Thus this causes a finite time for the current to fall to zero in one phase and transfer to another. As we have seen in Figure 2.11 that the presence of a large phase controlled rectifier will cause notches in the phase voltage. One case where these notches caused problems was in a concert hall. A new lift with a phase control was installed on the output of the same transformer supplying the microphone and stage lights. A simple dimmer circuit controlled the stage lights. This circuit measured the time from the zero crossing to determine the firing angle. When the lift was used, the firing angle for the lift controller changed and the notch moved along the waveform. When the notch neared the zero crossing of the phase voltage, there was a step change in the dimming level. The solution to this problem is often to provide the high power loads from a separate transformer. In this case there was additional inductance added at the lift motor such that the depth of notch seen by the dimmers was significantly reduced.

2.2.4 DC Offset in Loads

Consider again the distribution system shown in Figure 2.14. Let us assume that phase-a of Load-2 contains a half-bridge rectifier that draws dc current from the source. The output of the rectifier is connected to a 75 Ω resistor. The other two phases are unconnected. The feeder and remaining load parameters are as given in Example 2.3. The phase-a voltages and currents are shown in Figure 2.16. It can be seen that since the Load-2 current is dc, the source current also has a dc offset. The voltages both at Bus-2 and Bus-3 also have dc offsets. However, the voltage offset at Bus-2 is smaller compared to that of Bus-3. Further, the harmonic distortion at this bus is insignificant compared to that of Bus-3.

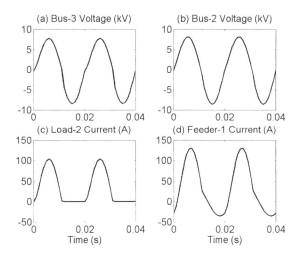

Figure 2.16. Effect of dc current in a distribution network

There are two main implications of the presence of a dc current in an electricity supply system. Usually a supply system is equipped with a transformer that changes the voltage levels in accordance with the need of the consumers. It was mentioned earlier that a dc current can offset the flux excursions in a distribution transformer. The positive flux excursion becomes heavily saturated while the negative excursion is well within the linear range. As a result the magnetic core of the transformer gets heavily saturated resulting in excessive heating.

The other aspect of the dc current is the earth path. The return path for a dc current can often involve current through the earth. This will sometimes involve the dc current passing through buried structures such as pipes or reinforced steel. The dc current greatly enhances corrosion of metallic structures as it carries the metallic ions in the direction of the current flow.

2.2.5 Unbalanced Loads

In a three-phase supply there is an expectation that the voltages in each phase will be equal in magnitude and are $120°$ phase shifted from each other. Now suppose Load-2 of Figure 2.14 is not balanced. The drawing of unbalanced current through supply impedance will mean that the supply voltage of the other two loads will also be unbalanced. For example consider the case in which Load-2 consists of three resistors of values 0.5 per unit, 1.0 per unit and 4.0 per unit in phases a, b and c respectively, while the two other loads and the feeder impedances remain unchanged. Then the voltages at Buses 2 and 3 are as shown in Figure 2.17. It can be seen that both these set of voltages are unbalanced due to the presence of the unbalanced loads. The degree of unbalance depends on the relative magnitude of the unbalanced currents drawn vis-à-vis that of the balanced currents drawn. The larger the unbalanced current, the larger is the unbalance.

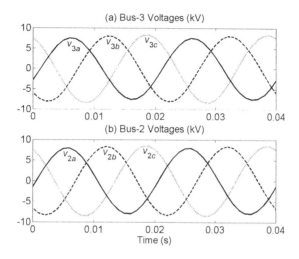

Figure 2.17. Unbalance in the bus voltages caused by unbalanced load

The voltage imbalance can be decomposed into a positive sequence voltage set, a negative sequence and a zero sequence voltage set. For induction motors, the positive sequence voltage set creates a positive torque that does the useful work. The negative sequence voltage set creates a flux rotating opposite to the rotor and creates a negative torque while the zero sequence voltage set may create current and extra losses but little effective torque. Thus the negative and zero sequence voltages generated due to voltage unbalance may give rise to extra losses and sometimes a torque reduction. Together these effects can contribute to overloading of induction motors.

Unbalanced loads in a three-phase system produce currents that give rise to negative phase sequence (NPS) voltages. The magnitude of the NPS voltage at a point of common coupling is usually limited by utilities because of the increased heating caused in three phase motors and generators. The permissible levels vary between countries but usually lies within the range of 1% to 2% [9-12]. The NPS voltage is defined in terms of the fundamental phase to neutral voltage phasors V_a, V_b and V_c as

$$V_{NPS} = \frac{\left(V_a + a^2 V_b + a V_c\right)}{\sqrt{3}} \qquad (2.2)$$

where $a = e^{j120°}$. In balanced systems this phasor summation would form a closed triangle and give $V_{NPS} = 0$.

Consider a three-phase induction motor having a single pole pair, the equivalent circuit of which is given in Figure 2.18. The positive sequence voltage creates a flux rotating in the positive sense at the fundamental frequency of 50 Hz. The rotor slip could be 2% for the rotor is moving in the positive direction at 49 Hz. A 2% NPS corresponds to a flux rotating in the reverse direction at a frequency of 99 Hz, i.e., at a slip of $s = 1.98$. Thus the model impedances are very close to slip of one, the condition for direct on-line starting. Typically start currents can be six times rated for 1.0 per unit input voltage. Thus a 2% NPS could give rise to a negative sequence current of 12% rated current. If the motor were already heavily loaded the additional current could give rise to overheating. This is in addition to the reverse torque on the rotor generated by the NPS flux.

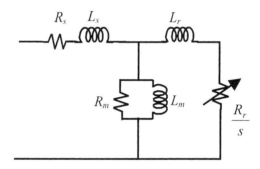

Figure 2.18. Equivalent circuit of an induction motor

The Australian Standard AS1359 [9] specifies that 3-phase machines should be designed for continuous voltage unbalance levels of 1.0%. This level is in agreement with the NEMA standard [10]. For the rail load in Queensland, the authority has previously allowed contributions from large unbalanced loads to the overall NPS at a point of common coupling (PCC) as follows [13]

- 2% NPS for 1-minute peak loads
- 1% NPS for 5 minute peak loads
- 0.7% NPS for 30 minute peak loads

The European practice is similar [11]. For example, in Germany the contributions from a single customer to the overall NPS at a point of

2. Characterization of Electric Power Quality

common coupling are constrained by the 10 minute geometric mean having to be less than 0.7%. This geometric mean is given by

$$NPS_{rms} = \sqrt{\frac{1}{T}\int_0^T nps(t)^2 dt} < 0.007 \qquad (2.3)$$

This represents the heating effect in a motor with a ten minute thermal time constant [14]. Compared to the levels indicated above, this criterion corresponds to an NPS_{rms} of 0.99% for five minutes and 2.21% for one minute [14].

The contribution, which an unbalanced load will make to the overall NPS voltage at a PCC, is readily calculated for fixed loads. Time varying loads appearing across different phase pairs make the prediction of NPS voltage levels more complex. Similarly the required rating of balancing plant, which ideally should utilize the unbalanced load absorption capability of the PCC, can become more difficult to assess.

Queensland Railways in Central Queensland in Australia operate an ac electrified railway. The railway is a heavy haul system comprising over 1000 km track and is used to transport coal from the inland mines to the export facilities on the east coast. Supply is provided from a 132 kV network via 13 railway substations, each of which has two or three 30 MVA single-phase, 132/50 kV transformers. Associated with each transformer is a 50 kV harmonic filter (HF) which may have a total rating of 4, 7 or 10 MVAr, depending on whether 3rd, 5th and 7th harmonic filter branches are included. Nine load balancing static var compensators (SVCs) are used to reduce the NPS voltages caused by the unbalanced loads, including the filters, to acceptable levels. The low fault levels (high source impedance) in most parts of the system used to supply the rail loads compound the problem of controlling NPS voltage levels.

An investigation into the capability of the railway supply substations was initiated after the rail authority, Queensland Railways, advised its intentions of significantly increasing the tonnage transported on the electrified system. The major concern was the containment of NPS voltage levels which is directly related to the ability of the nine load balancing SVCs to handle any increase in load. Figure 2.19 depicts one of the railway supply points of common coupling, which feeds two railway supply substations. Permanently connected 7 MVAr harmonic filters are installed on each of the 50 kV busbars and a 132 kV load balancing SVC is provided at one of the substations, Grantleigh. At Bouldercombe, the PCC, the fault level is approximately 1300 MVA, one of the highest among the railway supply points.

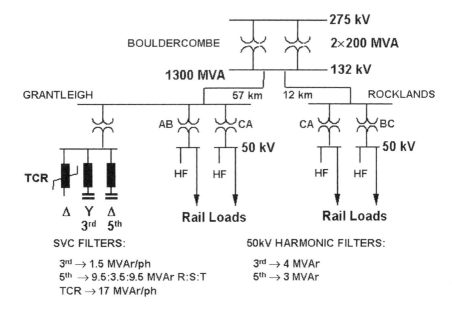

Figure 2.19. Two Queensland Railway substations supplied from a 132 kV bus

On site measurements of NPS levels, rail loads and SVC performance were made over a period of approximately a month. The rail loads were measured using single-phase power and reactive power transducers sampled at 15-second intervals. Figure 2.20 shows a typical single-phase traction load with a peak current of 450 A (50 kV busbar) corresponding to two loaded coal trains drawing full current simultaneously [15].

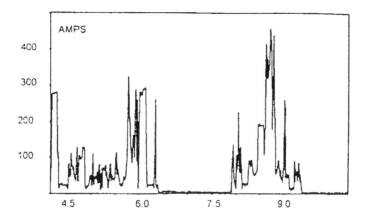

Figure 2.20. Typical load current supplied to a 50 kV traction load

2.Characterization of Electric Power Quality

With the load balancing SVC disabled, the NPS phasors at the point of common coupling, Bouldercombe 132 kV busbar, were found to lie on three trajectories at angles of approximately −60°, 60° and 180°. This result is expected since the power factors of the rail loads are similar, especially at high load. Excursions along these trajectories have been found to correlate well with single-phase trainloads in phase pairs AB, BC and CA respectively. Figure 2.21 shows the measured NPS excursions over a period of several hours. Note that the NPS measurements with high magnitude are aligned with the trajectories mentioned. The line segments connecting measurement points indicate time sequence.

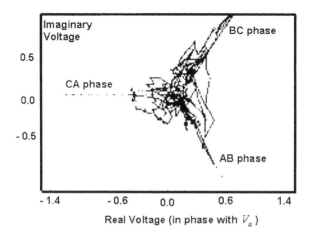

Figure 2.21. NPS voltage phasor measurement at Bouldercombe 132 kV busbar

The load character in phases AB and CA consists of comparatively short bursts of power corresponding to individual coal trains passing through the feeder sections. Phase BC has an additional component corresponding to more frequent but smaller loads due to local train traffic and shunting. This was observed to produce NPS voltage excursions along 60° trajectories, sometimes starting from part way down the − 60° or 180° trajectories of loads in the AB and CA phases. The additional loads are also responsible for the larger magnitudes of the NPS phasors in the BC phase supply. The generally low traffic density on the heavy haul rail system gives a low probability of coincident loads in adjacent sections, hence the lack of excursions in the region from − 60° to 180°.

To ascertain whether the NPS excursions were all caused by rail loading, the expected NPS currents due to rail loads were calculated using real and reactive power measurements at each railway supply transformer. The NPS current phasors showed excursions at angles of approximately 30°, 150° and

270° for loads in phases AB, BC, and CA respectively as shown in Figure 2.22. The currents have a phase shift of 90° with respect to the voltage due to the predominantly inductive nature of the source impedance. The NPS voltages were then determined by multiplying the NPS current with the NPS impedance where the source impedance was taken as the busbar short circuit level of 1300 MVA. Subtracting the measured NPS voltages and those calculated using the NPS current phasors and the estimated system NPS impedance leaves an NPS voltage phasor with a magnitude of approximately 0.15% and a steady phase angle. This residual NPS is thought to be due to transmission line unbalances or errors in voltage transformers and is the cause of the offsets from the origin in Figures 2.21 and 2.22. The rail load was found to be the dominant cause of NPS voltage excursions measured at this point of common coupling.

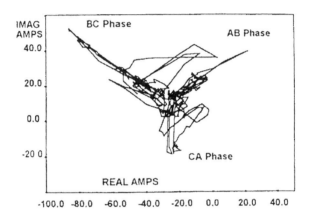

Figure 2.22. NPS Current phasor corresponding to Figure 2.20

2.2.6 Disturbance in Supply Voltage

As we have discussed before there can be various forms of disturbances in supply voltage such as interruption, distortion, overvoltage/undervoltage, Sag/Swell, flicker etc. These can have an adverse impact on the customers. For example, even a small duration voltage interruption can cause relay tripping, thereby completely stopping a process line. Many hours of production can be wasted through a few seconds of interruption. Even a short duration outage can cause defects in semiconductor processing. A sustained overvoltage can cause domestic lights to burn out faster and can put stress on capacitors. Voltage spikes or transient overvoltage can cause permanent damage on capacitors thereby burning power supply or other semiconductor components of computers, TVs, VCRs and household

2. Characterization of Electric Power Quality

appliances. Sustained undervoltage can cause motors to stall. Similarly a few cycle voltage sag can force motors to stop thereby ruining a process. Voltage flicker can be very annoying to the human eyes as it causes incandescent lamps to flicker. The impact of voltage disturbance on sensitive equipment is measured through the CBEMA curve discussed in the previous section.

Bonlac Foods Processing plant in Stanhope, Victoria, Australia processes diary milk into pasteurized milk, butter and cheese for high quality domestic and international consumption. The food processing plant is supplied by distributor Powercor Australia Ltd. Bonlac gets its supply from a 22 kV overhead line from Kyabram, with the incoming feeder to Kyabram being 66 kV. The number of faults in the Kyabram-Stanhope section rises during the summer due to storms and bird interference [16]. There are as many as 40 faults that occur in that section annually.

The plant equipment gets supply through six 22 kV/415 V transformers rated between 1 MVA and 1.5 MVA. The total load is approximately 5.25 MVA at 0.8 power factor. A large number of squirrel cage induction motors is used in this plant. These motors are used as evaporators/dryers or compressors. All these motors are sensitive to voltage dips, especially the motors running the sophisticated dryers. Each of the motor trips not only cause a loss of production but also a loss of raw material. Since this is a milk processing plant, the raw material cannot be recycled on the resumption of work as they may contaminate the new intake. Thus each voltage dip causes a huge loss of revenue.

In an interconnected distribution system, voltage disturbance can occur due to fault or badly behaved loads. For example in the distribution system shown in Figure 2.14, suppose Load-2 contains large induction motors. When the motor is started a large inrush current will flow, causing the Bus-3 voltage to drop in all three phases, thereby affecting Load-3. Also we have seen in Example 2.3 how a single-line-to-ground fault in one feeder causes the voltage to dip and rise in other feeders thereby affecting all loads connected to this feeder.

2.3 Conclusions

In this chapter we have discussed some of the problems that are facing the modern power system supply quality. There are a number of issues ranging from generation to transmission and distribution. In this book we concentrate on the last aspect. The aim of the book is to introduce the concepts of solid state power controllers that are utilized to compensate for some of the problems of distribution systems. The solid state solutions to these issues are covered in the later chapters. It is however to be noted that the solutions of all the problems facing modern power systems cannot solved using a few solid state controllers that are arbitrarily placed. The entire picture is very complicated. There are minor side issues that are also involved. We shall try to answer some of these questions as we go along.

2.4 References

[1] R. C. Dugan, M. F. McGranaghan and H. W. Beaty, *Electric Power Systems Quality*, McGraw-Hill, New York, 1996.
[2] Electrotek Concepts, "Power quality terms and definitions," *PQ Network Internet Site*: www.pqnet.electrotek.com/main/backgrnd, 1997.
[3] Electrotek Concepts, "Glossary of power quality terms," *PQ Network Internet Site*: www.pqnet.electrotek.com/main/backgrnd, 1997.
[4] D. Mueller and M. F. McGranaghan, "Effects of voltage sag in process industry applications," *PQ Network Internet Site*: www.pqnet.electrotek.com/main/backgrnd, 1997.
[5] T. Grebe, "Evaluation of utility capacitor switching transients," *PQ Network Internet Site*: www.pqnet.electrotek.com/main/backgrnd, 1997.
[6] N. Mohan, T. M. Undealand and W. P. Robbins, *Power Electronics: Converters, Applications and Design*, John Wiley, New York, 1989.
[7] W. McMurray, "Optimum snubbers for power semiconductors," *IEEE Industry Application Society Annual Meeting*, 1971.
[8] R. S. Thallam and G. T. Heydt, "Power acceptability and voltage sag indices in the three phase sense," *Panel Session on Power Quality – Voltage Sag Indices, IEEE Power Engineering Society Summer Meeting*, Seattle, 2000.
[9] Australian Standard AS1359.31 -1986, Rotating Electrical Machines - General Requirements, 1986.
[10] NEMA Standard MG1-14.34 (1980) and MG1-20.55 (1980).
[11] European Standard IEC 34-1: (1983), Rotating Electrical Machines, 1983.
[12] British Standard, BS4999, Specification of General Requirements for Rotating Electrical Machines, Part 101:1987, Specification for Rating and Performances, 1987.
[13] V. R. Sastry, D. V. Hill and C. J. Lee, "Static var compensators for balancing the single phase traction loads in central Queensland," *Journal Electrical & Electronic Engineering*, Australia - IE Aust & IREE Aust, Vol. 6, No. 3, 1986.
[14] G. Ledwich and T. A. George, "Using phasors to analyze power system negative phase sequence voltages caused by unbalanced loads," *IEEE Trans. Power Systems*, Vol. 9, No. 3, 1226-1232, 1994.
[15] P. Cummings, R. Kerr and J. Dunki-Jacobs, "Protection of induction motors against unbalanced voltage operation," *Proc. IEEE Industry Application Society Annual Meeting*, pp. 143-158, 1994.
[16] N. H. Woodley, A. Sundaram, B. Coulter and D. Morris, "Dynamic voltage restorer demonstration project experience," *12th Conf. Electric Power Supply Industry (CEPSI)*, Pattaya, Thailand, 1998.

Chapter 3

Analysis and Conventional Mitigation Methods

Power quality problems are not new in power systems, but the general customers' awareness of these problems has increased in the recent years. Modern technology such as computers and controls are largely responsible for the rise in the impacts of power quality but can also provide a tailor-made solution to these problems. Often these solutions are expensive, and in many cases, the cost has to be borne by the customer. Thus before the application of a power quality solution, the problem has to be analyzed in details and the cost to benefit ratio must also be calculated. The specifics of the analysis of power quality problems are an important issue and will be covered in this chapter.

Since power quality problems have existed for a long time, the conventional methods of mitigation of these problems also are quite well developed. For example, before the advent of active filters, passive filters based on inductors and capacitors were used and are still used in many power transmission and distribution applications. Some of these filters developed to high levels of sophistication and are even tuned to bypass specific harmonic frequencies. However, the use of passive elements at high power level makes these devices bulky. Moreover the passive filters have a fixed range of operation. Therefore before we introduce the custom power solution to power quality problems, it is important to discuss the conventional mitigation methods and highlight their deficiencies as well.

3.1 Analysis of Power Outages

The most common cause of an outage is equipment or component failure, e.g., loss of a generator, transformer or feeder due to faults. Sometimes utilities used scheduled outages to maintain the power equipment. Typical scheduled maintenance involves changing of transformer oil, replacement of a section of feeder conductors or changing of old and faulty switchgear or other equipment. During scheduled maintenance, a power distribution company may be able to cater to the large majority of the customers by channeling power through alternate feeders or supply transformers wherever available. However this may not always be possible. In general, such scheduled outages occur only occasionally and usually prior notice is given to customers that are affected by the outages.

It is the unscheduled outages that cause major problems to both utilities and customers alike. Such outages cause higher financial loss to the customers arising from loss of production in factories and assembly lines, rotting or contamination of edible materials in food processing plants, restaurants and even domestic households. The impact of even short outages in semiconductor plants can be very severe. It is therefore imperative that such outages are minimized. Amongst the unscheduled outages, some are caused by natural disasters and accidents like earthquakes, floods, blizzards, tornadoes, fires, arsons, terrorist activities etc. Even if some of these causes can be predicted, it is rather difficult to entirely prevent their impact on the power system. We shall therefore concentrate on the outages resulting from faults and equipment failures.

There are various reliability indices that define the response of the system to the outages. Below we define a few of them [1].

System Average Interruption Frequency Index (SAIFI): This defines the total number of customer interruption events that have occurred over a period of time (usually one year) divided by the total number of customers, i.e.,

$$SAIFI = \frac{\text{Total number of customer interruptions}}{\text{Total number of customers in the system}} \qquad (3.1)$$

This defines the average interruptions per customer over a year.

This defines the average number of interruptions per customer over a year.

Customer Average Interruption Frequency Index (CAIFI): This is defined as

$$CAIFI = \frac{\text{Number of customer interruptions}}{\text{Number of customers who had at least one interruption}} \qquad (3.2)$$

3. Analysis and Conventional Mitigation Methods

The index *SAIFI* is useful in that it gives the average interruptions per customer. The problem with this approach is that not all customers in the system face an equal amount of interruptions. For example *SAIFI* may produce an index of 1.5 in a year. However it may happen that only one quarter of the people suffered these interruptions. Then the average interruption for these customers is 6.0 and not 1.5. This aspect is addressed by the use of the index *CAIFI* which normalizes the number of interruptions with respect to the total number of customers who have faced interruptions. The numerical value of *CAIFI* will be greater than or equal to that of *SAIFI*.

It is interesting to note that a comparison of these two indices can give us an insight into the system. For example if the relative difference between these two indices is negligible, then it can be concluded that the interruptions have affected most groups of customers equally. If, on the other hand, there is a large difference between these indices, then it means that the interruptions have affected some groups of customers more than the others. This can be due to poor grounding, poor design or poor maintenance. Further investigation will then be required to determine and rectify the cause.

System Average Interruption Duration Index (SAIDI): This defines the average duration of all interruptions per customer, i.e.,

$$SAIDI = \frac{\text{Sum total of the duration of all customer interruptions}}{\text{Total number of customers in the system}} \quad (3.3)$$

In this index the sum total of the duration of interruptions of all customers are normalized with respect to the total number of customers.

Customer Average Interruption Duration Index (CAIDI): The total interruption duration over a year is averaged amongst the customers, who had at least one interruption, i.e.,

$$CAIDI = \frac{\text{Sum total of the duration of all customer interruptions}}{\text{Number of customers with at least one interruption}} \quad (3.4)$$

As in the case of *SAIFI* and *CAIFI*, a large difference between *SAIDI* and *CAIDI* will indicate that the outages are concentrated on a limited set of customers and hence further investigation will be required.

Momentary Average Interruption Frequency Index (MAIFI): This index deals with momentary or short duration interruptions. In general the utilities do not treat the short duration interruptions as outages and hence momentary interruptions are not classified under *SAIFI* or *CAIFI*. The momentary index is computed as

$$MAIFI = \frac{\text{Number of customer momentary interruptions}}{\text{Total number of customers}} \quad (3.5)$$

We demonstrate the calculation of the interruptions with the help of the following example. This follows the guidelines provided in [2].

Example 3.1: Consider a distribution system with 100,000 customers. These customers are served from six different buses. These buses and the number of customers per bus are listed in Table 3.1. The actual configuration of the distribution system is not important. The system interruption data over a year is listed in Table 3.2. Note from this table that for the interruption case 2, 10,000 customers of bus 2 and 2,000 of bus 4 are affected for different duration. Again in the interruption case 5 that involves bus 4, the same 2,000 customers involved in case 2 are affected again. In addition another 3,000 customers are also affected for the interruption case 5.

Table 3.1. Distribution system data

Bus number	Number of customers served
1	30,000
2	25,000
3	20,000
4	12,000
5	8,000
6	5,000

Table 3.2. Customer interruption data

Interruption case	Bus number	Affected customers	Duration (hours)
1	1	15,000	2.0
2	2	10,000	1.5
	4	2,000	1.0
3	6	5,000	4.0
4	5	3,000	0.5
5	4	5,000	1.0

Therefore the total number of customer interruptions, calculated from the data given in Table 3.2 is

$$\text{Total interruptions} = (15 + 10 + 2 + 5 + 3 + 5) \times 10^3 = 40,000$$

Hence,

3. Analysis and Conventional Mitigation Methods

$$SAIFI = \frac{40,000}{100,000} = 0.4$$

To compute *CAIFI* we note that the total number of customers affected by the interruptions is 38,000. Therefore,

$$CAIFI = \frac{40,000}{38,000} = 1.05$$

Note that *CAIFI* has a numerical value that is greater than 1. This implies that some customers have undergone more number of outages than the others per year.

To compute *SAIDI* and *CAIDI* we have to translate the interruption data given in Table 3.2 into customer-minutes lost due to interruptions. This is shown in Table 3.3. From this table we compute the customer-minutes lost due to interruptions as

$$\text{Customer - minutes lost} = (1.8 + 0.9 + 0.12 + 1.2 + 0.09 + 0.3) \times 10^6$$
$$= 4,410,000$$

Therefore

$$SAIDI = \frac{4,410,000}{100,000} = 44.1 \text{ minutes}$$

$$CAIDI = \frac{4,410,000}{38,000} = 116.05 \text{ minutes}$$

Table 3.3. Customer-minutes lost due to interruptions

Interruption case	Bus number	Affected customers	Duration (hours)	Customer-minutes
1	1	15,000	2.0	1,800,000
2	2	10,000	1.5	900,000
	4	2,000	1.0	120,000
3	6	5,000	4.0	1,200,000
4	5	3,000	0.5	90,000
5	4	5,000	1.0	300,000

ΔΔΔ

Note from the above example that

$$SAIFI \leq CAIFI, \quad SAIDI \leq CAIDI, \quad CAIFI \geq 1 \tag{3.6}$$

Also note that

$$\frac{CAIDI}{SAIDI} = \frac{CAIFI}{SAIFI} \tag{3.7}$$

= Fraction of customers who had at least one outage

The frequency indices like *SAIFI*, *CAIFI* and *MAIFI* tell us how often faults occur. They give us an indication about system equipment and network layout. The regulator of utility can declare a maximum limit on any of these indices as the key performance measure and the utility can respond by rescheduling their maintenance procedure to be within the maximum limit. Other approaches are to use live line work to limit the outages experienced. The duration indices like *SAIDI* and *CAIDI*, on the other hand, are functions of the organization ability of the utility to limit the faulted section to the smallest number of customers and the ability to control the repair time. These indices can be used to identify when it is critical to reschedule the repair procedures of the utility such that the load curtailment can be kept at the minimum.

3.2   Analysis of Unbalance

Historically unbalance in a three-phase ac system has always been treated through symmetrical components. In this approach, a set of unbalanced ac voltage or current phasors is converted to three balanced phasors. Also an unbalanced ac network can be decomposed into three sequence networks.

3.2.1   Symmetrical Components of Phasor Quantities

Symmetrical components are used to analyze unbalanced conditions in three-phase circuits in the steady state. It is well known that a set of three unbalanced phasors representing either three-phase voltages or three-phase currents can be resolved into the following three sets of three balanced phasors:

− Positive sequence: These are a set of equal magnitude three phase vectors that are displaced from each other by 120° and have the same phase sequence as the original phasors. The positive sequence components of

3. Analysis and Conventional Mitigation Methods

the voltage phasors V_a, V_b and V_c are usually denoted by V_{a1}, V_{b1} and V_{c1} respectively. The currents are also defined similarly.

- Negative sequence: These are a set of equal magnitude three phase vectors that are displaced from each other by 120° and have the opposite phase sequence to the original phasors. The negative sequence components of the voltage phasors V_a, V_b and V_c are usually denoted by V_{a2}, V_{b2} and V_{c2} respectively.
- Zero sequence: These are a set of equal magnitude three phase vectors that are exactly in phase with each other. The zero sequence voltage components are usually denoted by V_{a0}, V_{b0} and V_{c0}.

Symmetrical components are defined in terms of the operator a that is given as

$$a = e^{j120°} \tag{3.8}$$

We can then write

$$a^2 = e^{j240°}, \ a^3 = 1, \ a^4 = a \text{ and } 1 + a + a^2 = 0$$

Let us define the following vectors

$$V_{abc}^T = \begin{bmatrix} V_a & V_b & V_c \end{bmatrix} \text{ and } V_{a012}^T = \begin{bmatrix} V_{a0} & V_{a1} & V_{a2} \end{bmatrix}$$

Then transformation from abc to 012-plane is given by

$$V_{a012} = \begin{bmatrix} V_{a0} \\ V_{a1} \\ V_{a2} \end{bmatrix} = \frac{1}{K} \begin{bmatrix} 1 & 1 & 1 \\ 1 & a & a^2 \\ 1 & a^2 & a \end{bmatrix} \begin{bmatrix} V_a \\ V_b \\ V_c \end{bmatrix} = PV_{abc} \tag{3.9}$$

where K is a constant that is chosen either 1/3 or 1/√3. Note that we can also transform currents using a similar transform, i.e.,

$$I_{a012} = \begin{bmatrix} I_{a0} \\ I_{a1} \\ I_{a2} \end{bmatrix} = \frac{1}{K} \begin{bmatrix} 1 & 1 & 1 \\ 1 & a & a^2 \\ 1 & a^2 & a \end{bmatrix} \begin{bmatrix} I_a \\ I_b \\ I_c \end{bmatrix} = PI_{abc} \tag{3.10}$$

The three-phase power in the original unbalanced system is given by

$$P_{abc} + jQ_{abc} = V_a I_a^* + V_b I_b^* + V_c I_c^* \qquad (3.11)$$

where I^* is the complex conjugate of the vector I. Now from (3.9) and (3.10) we get

$$V_{a0} I_{a0}^* + V_{a1} I_{a1}^* + V_{a2} I_{a2}^* = \frac{3}{K^2}\left[V_a I_a^* + V_b I_b^* + V_c I_c^*\right] \qquad (3.12)$$

From (3.12) there are two ways of choosing the value of K. If the three-phase complex power is to be equated to the sum of the complex powers in the transformed system, i.e.,

$$P_{abc} + jQ_{abc} = \sum_{i=a,b,c}\left(V_{i0} I_{i0}^* + V_{i1} I_{i1}^* + V_{i2} I_{i2}^*\right)$$

then the value of K must be equal to 3. The transformation matrix and its inverse is then given by

$$P = \frac{1}{3}\begin{bmatrix} 1 & 1 & 1 \\ 1 & a & a^2 \\ 1 & a^2 & a \end{bmatrix}, \quad P^{-1} = \begin{bmatrix} 1 & 1 & 1 \\ 1 & a^2 & a \\ 1 & a & a^2 \end{bmatrix} \qquad (3.13)$$

Alternatively if we choose K to be equal to $\sqrt{3}$, both sides of (3.12) become equal, i.e.,

$$P_{abc} + jQ_{abc} = V_{a0} I_{a0}^* + V_{a1} I_{a1}^* + V_{a2} I_{a2}^* = V_{a012}^T I_{a012}^* \qquad (3.14)$$

The transformation matrix and its inverse is then given by

$$P = \frac{1}{\sqrt{3}}\begin{bmatrix} 1 & 1 & 1 \\ 1 & a & a^2 \\ 1 & a^2 & a \end{bmatrix}, \quad P^{-1} = \frac{1}{\sqrt{3}}\begin{bmatrix} 1 & 1 & 1 \\ 1 & a^2 & a \\ 1 & a & a^2 \end{bmatrix} \qquad (3.15)$$

In the above equation $P^T P^* = I_3$, where I_3 is a (3×3) identity matrix. This implies that the matrix P is unitary. Further this transformation has the advantage that the three-phase power can be computed based only on the phase-a of the sequence components.

3. Analysis and Conventional Mitigation Methods

Example 3.2: Let us consider a three-phase balanced source supplying an unbalanced load. The supply voltages and the load currents are given in per unit by

$$V_a = 1.0\angle 0°, \quad V_b = 1.0\angle -120°, \quad V_c = 1.0\angle 120°, \quad I_a = 0.55\angle -30°,$$
$$I_b = 0.45\angle -160° \text{ and } I_c = 0.60\angle 95°$$

The complex power is then given by

$$P_{abc} + jQ_{abc} = 1.3648 + j0.8178 \text{ per unit}$$

Let us now investigate the power in the sequence circuits. The zero-sequence power can be computed as

$$P_0 + Q_0 = V_{a0}I_{a0}^* = \frac{1}{3}(V_a + V_b + V_c)(I_a + I_b + I_c)^*$$

Since the supply voltage is balanced, $V_a + V_b + V_c = 0$ and hence both real and reactive powers in the zero-sequence are zero. As $1 + a + a^2 = 0$, the power in the negative sequence

$$P_2 + Q_2 = V_{a2}I_{a2}^* = \frac{1}{3}(V_a + a^2V_b + aV_c)(I_a^* + aI_b^* + a^2I_c^*) = 0$$

It is then needless to say that the total power in the positive-sequence circuit is equal to the power in the three-phase circuit, i.e.,

$$P_1 + jQ_1 = V_{a1}I_{a1}^* = 1.3648 + j0.8178 \text{ per unit}$$

ΔΔΔ

It can thus be seen that powers in the zero and negative sequence circuits are zero when the supply voltage is balanced but the current is not. Similarly, we can show that the power in these sequence components will be zero when the current is balanced but the supply voltage is unbalanced. Let us now consider what happens when both the voltages and currents are unbalanced.

Example 3.3: Let us consider a system in which the supply voltages are unbalanced and are given in per unit by

$V_a = 1.05\angle 0°$, $V_b = 1.0\angle -130°$ and $V_c = 0.95\angle 120°$

The loads are such that the load currents are the same as that given in Example 3.2. The complex power is then

$$P_{abc} + jQ_{abc} = 1.4064 + j0.7546 \text{ per unit}$$

The zero, positive and negative sequence powers are then

$$P_0 + jQ_0 = 0.0032 + j0.0038 \text{ per unit}$$
$$P_1 + jQ_1 = 1.4052 + j0.7347 \text{ per unit}$$
$$P_2 + jQ_2 = -0.0020 + j0.0161 \text{ per unit}$$

It can be seen that

$$P_{abc} + jQ_{abc} = P_{a012} + jQ_{a012} = P_0 + jQ_0 + P_1 + jQ_1 + P_2 + jQ_2$$

△△△

To analyze an unbalanced network, it is decomposed into three sequence networks. These three sequence networks are then combined in a particular fashion to determine the fault currents. For example, for a single-line to ground fault, the Thevenin equivalent of the three networks are placed in series. These details are given in any textbook on power system (e.g., [3]). Also note that any transformer connection has an important role to play in the constitution of the zero-sequence network. The zero-sequence network of a grounded Y-Y transformer is different than that of an ungrounded Y-Y transformer or a Δ-Δ transformer. The details of the zero-sequence components of different transformer connections are also given in [3].

3.2.2 Instantaneous Symmetrical Components

The symmetrical component transformation matrix can also be applied to instantaneous voltages and currents [4]. Through this transformation we can convert three-phase instantaneous quantities into a zero-sequence component and two phasor components. Let the instantaneous three-phase currents be given by I_a, I_b and I_c. We can then define the instantaneous symmetrical components as

3. Analysis and Conventional Mitigation Methods

$$\begin{bmatrix} i_{a0} \\ i_{a1} \\ i_{a2} \end{bmatrix} = \frac{1}{\sqrt{3}} \begin{bmatrix} 1 & 1 & 1 \\ 1 & a & a^2 \\ 1 & a^2 & a \end{bmatrix} \begin{bmatrix} i_a \\ i_b \\ i_c \end{bmatrix} \qquad (3.16)$$

where I_{a0} is the zero-sequence and I_{a1} and I_{a2} are two vectors that are complex conjugates of each other. A similar transformation can also be defined for voltages. Note that through the instantaneous symmetrical components we can analyze waveforms that are distorted, unbalanced or purely fundamental. To illustrate the idea, let us consider the following example.

Example 3.4: Let us consider a balanced three-phase supply voltage given in per unit by

$$v_a = \sin \omega t, \ v_b = \sin(\omega t - 120°) \text{ and } v_c = \sin(\omega t + 120°)$$

where $\omega = 100\pi$ rad/s. First let us assume that it supplies a balanced load. The load currents are given in per unit by

$$i_a = 0.5 \sin(\omega t - 30°), \ i_b = 0.5 \sin(\omega t - 150°) \text{ and } i_c = 0.5 \sin(\omega t + 90°)$$

The vectors v_{a1} and I_{a1} are computed for one cycle and the loci of the tip of these vectors are plotted in Figure 3.1. The starting point of the vectors at $t = 0$ are marked by a cross ($\times$) and their direction of rotation as t increases are also indicated in the figure. It can be seen that these loci form a closed path. Since both the supply voltage and load currents are balanced, the paths traced by them form circles. The magnitude of the circles are however different. Note that since the vector v_{a2} (I_{a2}) is the complex conjugate of v_{a1} (I_{a1}), the path traced by v_{a2} (I_{a2}) will also be a circle that overlaps the path traced by v_{a1} (I_{a1}), even though these two vectors move in the opposite directions.

To compute the magnitude of this vector let us assume that the peak of the balanced voltages is given by V_m. Then the vector v_{a1} is given by

$$\begin{aligned} v_{a1} &= \frac{1}{\sqrt{3}} \{v_a + av_b + a^2 v_c\} \\ &= \frac{1}{\sqrt{3}} \left\{ v_a - \frac{1}{2}(v_b + v_c) + j\frac{\sqrt{3}}{2}(v_b - v_c) \right\} \end{aligned} \qquad (3.17)$$

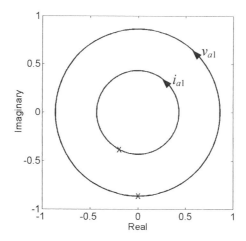

Figure 3.1. Loci of the positive sequence vectors under balanced condition

Since the voltages are balanced, $v_a + v_b + v_c = 0$. We then have

$$v_a - \frac{1}{2}(v_b + v_c) = \frac{3}{2}v_a = \frac{3V_m}{2}\sin \omega t$$

$$v_b - v_c = V_m\{\sin(\omega t - 120°) - \sin(\omega t + 120°)\}$$

$$= -2V_m \cos \omega t \sin 120° = -\sqrt{3}\,V_m \cos \omega t$$

Substituting the above two equations in (3.17) we get

$$v_{a1} = \frac{V_m}{\sqrt{3}}\left\{\frac{3}{2}\sin \omega t - j\frac{3}{2}\cos \omega t\right\} \tag{3.18}$$

The magnitude of the voltage vector v_{a1} is given from (3.18) as

$$|v_{a1}| = |v_{a2}| = \frac{\sqrt{3}\,V_m}{2} \tag{3.19}$$

The magnitude of the voltage vector of Figure 3.1 is then √3/2, i.e., 0.866 and that of the current vector is 0.433. Also note that the vector I_{a1} lags the vector v_{a1} by 30° for every value of t. This implies that the phase angle difference between these two vectors is the power factor angle.

Let us now assume that the load currents are unbalanced and are given in per unit as

3. Analysis and Conventional Mitigation Methods

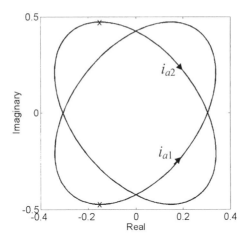

Figure 3.2. Loci of positive and negative sequence vectors of unbalanced currents

$i_a = 0.5 \sin \omega t$, $i_b = 0.2 \sin(\omega t - 100°)$ and $i_c = 0.75 \sin(\omega t + 90°)$

The loci of the vectors I_{a1} and I_{a2} are plotted in Figure 3.2. In this figure, the direction of rotation of the loci is also indicated. Since the currents are unbalanced, the paths traced by I_{a1} and I_{a2} are ellipses. Moreover, the vector I_{a1} rotates in the counterclockwise direction while the vector I_{a2} rotates in the clockwise direction. The starting point of each of these vectors is indicated by a cross.

ΔΔΔ

3.2.3   Instantaneous Real and Reactive Powers

In this section we shall define the real and reactive powers in a three-phase system in terms of the instantaneous voltages and currents. Let us define the instantaneous voltage and current vectors for the three-phase instantaneous quantities as

$$v_{abc}^T = [v_a \quad v_b \quad v_c] \text{ and } i_{abc}^T = [i_a \quad i_b \quad i_c]$$

The instantaneous active (real) power is defined as the dot (scalar) product of these two vectors, i.e.,

$$p = v_{abc} \cdot i_{abc} = v_a i_a + v_b i_b + v_c i_c \qquad (3.20)$$

The instantaneous reactive power is a vector that is defined as the cross product of the vectors v_{abc} and I_{abc} [5], i.e.,

$$q = v_{abc} \times i_{abc} = \begin{bmatrix} q_a \\ q_b \\ q_c \end{bmatrix} = \begin{bmatrix} \begin{vmatrix} v_b & v_c \\ i_b & i_c \end{vmatrix} \\ \begin{vmatrix} v_c & v_a \\ i_c & i_a \end{vmatrix} \\ \begin{vmatrix} v_a & v_b \\ i_a & i_b \end{vmatrix} \end{bmatrix} \quad (3.21)$$

The norm of the vector

$$\|q\| = \sqrt{q_a^2 + q_b^2 + q_c^2} \quad (3.22)$$

may sometimes be used as the scalar representing the instantaneous reactive power. Alternatively, the algebraic sum

$$q_{sum} = \frac{q_a + q_b + q_c}{\sqrt{3}} \quad (3.23)$$

can also be used as a scalar representing the total reactive power that circulates in the three phases. An advantage of the representation of (3.23) is that it can indicate the polarity of the instantaneous reactive power unlike $\|q\|$, which is always positive.

The instantaneous active current vector is defined in terms of the power p as

$$i_p = \begin{bmatrix} i_{ap} \\ i_{bp} \\ i_{cp} \end{bmatrix} = \frac{p}{v_{abc} \cdot v_{abc}} v_{abc} \quad (3.24)$$

Similarly, the instantaneous reactive current vector is defined in term of the instantaneous reactive power vector q as

3. Analysis and Conventional Mitigation Methods

$$i_q = \begin{bmatrix} i_{aq} \\ i_{bq} \\ i_{cq} \end{bmatrix} = \frac{q \times v_{abc}}{v_{abc} \cdot v_{abc}} \qquad (3.25)$$

The total current vector is the sum of the active current vector and the reactive current vector, i.e.,

$$i_{abc} = i_p + i_q \qquad (3.26)$$

Further note that $v_{abc} \cdot I_q = 0$, i.e., I_q is orthogonal to v_{abc}. Also, $v_{abc} \times I_p = 0$, i.e., I_p is parallel to v_{abc}.

The instantaneous apparent power s is a scalar defined as

$$s = \|v_{abc}\| \|i_{abc}\| \qquad (3.27)$$

where $\|v_{abc}\|$ and $\|I_{abc}\|$ are the instantaneous vector norms defined as

$$\|v_{abc}\| = \sqrt{v_a^2 + v_b^2 + v_c^2}, \quad \|i_{abc}\| = \sqrt{i_a^2 + i_b^2 + i_c^2}$$

The instantaneous power factor λ can then be defined as the ratio of active and apparent powers, i.e.,

$$\lambda = \frac{p}{s} \qquad (3.28)$$

Let us now define a vector containing the instantaneous symmetrical component vectors as

$$v_{a012}^T = \begin{bmatrix} v_{a0} & v_{a1} & v_{a2} \end{bmatrix} \quad \text{and} \quad i_{a012}^T = \begin{bmatrix} i_{a0} & i_{a1} & i_{a2} \end{bmatrix}$$

The instantaneous voltage and current vectors can then be written as

$$v_{abc} = P^{-1} v_{a012} \quad \text{and} \quad i_{abc} = P^{-1} i_{a012} \qquad (3.29)$$

where the inverse of the matrix P is given in (3.15). Combining (3.20) and (3.29) we get

$$p = v_{a012}^T P^{-T} P^{-1} i_{a012} = v_{a012}^T \begin{bmatrix} 1 & 0 & 0 \\ 0 & 0 & 1 \\ 0 & 1 & 0 \end{bmatrix} i_{a012} \quad (3.30)$$

$$= v_{a0} i_{a0} + v_{a2} i_{a1} + v_{a1} i_{a2}$$

Since v_{a2} and v_{a1} are complex conjugate of each other and also I_{a2} and I_{a1} are complex conjugate of each other, (3.30) can be written as

$$p = v_{a0} i_{a0} + 2 \operatorname{Re}\{v_{a1} i_{a1}^*\} \quad (3.31)$$

where Re denotes the real part.

In a similar way the reactive power is given by [6]

$$q = \begin{bmatrix} q_a \\ q_b \\ q_c \end{bmatrix} = -\frac{2}{\sqrt{3}} \operatorname{Im} \begin{bmatrix} v_{a1} i_{a1}^* \\ v_{a1} i_{a1}^* \\ v_{a1} i_{a1}^* \end{bmatrix} - \frac{2 v_{a0}}{\sqrt{3}} \operatorname{Im} \begin{bmatrix} i_{a1} \\ i_{a1} e^{-j2\pi/3} \\ i_{a1} e^{+j2\pi/3} \end{bmatrix}$$

$$+ \frac{2 i_{a0}}{\sqrt{3}} \operatorname{Im} \begin{bmatrix} v_{a1} \\ v_{a1} e^{-j2\pi/3} \\ v_{a1} e^{+j2\pi/3} \end{bmatrix} \quad (3.32)$$

where Im denotes the imaginary parts. From the above equation the instantaneous scalar reactive power is given by

$$q_{sum} = \frac{q_a + q_b + q_c}{\sqrt{3}} = -2 \operatorname{Im}\{v_{a1} i_{a1}^*\} \quad (3.33)$$

It can be seen from (3.33) that the zero sequence components do not contribute to q_{sum}. Also the last two terms of (3.32) do not contribute to q_{sum} and only the summation of the first term of (3.32) appears on the right hand side of (3.33).

In general both p and q_{sum} have two components. They are given by

$$\begin{aligned} p &= p_{av} + p_{osc} \\ q_{sum} &= q_{sum,av} + q_{sum,osc} \end{aligned} \quad (3.34)$$

In the above equation, the subscript *av* denotes the average (dc) value and the subscript *osc* denotes the oscillating component that has an oscillation

3. Analysis and Conventional Mitigation Methods

frequency of 100 Hz when the system frequency is 50 Hz with no distortion. In a balanced circuit the oscillating components are zero.

Example 3.5: Let us consider a set of balanced supply voltages that are given in per unit by

$$v_a = \sqrt{2}\sin\omega t,\ v_b = \sqrt{2}\sin(\omega t - 120°)\ \text{and}\ v_c = \sqrt{2}\sin(\omega t + 120°)$$

where $\omega = 100\pi$ rad/s. The source supplies an unbalanced load. The steady state load currents are given in per unit as

$$i_a = \frac{\sqrt{2}}{2}\sin(\omega t - 30°),\ i_b = \frac{3\sqrt{2}}{4}\sin(\omega t - 150°),\ i_c = \frac{\sqrt{2}}{4}\sin(\omega t + 90°)$$

The real power computed from (3.31) and the total reactive power computed from (3.33) are plotted in Fig. 3.3. It can be seen that these quantities have a frequency of 100 Hz. Their mean values are also plotted in this figure. The average real power supplied to the load is 1.5 cos 30° = 1.299 per unit. Similarly the average reactive power required by the load is equal to 1.5 sin (– 30°) = – 0.75 per unit. It can be seen that the mean values given in Figure 3.3 are the same as these values. We can therefore conclude that q_{sum} is indeed the scalar instantaneous reactive power.

△△△

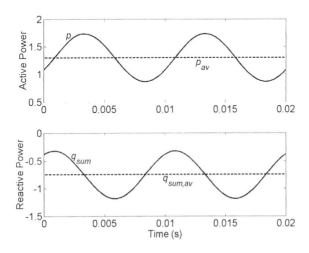

Figure 3.3. Instantaneous active and reactive powers in an unbalanced circuit

3.3 Analysis of Distortion

The main causes of voltage and current waveform distortions are harmonics, notching and interharmonics. We have discussed harmonics and notching in Chapter 2. Interharmonics are defined as the frequencies that are not integer multiples of the fundamental frequency but are present in the voltage or current waveforms [7]. These can appear as discrete frequencies or as wide-band spectrum. Let the fundamental frequency of a signal be f_0. Then the terms such as harmonics, dc offset, sub-harmonics and interharmonics have the frequency domain interpretations shown in Table 3.4. In this table the term f denotes a particular frequency component of the signal. The term sub-harmonic is loosely called as interharmonics with frequency component less than the fundamental frequency.

Table 3.4. Interpretations of various harmonic terms

Term	A frequency component
Harmonics	$f = n \times f_0$, for an integer $n > 0$
DC offset	$f = n \times f_0$ for $n = 0$
Interharmonics	$f = n \times f_0$ for a non-integer $n > 0$
Sub-harmonics	$0 < f < f_0$

The rise in the use of power electronic loads and the increasing use of power factor correction capacitors is causing a general rise in the level of harmonics with particular locations exhibiting strong effects on other customers. For many purposes the distorting loads can be modeled as harmonic current sources. When there is a finite source impedance at the nth harmonic $z(n\omega_0)$ then the nth harmonic current flowing to the source will generate a harmonic voltage at the connection point of common coupling resulting in voltage distortion that can affect other customers.

The main impacts can be summarized as

- Increased losses
- Reduced equipment life
- Interference with protection control and communication circuits
- Interference with customer equipment

The increased losses can be in induction motors where the harmonic voltages can create significant harmonic currents without providing a commensurate increase in useful torque. These increased losses may not cause immediate equipment failure but can result in increased temperature in equipment which can be reliably mapped to a reduced life particularly for electrical insulation. Some control equipment presumes the existence of sinusoidal voltage supplies and takes its timing reference from the zero

3. Analysis and Conventional Mitigation Methods

crossing of the supply voltage. When there are variable distorting loads present this can give rise to a variation of the zero crossing and hence the control output. An example of this was a newly installed dc drive on a lift connected from the same transformer as the thyristor controlled lighting for the main stage. The distortion from the lift drive caused changes in the voltage supplied to the rest of the loads resulting in significant flicker of the stage lighting if the lift were used during a performance.

As mentioned earlier the harmonics in power systems are mainly caused by power electronic loads like UPS adjustable speed drives etc. Most of these produce harmonics and some of them produce dc offsets. The main cause of interharmonics is the cycloconverters [8]. These converters are used extensively in rolling mills and linear motor drives in cement and mining industries and their maximum size can be as high as 20 MW. The arc loads like arc furnace, welding machines and arc lamps also produce interharmonics. These are also the most common source of voltage flickers that can be construed as fluctuations due to low frequency interharmonic components. The other sources of interharmonics are power line carrier signals, induction motors and integral cycle control that are extensively used to control street lighting and hot water or other controlled customer loads [8].

There are many effects of interharmonics. Similar to the harmonic currents they also cause overheating. The other effects are flicker in TV picture tubes, torsional oscillations in a turbine-generator shaft, communication interference, ripple control (power line carrier) interference and CT saturation [8].

It is well known that the distortions in the voltage or current from the fundamental frequency sinewave can be represented as a superposition of all the harmonic frequency sinewaves on the fundamental sinewave. The harmonic component values are separated from the fundamental frequency components using the Fourier analysis. Fourier series is used to separate the frequency components of periodic but non-sinusoidal waveforms. Through the use of this series, the fundamental component, the dc component and the harmonic components that are integer multiples can be separated.

Using the exponential form of Fourier series, a periodic signal $x(t)$ can be expressed as [9]

$$x(t) = \sum_{k=-\infty}^{k=\infty} c_k e^{jk\omega t} \qquad (3.35)$$

where ω is the fundamental frequency in rad/s and the coefficients c_k, $k = -\infty, \ldots, \infty$ are given by

$$c_k = \frac{1}{T_0} \int_{T_0} x(t) e^{-jk\omega t} dt \qquad (3.36)$$

To illustrate, let us consider a signal given by

$$x(t) = 0.3 + 1.5 \sin \omega t + 0.5 \sin 3\omega t + 0.3 \sin 5\omega t \qquad (3.37)$$

where $\omega = 100\pi$. Let us choose T_0 to be 20 ms, i.e., one cycle. Then

$$c_0 = \frac{1}{2\pi} \int_0^{2\pi} x(t) d\omega t = 0.3$$

Note that the last three terms of (3.37) do not contribute to the average in determining c_0.

In order to determine the kth harmonic, the first step is to replace the sinusoidal terms in (3.37) by an exponential form, i.e.,

$$\sin k\omega t = \frac{e^{jk\omega t} - e^{-jk\omega t}}{2j} \qquad (3.38)$$

This form, when substituted in (3.36), results in one constant term at frequency $k\omega$ and exponentials in the integrand. Clearly, only the constant term contributes to c_k. Therefore

$$c_k = -c_{-k} = \frac{1}{2} \frac{M_k}{j} \qquad (3.39)$$

where M_k is the magnitude of the kth harmonic component. Then

$$c_1 = -c_{-1} = \frac{1.5}{j2}, \quad c_3 = -c_{-3} = \frac{0.5}{j6}, \quad c_5 = -c_{-5} = \frac{0.3}{j10}$$

Therefore from (3.35) we get

$$x(t) = 0.3 + \frac{1.5}{j2}\left(e^{j\omega t} - e^{-j\omega t}\right) + \frac{0.5}{j2}\left(e^{j3\omega t} - e^{-j3\omega t}\right) + \frac{0.3}{j2}\left(e^{j5\omega t} - e^{-j5\omega t}\right)$$

3. Analysis and Conventional Mitigation Methods

For a general function it can be shown that c_{-k} is the complex conjugate of c_k [9]. This is the original $x(t)$ given in (3.37). Note that $x(t)$ did not have a dc term, the integration interval T_0 can be reduced to 10 ms, i.e., half a cycle. Further note that the sub harmonics that have frequency components ω/n for integer n can also be extracted using the series (3.35) by choosing T_0 to be integer multiples of 20 ms.

The complex Fourier series given in (3.35) is related to the usual form as given below

$$x(t) = \frac{a_0}{T} + \frac{2}{T}\sum_{k=1}^{\infty}\{a_k \cos\omega t + b_k \sin\omega t\} \tag{3.40}$$

$$c_k = a_k - jb_k \tag{3.41}$$

The sine and cosine terms in (3.40) can be combined into a convenient single sinusoidal term by modifying (3.36) as

$$c'_k = jc_k = \frac{1}{T_0}\int_{T_0} x(t)e^{-jk(\omega t - \pi/2)}dt \tag{3.42}$$

The amplitude and phase of c'_k is the amplitude and phase of the combined sinusoidal term.

Fourier transform can be applied to waveforms that may or may not be periodic. For a non-periodic waveform this will produce a frequency spectrum that is continuous with no fundamental component. For periodic waveforms the application of Fourier transform will result in a spectrum containing the fundamental and the harmonic components contained in the signal. Fourier transforms are numerically evaluated on a digital computer using numerical methods like discrete Fourier transform (DFT) or fast Fourier transform (FFT).

Interharmonics have the potential to interfere with a power system but are rather difficult to detect even using FFT. For example consider the voltage waveform

$$\begin{aligned}v(t) = {}&0.2 + \sin(\omega t) + 0.6\sin(40\pi t) + 0.3\sin(222\pi t) \\ &+ 0.4\sin(246\pi t) + 0.2\sin(288\pi t) + 0.5\sin(5\omega t)\end{aligned} \tag{3.43}$$

where $\omega = 100\pi$. This implies that the waveform, in addition to the fundamental, contains dc, 5th harmonic, a sub-harmonic at 20 Hz and interharmonics at 111 Hz, 123 Hz and 144 Hz. The waveform is shown in Figure 3.4 (a). We now use an FFT algorithm to find the spectrum of this waveform. This is shown in Figure 3.4 (b). It can be seen that the detection of the dc and the interharmonic components are rather inaccurate here. In fact the proper use of FFT to detect these components is rather an art. In [8] an example is given in which a Hanning window is used to limit the time duration over which the waveform is observed and an FFT algorithm is used subsequently with a four-fold zero padding. This example is repeated in [10] in which the FFT and windowing algorithms are also discussed and shows how the appropriate values can be extracted.

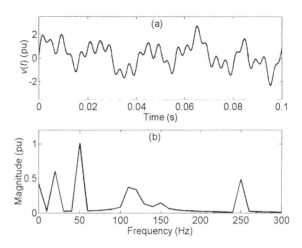

Figure 3.4. (a) Waveform containing interharmonics and (b) its spectrum

3.3.1 On-line Extraction of Fundamental Sequence Components from Measured Samples

The sequence components are sinusoidal steady state quantities. They can thus be obtained from the measurements of peak values and the phase angles once the voltages or currents are in steady state. These quantities however are required in many applications for on-line synthesis of voltages or currents. In this section we shall present an algorithm to obtain these components through sampling and averaging the instantaneous voltages or currents.

Let us define a set of three unbalanced voltage phasor terms as

$$V_a = \frac{V_{ma}}{\sqrt{2}} \angle \phi_a, \ V_b = \frac{V_{mb}}{\sqrt{2}} \angle \phi_b \text{ and } V_c = \frac{V_{mc}}{\sqrt{2}} \angle \phi_c$$

3. Analysis and Conventional Mitigation Methods

Their instantaneous components are then

$$v_a = V_{ma}\sin(\omega t + \phi_a), \quad v_b = V_{mb}\sin(\omega t + \phi_b) \text{ and } v_c = V_{mc}\sin(\omega t + \phi_c)$$

The phasor symmetrical components of these quantities can then be written from (3.9) as

$$\begin{bmatrix} V_{a0} \\ V_{a1} \\ V_{a2} \end{bmatrix} = \frac{1}{\sqrt{3}}\begin{bmatrix} 1 & 1 & 1 \\ 1 & a & a^2 \\ 1 & a^2 & a \end{bmatrix}\begin{bmatrix} V_a \\ V_b \\ V_c \end{bmatrix} = \frac{1}{\sqrt{6}}\begin{bmatrix} V_{ma}e^{j\phi_a} + V_{mb}e^{j\phi_b} + V_{mc}e^{j\phi_c} \\ V_{ma}e^{j\phi_a} + aV_{mb}e^{j\phi_b} + a^2V_{mc}e^{j\phi_c} \\ V_{ma}e^{j\phi_a} + a^2V_{mb}e^{j\phi_b} + aV_{mc}e^{j\phi_c} \end{bmatrix}$$

(3.44)

We shall now use the instantaneous symmetrical component transformation given in (3.16) on the instantaneous quantities and use (3.42) to obtain the relations given in (3.44) [11]. From (3.42) it can be seen that the fundamental component of a signal $x(t)$ is given by

$$c'_1 = \frac{1}{T_0}\int_{T_0} x(t)e^{-j(\omega t - \pi/2)}dt \tag{3.45}$$

Let us first consider the zero-sequence. From (3.16) we get

$$v_{a0} = \frac{1}{\sqrt{3}}\{v_a + v_b + v_c\}$$
$$= \frac{1}{\sqrt{3}}\{V_{ma}\sin(\omega t + \phi_a) + V_{mb}\sin(\omega t + \phi_b) + V_{mc}\sin(\omega t + \phi_c)\} \tag{3.46}$$

We now use (3.45) with $x(t) = v_{a0}$. Then we have

$$c_{a0} = \frac{1}{T_0\sqrt{3}}\int_0^{T_0}\{V_{ma}\sin(\omega t + \phi_a) + V_{mb}\sin(\omega t + \phi_b) + V_{mc}\sin(\omega t + \phi_c)\}(\sin\omega t + j\cos\omega t)dt \tag{3.47}$$

The integrand of (3.47) can be expanded as

$$\frac{1}{2}[V_{ma}\{\cos\phi_a - \cos(2\omega t + \phi_a)\} + V_{mb}\{\cos\phi_b - \cos(2\omega t + \phi_b)\}$$

$$+ V_{mc}\{\cos\phi_c - \cos(2\omega t + \phi_c)\}] + \frac{j}{2}[V_{ma}\{\sin\phi_a + \sin(2\omega t + \phi_a)\}$$

$$+ V_{mb}\{\sin\phi_b - \sin(2\omega t + \phi_b)\} + V_{mc}\{\sin\phi_c - \sin(2\omega t + \phi_c)\}]$$

Now let us choose T_0 such that the integrals of the double frequency terms are zero. We then get

$$c_{a0} = \frac{1}{T_0 2\sqrt{3}} \int_0^{T_0} \{V_{ma}e^{j\phi_a} + V_{mb}e^{j\phi_b} + V_{mc}e^{j\phi_c}\} dt$$

$$= \frac{1}{2\sqrt{3}} \{V_{ma}e^{j\phi_a} + V_{mb}e^{j\phi_b} + V_{mc}e^{j\phi_c}\}$$

(3.48)

Comparing the first row of (3.44) with (3.48) we can write

$$V_{a0} = \sqrt{2}\, c_{a0} \qquad (3.49)$$

Let us now consider the positive sequence. From (3.44) we get

$$v_{a1} = \frac{1}{\sqrt{3}}\{v_a + av_b + a^2 v_c\}$$

$$= \frac{1}{\sqrt{3}}\{V_{ma}\sin(\omega t + \phi_a) + aV_{mb}\sin(\omega t + \phi_b) + a^2 V_{mc}\sin(\omega t + \phi_c)\}$$

(3.50)

Again using (3.45) the fundamental component of the above equation is

$$c_{a1} = \frac{1}{T_0\sqrt{3}} \int_0^{T_0} \{V_{ma}\sin(\omega t + \phi_a) + aV_{mb}\sin(\omega t + \phi_b)$$

$$+ a^2 V_{mc}\sin(\omega t + \phi_c)\}(\sin\omega t + j\cos\omega t)dt$$

(3.51)

The integrand of (3.51) can be expanded as

$$\frac{1}{2}[V_{ma}\{\cos\phi_a - \cos(2\omega t + \phi_a)\} + aV_{mb}\{\cos\phi_b - \cos(2\omega t + \phi_b)\}$$
$$+ a^2 V_{mc}\{\cos\phi_c - \cos(2\omega t + \phi_c)\}] + \frac{j}{2}[V_{ma}\{\sin\phi_a + \sin(2\omega t + \phi_a)\}$$
$$+ aV_{mb}\{\sin\phi_b - \sin(2\omega t + \phi_b)\} + a^2 V_{mc}\{\sin\phi_c - \sin(2\omega t + \phi_c)\}]$$

Again since the integrals of the double frequency terms are zero, we have

$$c_{a1} = \frac{1}{T_0 2\sqrt{3}} \int_0^{T_0} \{V_{ma}e^{j\phi_a} + aV_{mb}e^{j\phi_b} + a^2 V_{mc}e^{j\phi_c}\} dt$$
$$= \frac{1}{2\sqrt{3}} \{V_{ma}e^{j\phi_a} + aV_{mb}e^{j\phi_b} + a^2 V_{mc}e^{j\phi_c}\} \tag{3.52}$$

Comparing the second row of (3.44) with (3.52) we get

$$V_{a1} = \sqrt{2}\, c_{a1} \tag{3.53}$$

Proceeding in the same way as above we get show that

$$V_{a2} = \sqrt{2}\, c_{a2} \tag{3.54}$$

where

$$c_{a2} = \frac{1}{T_0} \int_{T_0} v_{a2} e^{-j(\omega t - \pi/2)} dt$$
$$= \frac{1}{2\sqrt{3}} \{V_{ma}e^{j\phi_a} + a^2 V_{mb}e^{j\phi_b} + aV_{mc}e^{j\phi_c}\} \tag{3.55}$$

We can then summarize the following from (3.49), (3.53) and (3.54)

$$V_{a0} = \sqrt{2}\, c_{a0} = \frac{\sqrt{2}}{T_0} \int_{T_0} v_{a0} e^{-j(\omega t - \pi/2)} dt$$
$$V_{a1} = \sqrt{2}\, c_{a1} = \frac{\sqrt{2}}{T_0} \int_{T_0} v_{a1} e^{-j(\omega t - \pi/2)} dt \tag{3.56}$$
$$V_{a2} = \sqrt{2}\, c_{a2} = \frac{\sqrt{2}}{T_0} \int_{T_0} v_{a2} e^{-j(\omega t - \pi/2)} dt$$

Note that the sequence component extraction algorithm of (3.56) use the Fourier integral. Therefore, it can also extract the fundamental of the sequence components even when the system contains harmonics or dc offset. The choice of the time T_0 becomes very crucial in such cases. We shall demonstrate the idea with the help of the following examples.

Example 3.6: Let us consider the unbalanced three-phase voltages that are given in the phasor form as

$$V_a = 1\angle 0°, \; V_b = 1.2\angle -110° \text{ and } V_c = 0.85\angle 100°$$

Their symmetrical component transformation (3.9) leads to the following

$$\begin{bmatrix} V_{a0} \\ V_{a1} \\ V_{a2} \end{bmatrix} = \begin{bmatrix} 0.2552 - j0.1677 \\ 1.7208 - j0.0475 \\ -0.2439 + j0.2153 \end{bmatrix} = \begin{bmatrix} 0.3054\angle -33.32° \\ 1.7215\angle -1.58° \\ 0.3253\angle 138.57° \end{bmatrix}$$

The instantaneous voltages corresponding to the above voltage phasors are

$$v_a = \sqrt{2}\sin\omega t, \quad v_b = 1.2 \times \sqrt{2}\sin(\omega t - 110°)$$
$$v_c = 0.85 \times \sqrt{2}\sin(\omega t + 100°)$$

with $\omega = 100\pi$. The coefficients c_{a0}, c_{a1} and c_{a2} are then given by

$$\begin{bmatrix} c_{a0} \\ c_{a1} \\ c_{a2} \end{bmatrix} = \begin{bmatrix} 0.1804 - j0.1186 \\ 1.2168 - j0.0336 \\ -0.1725 + j0.1522 \end{bmatrix} = \begin{bmatrix} 0.2159\angle -33.32° \\ 1.2173\angle -1.58° \\ 0.2300\angle 138.57° \end{bmatrix}$$

It can be seen that by multiplying the coefficients by $\sqrt{2}$ we can get the phasor form of symmetrical components calculated above. Note that the above coefficients do not require the use of instantaneous samples. They can be obtained directly from magnitude and phase as seen from (3.48), (3.52) and (3.55). The next example illustrates the use of instantaneous samples for the case when the voltages are distorted.

△△△

3. Analysis and Conventional Mitigation Methods

Example 3.7: Let us now add 5th and 7th harmonic components to the fundamental voltages with magnitudes that are inversely proportional to their harmonic numbers. In addition, a 50% dc component has been added to the phase-a voltage. Therefore the three voltages are given by the following instantaneous equations

$$v_a = 0.5 + \sqrt{2}\left\{\sin \omega t + \frac{1}{5}\sin 5\omega t + \frac{1}{7}\sin 7\omega t\right\}$$

$$v_b = 1.2 \times \sqrt{2}\left\{\sin(\omega t - 110°) + \frac{1}{5}\sin 5(\omega t - 110°) + \frac{1}{7}\sin 7(\omega t - 110°)\right\}$$

$$v_c = 0.85 \times \sqrt{2}\left\{\sin(\omega t + 100°) + \frac{1}{5}\sin 5(\omega t + 100°) + \frac{1}{7}\sin 7(\omega t + 100°)\right\}$$

We now extract the fundamental rms sequence components using (3.56) with the period T_0 being equal to 10 ms (i.e., half the time required by a 50 Hz cycle). Thus we obtain the sequence components based on the samples of the previous half cycle at every 10 ms. We can then obtain the fundamental rms voltage of each phase through inverse symmetrical component transform. The instantaneous fundamental voltages can be reconstructed during every half cycle based on the latest estimates. This is shown in Figure 3.5 in which the actual and the extracted fundamental voltages are plotted. The effect of filtering out the dc component is clearly seen in phase-a.

ΔΔΔ

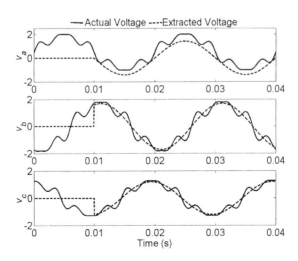

Figure 3.5. Fundamental reconstruction using half-cycle averaging

The procedure described above is based on the samples of the previous half cycle. It has a 10 ms delay in estimation and a maximum delay of another 10 ms during reconstruction. Therefore it is not suitable for on-line tracking of fundamental components. The algorithm can be improved by using moving average in the calculation of the integrals in (3.56). The following example illustrates this.

Example 3.8: Consider the integrals of (3.56). Suppose the integration interval of 10 ms is divided into N equal sub-intervals and samples of the phase voltages are obtained at each of these sample points to compute the integrand values of (3.56). Then the integrations are performed by computing the average values. Now note that the integration can be performed between any two points in the power cycles that are 10 ms apart. This is the basis of moving average filtering. In this the values of V_{a0}, V_{a1} and V_{a2} are computed as per (3.56) at each sampling instant with the data of the last N integrand sample values. Since this is a continuous process, the settling time is just half cycle for any change in the voltages.

To illustrate the idea, let us assume the same voltages that are given in Example 3.7. At the end of 2 cycles, the voltages are changed to

$$v_a = 0.5 \times \sqrt{2} \left\{ \sin \omega t + \frac{1}{5} \sin 5\omega t + \frac{1}{7} \sin 7\omega t \right\}$$

$$v_b = 0.7 \times \sqrt{2} \left\{ \sin(\omega t - 110°) + \frac{1}{5} \sin 5(\omega t - 110°) + \frac{1}{7} \sin 7(\omega t - 110°) \right\}$$

$$v_c = 1.3 \times \sqrt{2} \left\{ \sin(\omega t + 100°) + \frac{1}{5} \sin 5(\omega t + 100°) + \frac{1}{7} \sin 7(\omega t + 100°) \right\}$$

The actual and reconstructed voltages are shown in Figure 3.6. It can be seen that the reconstructed waveforms settle within half a cycle after the voltages are changed. Also note that for the first half cycle the waveforms are not reconstructed as during this time measurements are gathered.

∆∆∆

Interharmonics are of potential concern to a power system. It has been mentioned before that they cannot be easily detected using FFT. Also the fundamental component extraction process discussed above will fail in their presence. The following example illustrates the idea.

Example 3.9: Let us consider a set of balanced voltage waveforms that are laced with interharmonics. The voltages are given by

3. Analysis and Conventional Mitigation Methods

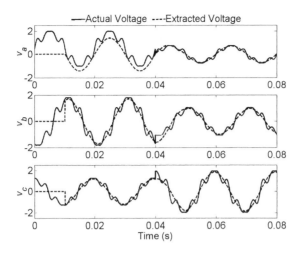

Figure 3.6. Fundamental reconstruction using moving average filtering

$$v_a = \sqrt{2}\{\sin \omega t + 0.3\sin 2.4\omega t + 0.5\sin 3.8\omega t\}$$
$$v_b = \sqrt{2}\{\sin(\omega t - 120°) + 0.3\sin 2.4(\omega t - 120°) + 0.5\sin 3.8(\omega t - 120°)\}$$
$$v_c = \sqrt{2}\{\sin(\omega t + 120°) + 0.3\sin 2.4(\omega t + 120°) + 0.5\sin 3.8(\omega t + 120°)\}$$

The fundamental sequence component extraction algorithm of (3.56) is now applied in which the integral is computed through moving average filtering. The averaging window is taken to be 10 ms. The results are shown in Figure 3.7. It can be seen that none of the reconstructed waveforms are sinusoidal.

The problem with interharmonics is that they are not integer multiples of the fundamental frequency. Therefore the integrals of the resulting frequency components do not vanish in half a cycle. In fact the waveforms given above have frequency components of 50 Hz, 120 Hz and 190 Hz. The least common multiplier of these frequencies is 11400 Hz, i.e., 228 times a 50 Hz cycle. Therefore the averaging time for this case must be chosen to be 114 cycles, i.e., 2.28 s. Clearly this will result in a significant delay in obtaining the average following any change. Furthermore, the interharmonic frequency components are not known a priori. Therefore it is also not possible to choose a suitable averaging time that will eliminate all these frequencies to produce the fundamental.

ΔΔΔ

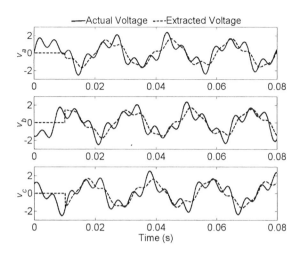

Figure 3.7. Fundamental reconstruction in the presence of interharmonics

3.3.2 Harmonic Indices

The amplitude of harmonic contents in a current or voltage signal is characterized by the total harmonic distortion (*THD*). The percentage total harmonic distortion of a voltage signal with the rms value of V_1 and rms harmonic contents of V_n, $n = 2, ..., \infty$ is given in (2.1). In per unit form the *THD* is given by

$$THD = \frac{\sqrt{\sum_{n=2}^{\infty} V_n^2}}{V_1} \qquad (3.57)$$

The problem with this approach is that the *THD* become infinity if no fundamental is present. One way to avoid this ambiguity is to use an alternate definition that represents the harmonic distortion. This is called the distortion index (*DIN*) and is defined as [12]

$$DIN = \frac{\sqrt{\sum_{n=2}^{\infty} V_n^2}}{\sqrt{\sum_{n=1}^{\infty} V_n^2}} \qquad (3.58)$$

3. Analysis and Conventional Mitigation Methods

It can be seen that the *THD* and *DIN* are interrelated by the following equations

$$DIN = \frac{THD}{\sqrt{1+THD^2}} \qquad (3.59)$$

$$THD = \frac{DIN}{\sqrt{1-DIN^2}} \qquad (3.60)$$

The following example illustrates their relations.

Example 3.10: The waveform of Figure 2.10 (a) contains 50 Hz fundamental, plus 3rd, 5th, 7th, 9th and 11th harmonics with their magnitudes being reciprocal of their harmonic numbers. In Chapter 2 its *THD* is calculated as 43.83%, i.e., 0.4383 per unit. The *DIN* is then given by

$$DIN = \frac{\sqrt{\left(\frac{1}{3}\right)^2+\left(\frac{1}{5}\right)^2+\left(\frac{1}{7}\right)^2+\left(\frac{1}{9}\right)^2+\left(\frac{1}{11}\right)^2}}{\sqrt{1+\left(\frac{1}{3}\right)^2+\left(\frac{1}{5}\right)^2+\left(\frac{1}{7}\right)^2+\left(\frac{1}{9}\right)^2+\left(\frac{1}{11}\right)^2}} = 0.4015$$

It can be seen that the values of *THD* and *DIN* obtained are in accordance to (3.59) and (3.60).

Let us now consider a waveform that, in addition to 1.0 per unit fundamental, has a 7th harmonic whose magnitude is 1/7. The *THD* and *DIN* for this waveform respectively are 0.1429 and 0.1414. It can be seen that these values are very close to each other. Through the Taylor's series expansion of (3.59) and (3.60) it has been shown in [12] that the values of *THD* and *DIN* are almost the same when the harmonic distortion is low.

ΔΔΔ

Since both these indices quantify the harmonic distortion their use is a matter of preference. For example the IEEE prefers the use of *THD* while the European bodies like IEC prefer *DIN*. The detection of *THD* in the presence of interharmonics in not straightforward. We can rewrite (3.57) as

$$THD = \frac{\sqrt{V_{rms}^2 - V_1^2}}{V_1} \tag{3.61}$$

This equation remains valid even in the presence of interharmonics. However any waveform in the presence of interharmonics is not strictly periodic especially if the harmonic frequency is irrational. Thus the evaluation of (3.61) depends on finding the approximate value of the rms voltage V_{rms} of the non-periodic waveform. However if a large number of cycles is considered over which the waveform is nearly periodic for the calculation of the rms value, then this will approximately be equal to the true rms value. Then (3.61) will give an almost correct indication of the total harmonic distortion.

Power factor (*PF*) is a well known quantity to all electrical engineers. It is defined as

$$PF = \frac{P}{V_{rms} I_{rms}} \tag{3.62}$$

This quantity cannot however be used in a case when the voltage and current waveforms are not sinusoidal. In such a case a quantity like the displacement factor (*DF*) may be of interest. This is defined by [12]

$$DF = \cos\phi_1 = \frac{P_1}{|V_1||I_1|} \tag{3.63}$$

where the subscript 1 refers to the fundamental components only. It can be seen that the displacement factor is equal to the power factor for the sinusoidal case. However for non-sinusoidal cases, the following inequality holds [12]

$$PF \leq DF$$

3.4 Analysis of Voltage Sag

Voltage dips are experienced when other customers share a common supply impedance with an over current event on the supply system or in customer premises. There are several aspects to this problem. Let us consider them in turn.

At the low voltage supply a frequent cause of complaint is dips caused by motor starts in neighboring premises. Direct on line starts of certain classes of motor in appliances such as air conditioners can provide significant voltage drops for fractions of a second. The severity of the dip is high when a low voltage transformer is rated to supply a very few customers. A higher rating transformer feeding a larger number of customers reduces the depth of the voltage dip but affects more customers.

Faults in low voltage supply to customers have a very similar effect as the motor starts. The duration of the disturbance is now determined by the fuse characteristic in the customer premises. A fault on a higher voltage supply to customers such as an 11kV connection can affect a much larger number of customers. Once again the grading of the protection between the local supply on customer premises and the feeder protection should mean that the duration is determined by customer equipment.

Faults on the feeder can be initiated by

- lightning strike,
- trees or branches falling on conductors,
- animals across lines,
- wind causing conductors to clash together and
- digging equipment breaking cables.

The statistics of overhead lines indicate that

- 70% of the faults are single line to ground.
- 15% of the faults are double line to ground.
- 10% of the faults are line to line.
- 5% of the faults are three phase faults.

The low impedance faults are referred to as bolted faults indicating that the faulted conductors are effectively bolted together. A much more common effect is where the fault has some finite impedance. When a line falls on sandy soil or there is a significant distance for an arc to jump, then the characteristic may have a constant voltage characteristic. In each half cycle there is no conduction until the voltage has risen sufficiently to cause conduction and this continues until the voltage falls below this level. During this time the voltage at the fault point is roughly constant. The process repeats itself on each half cycle. There is a degree of variability of the separations and arc path, which means that the fault voltage may vary in each half cycle and may demonstrate restriking.

There are many loads that cannot tolerate voltage sags like adjustable speed drives, computers, programmable logic controllers etc. The ASDs are used in many process industries. If these drives freeze due to voltage sags, then that can result in the termination of the process in which these drives are employed. These may result in the fall of the quality of the product or simply in the loss of raw material that cannot be used again. Therefore these undesirable events must be avoided.

The voltage sag events must be qualified for the calculation of an index. Usually a sag is said to have occurred if the rms voltage in any one phase has fallen below 75% of the nominal value [13]. A sag below 10% of the nominal value is considered as an interruption and this has to be qualified using the outage indices discussed earlier in the chapter. Many methods have been proposed for the development of a composite index of all voltage sag events during a defined period. Below we shall discuss a few of them.

3.4.1 Detroit Edison Sag Score

The Detroit Edison sag score (*SS*) is defined as [13]

$$SS = \frac{V_A + V_B + V_C}{3} \qquad (3.64)$$

where V_A, V_B and V_C are the rms values of the phase voltages in per unit. Even though this method is very simple to use, it does not consider the duration of the voltage sag. We have seen from the voltage tolerance level curves (e.g. CBEMA curve) in Chapter 2 that the time duration of any sag or swell event is an important parameter to quantify the impact of these events on a particular load. Therefore despite its simplicity the sag score cannot indicate the damage a sag event may have on a particular load.

3.4.2 Voltage Sag Energy

The voltage sag energy is defined as [14]

$$E_{VS} = \int_0^T \left\{ 1 - \frac{V(t)}{V_{nom}} \right\}^2 dt \qquad (3.65)$$

where V is the magnitude of the voltage and V_{nom} is the nominal voltage and T duration of the sag.

3.4.3 Voltage Sag Lost Energy Index (VSLEI)

This index gives the lost energy during a sag event as [13]

3. Analysis and Conventional Mitigation Methods

$$W = \left\{1 - \frac{V}{V_{nom}}\right\}^{3.14} \times T \qquad (3.66)$$

where V is the phase voltage in per unit of nominal voltage during a sag event and T sag duration in milliseconds. Note that factor 3.14 associated with the power of voltage is derived from the CBEMA curve. This is done by applying the least squares curve fitting of the log plot of the CBEMA curve [13]. For a three-phase sag event, (3.66) can be modified to include the unequal duration sag of all the phases as

$$W = \left\{1 - \frac{V_a}{V_{nom}}\right\}^{3.14} \times T_a + \left\{1 - \frac{V_b}{V_{nom}}\right\}^{3.14} \times T_b + \left\{1 - \frac{V_c}{V_{nom}}\right\}^{3.14} \times T_c \quad (3.67)$$

where V_a, V_b and V_c are the voltage of the three phases and T_a, T_b and T_c are their respective duration of sag events. Consider the following example.

Example 3.11: Let us consider a set of three voltages that are phase displaced by 120°. The voltages start with a peak value of $\sqrt{2}$ per unit when an unbalanced sag occurs. The instant of occurrence of the sag in the three phases is not the same, neither is the time when the sag disappears. Also the magnitudes of the sag in the three phases are not the same. The peak values of the voltage in phases a, b and c during the sag are $\sqrt{2} \times 0.72$ per unit, $\sqrt{2} \times 0.9$ per unit and $\sqrt{2} \times 0.65$ per unit respectively. The voltages are shown in Figure 3.8 (a).

To compute VSLEI (W) from (3.67) we require the rms value of the phase voltage and the duration of the sag. The rms value can be computed from

$$V_{rms} = \sqrt{\frac{1}{T_0} \int_0^{T_0} v^2(t)\, dt} \qquad (3.68)$$

where T_0 is the time period over which the integral is computed. Note that the rms value can be calculated by obtaining the moving average of the square of the instantaneous voltage in which the time window T_0 can be selected as 10 ms (half a cycle). As we have mentioned before this enables the rms measurement to settle to the steady state value exactly half a cycle following any change in the voltage. The moving averaged rms values of the three phase voltage are shown in Figure 3.8 (b-d).

Notice that rms voltage starts changing as soon as the sag occurs. However it takes 10 ms to reach the steady state value. We can therefore take minimal value as the sag voltage. We assume that the inception of the sag is the instant at which the moving rms voltage starts falling from the steady state value of 1.0 per unit. Also the sag ends at the instant in which the rms value starts climbing towards the steady state value of 1.0 per unit. The time period of the sag is then the time difference between these two instants.

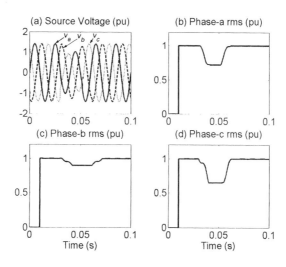

Figure 3.8. Voltage sag and its moving averaged rms measurement

Using the above simple logic the time duration of the sag is computed as 20.5 ms, 29.6 ms and 22.7 ms. Therefore we get the following from (3.67)

$$W = (1 - 0.72)^{3.14} \times 20.5 + (1 - 0.9)^{3.14} \times 29.6 + (1 - 0.65)^{3.14} \times 22.7$$
$$= 1.2382$$

Note that the complexities involved with the calculation of the duration of the sag make the Detroit Edison sag score easier to use. In this case the score will be one third of the sum of 0.72, 0.9 and 0.65, i.e., 0.7567.

△△△

3.5 Analysis of Voltage Flicker

Voltage flicker is a problem of human perception. Usually the deviation in the flickering voltage is much less than the threshold of susceptibility of

3. Analysis and Conventional Mitigation Methods

the electrical equipment. It is therefore unlikely that such voltage can cause damage to any equipment. However a flickering voltage will cause a continuous deviation in the light intensity of incandescent lamps. This variation is perceived as disturbing by customers particularly in the range of 3 to 15 times per second.

The main cause of voltage flicker is the arc loads like arc furnace, arc welder and arc lamp. Also starting of large motors can cause the voltage to flicker, as this will require a large inrush current during the starting. Electric arc furnaces are time varying and non-linear loads that generate random perturbation at the point of common coupling with the electric utility [15]. Such furnaces are usually used for steel melting and these units can be very big in size (100 MW or above). Usually a melting cycle can be divided into three distinct steps – drilling period, melting period and reheating period [15]. During the drilling period, reduced voltage and power are required for a short duration to reduce the material to scrap. The arc length varies irregularly during this period. The full voltage and power is given during the melting period for melting the scrap. The time required for melting is relatively large. Lower voltage and power is required during the reheat process when the arc lengths are shorter. Due to uneven arc length during the entire melting cycle, the bus voltage fluctuates continuously during the melting cycle.

A typical fluctuating voltage is shown in Figure 3.9 (a). This is the voltage across a non-linear and time varying load. The load current is shown in Figure 3.9 (b). It can be seen that load current continuously varies and the load voltage continuously fluctuates.

The IEC standard 61000-3-7 [16] presents a three-step procedure for evaluating loads. The first step is an "automatic acceptance" procedure that can be applied to assess the impact of a potential customer without detailed analysis. Table 3.5 shows the criteria for medium voltage (MV) connections for 1 kV < MV < 35kV which specifies the maximum allowable ratio of load power variation ΔS to the available short circuit power S_{SC} as a function of the fluctuation rate. Fluctuating loads connected directly to a high voltage (HV) supply for 35 kV < HV < 230 kV can be accepted without further study provided the ratio $S_{max}/S_{SC} < 0.1\%$ where S_{max} is the maximum load power.

The consequence of these flicker standards is to encourage connection of potentially disturbing loads to higher voltage supplies. For large arc furnaces there is a limit to the extent that the connection point selection can solve all the problems. Reactive voltage controllers may then be required to correct the voltage within the specified limits.

Table 3.5. Maximum possible load variations for automatic acceptance of MV loads

Number of variations per minute (R)	$\Delta S/S_{SC}$
$R > 200$	0.1
$10 < R < 200$	0.2
$R < 10$	0.4

Power quality problems are as old as power distribution through feeders. Therefore at least the partial mitigation of these problems existed even before the advent of power electronic controllers. We shall call these conventional mitigation techniques. We shall now discuss some of these techniques. However it must be appreciated that these conventional techniques are not flexible and in no way can match the performance of the power electronic controllers.

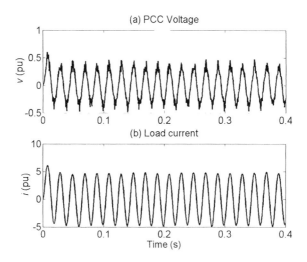

Figure 3.9. Voltage flicker due to time varying, non-linear load

3.6 Reduced Duration and Customer Impact of Outages

The duration of an outage and the number of customers involved are key parameters to address when seeking to improve power quality. The conventional approach to reduce impact of faults is by the use of reclosers and provision of alternate paths of supply. When a fault occurs on a radial feeder the feeder breaker needs to open for safety and to prevent equipment damage. This leaves every customer on that feeder without supply. Up to 70% of faults on overhead lines are transient in nature and if a circuit breaker recloses there is a good chance that the lines that clashed have separated or the branch that fell on the lines has fallen to ground. A common practice in Australia is to have 2 attempts to reclose first at a short interval

3. Analysis and Conventional Mitigation Methods

less than 5 seconds and next at an interval up to 30 seconds. If this fails then the circuit is locked out until manual line inspection can occur. This use of reclosing can significantly reduce the customer minutes lost for a distribution company but its customers will still experience an outage which will stop their computers, fax machines and will have clocks on ovens, microwaves, stereos, VCRs, clock radios all flashing.

For permanent faults in urban areas air break switches (ABS) and Normally Open Tie Switches (NOTS) are the main tools to restore supply. The ABS are shown as 'X' on Figure 3.10. The NOTS connecting to feeders from the same substation are shown in lower case while NOTS to feeders from other substations (not shown in the figure) are shown as Capitals.

The aim is to determine the faulted segment and open air break switches on either side of the fault. The main circuit breaker can now be reclosed and the supply restored to the upstream side. In many cases a connection to another intact feeder from the same or nearby substation has been designed into the system, thus supply can be restored to customers downstream of the faulted section. Once the faulted section is repaired it usually requires another break in supply while the original configuration is restored.

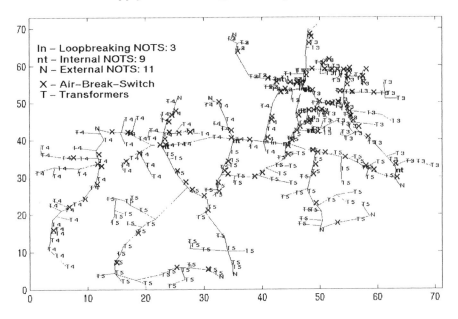

Figure 3.10. System Topology diagram for an 11kV Net

3.7 Classical Load Balancing Problem

The traditional approach to load balancing is to connect equal nominal loads to each phase. Normally the load diversity is sufficient so that severe imbalance will be very infrequent. For larger loads, diversity is not sufficient and connection agreements limit the size of single-phase loads that can be connected without negotiation. Where load imbalance is unavoidable, the point of common coupling can be reinforced by a stronger supply, or the offending load more is made isolated by insertion of a separate high voltage connection via transformer.

Below we discuss the classical load balancing problem that has been first introduced by Charles Steinmetz [17]. We shall discuss the open-loop approach first and then present how this can be implemented in closed-loop through thyristor-based controllers.

3.7.1 Open-Loop Balancing

The load here is represented as a Δ-connected set of admitances, Y_{ab}, Y_{bc} and Y_{ca} as shown in Figure 3.11. Let us decompose the load admittance into real and imaginary terms, such that

$$Y_{ab} = G_{ab} + jB_{ab}$$
$$Y_{bc} = G_{bc} + jB_{bc} \qquad (3.69)$$
$$Y_{ca} = G_{ca} + jB_{ca}$$

The basic idea of load compensation is to connect three purely reactive elements (susceptances) B_{yab}, B_{ybc} and B_{yca} in parallel with the load (see Figure 3.11). Let us first compensate for the reactive portion of the load such that it becomes unity power factor. To do that let us define

$$B_{yab} = B_{yab}(1) + B_{yab}(2)$$
$$B_{ybc} = B_{ybc}(1) + B_{ybc}(2) \qquad (3.70)$$
$$B_{yca} = B_{yca}(1) + B_{yca}(2)$$

where

$$B_{yab}(1) = -B_{ab}$$
$$B_{ybc}(1) = -B_{bc} \qquad (3.71)$$
$$B_{yca}(1) = -B_{ca}$$

The overall load is now real. However, it still is unbalanced. Thus to balance it we must determine the components $B_{yab}(2)$, $B_{ybc}(2)$ and $B_{yca}(2)$ of the compensator.

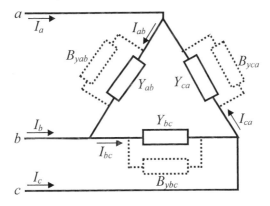

Figure 3.11. Scheme for balancing an unbalanced Δ-connected load

Since the load is Δ-connected, the zero-sequence current is zero. We must therefore choose the susceptances $B_{yab}(2)$, $B_{ybc}(2)$ and $B_{yca}(2)$ in such a way that the negative sequence current through the load is forced to zero. The compensation algorithm depends on the exact voltage applied. For this discussion the voltage is assumed to be pure positive sequence and has line-to-line rms value of 1.0 per unit. The sequence currents I_{a0}, I_{a1} and I_{a2} can then be obtained using (3.10). Now from Figure 3.11, we have

$$I_a = I_{ab} - I_{ca}, \; I_b = I_{bc} - I_{ab} \text{ and } I_c = I_{ca} - I_{bc}$$

The negative sequence of the line current can then be found in terms of the phase currents as

$$I_{a2} = \frac{1}{\sqrt{3}}(1-a^2)\{I_{ab} + a^2 I_{bc} + aI_{ca}\} \qquad (3.72)$$

The reactive components for balancing are connected in Δ. Let us assume $V_{ab} = 1$, $V_{bc} = a^2$ and $V_{ca} = a$. Then the negative-sequence component for the line currents that is flowing through the added compensation of $B_{yab}(2)$, $B_{ybc}(2)$ and $B_{yca}(2)$ can be found as

$$\begin{aligned}I_{a2} &= \frac{1}{\sqrt{3}}(1-a^2)\{jV_{ab}B_{yab}(2) + ja^2 V_{bc}B_{ybc}(2) + jaV_{ca}B_{yca}(2)\} \\ &= \frac{1}{\sqrt{3}}(1-a^2)\{jB_{yab}(2) + jaB_{ybc}(2) + ja^2 B_{yca}(2)\}\end{aligned} \qquad (3.73)$$

This must cancel the negative sequence current due to the load conductances. Thus

$$\frac{1}{\sqrt{3}}(1-a^2)\{jB_{yab}(2)+jaB_{ybc}(2)+ja^2B_{yca}(2)\}$$
$$+\frac{1}{\sqrt{3}}(1-a^2)\{G_{ab}+aG_{bc}+a^2G_{ca}\}=0 \qquad (3.74)$$

The positive sequence of the balancing compensator yields

$$I_{a1}=\frac{1}{\sqrt{3}}(1-a)\{jB_{yab}(2)+jB_{ybc}(2)+jB_{yca}(2)\}$$

As we want the compensated load to be unity power factor, the imaginary component of the above equation must be zero. Therefore,

$$\{jB_{yab}(2)+jB_{ybc}(2)+jB_{yca}(2)\}=0 \qquad (3.75)$$

Separating the real and imaginary parts of (3.74) and combining with (3.75) we get

$$\begin{bmatrix} 0 & -\sqrt{3}/2 & \sqrt{3}/2 \\ 1 & -1/2 & -1/2 \\ 1 & 1 & 1 \end{bmatrix} \begin{bmatrix} B_{yab}(2) \\ B_{ybc}(2) \\ B_{yca}(2) \end{bmatrix} = \begin{bmatrix} -1 & 1/2 & 1/2 \\ 0 & -\sqrt{3}/2 & \sqrt{3}/2 \\ 0 & 0 & 0 \end{bmatrix} \begin{bmatrix} G_{ab} \\ G_{bc} \\ G_{ca} \end{bmatrix} \qquad (3.76)$$

Solving the above equation we get

$$\begin{bmatrix} B_{yab}(2) \\ B_{ybc}(2) \\ B_{yca}(2) \end{bmatrix} = \frac{1}{\sqrt{3}} \begin{bmatrix} 0 & -1 & 1 \\ 1 & 0 & -1 \\ -1 & 1 & 0 \end{bmatrix} \begin{bmatrix} G_{ab} \\ G_{bc} \\ G_{ca} \end{bmatrix} \qquad (3.77)$$

Combining (3.71) and (3.77) and substituting in (3.70) we get the susceptances that will balance the unbalanced load as

$$\begin{bmatrix} B_{yab} \\ B_{ybc} \\ B_{yca} \end{bmatrix} = -\begin{bmatrix} B_{ab} \\ B_{bc} \\ B_{ca} \end{bmatrix} + \frac{1}{\sqrt{3}} \begin{bmatrix} 0 & -1 & 1 \\ 1 & 0 & -1 \\ -1 & 1 & 0 \end{bmatrix} \begin{bmatrix} G_{ab} \\ G_{bc} \\ G_{ca} \end{bmatrix} \qquad (3.78)$$

3. Analysis and Conventional Mitigation Methods

Example 3.12: Let us consider the following unbalanced Δ-connected load

$$Z_{ab} = 6.0 + j3.0 \text{ per unit}$$
$$Z_{bc} = 3.0 + j1.5 \text{ per unit}$$
$$Z_{ca} = 7.5 + j1.5 \text{ per unit}$$

The admittances are then

$$Y_{ab} = 0.1333 - j0.0667 \text{ per unit}$$
$$Y_{bc} = 0.2667 - j0.1333 \text{ per unit}$$
$$Y_{ca} = 0.1282 - j0.0256 \text{ per unit}$$

The compensator susceptances will then be

$$B_{yab} = 0.0667 + (0.1282 - 0.2667)/\sqrt{3} = -0.0133 \text{ per unit}$$
$$B_{ybc} = 0.1333 + (0.1333 - 0.1282)/\sqrt{3} = 0.1363 \text{ per unit}$$
$$B_{yca} = 0.0256 + (0.2667 - 0.1333)/\sqrt{3} = 0.1025 \text{ per unit}$$

The cumulative load and compensator (we refer it as compensated) admittances are then

$$Y_{comp}^{ab} = Y_{ab} + jB_{yab} = 0.1333 - j0.0799 \text{ per unit}$$
$$Y_{comp}^{bc} = Y_{bc} + jB_{ybc} = 0.2667 + j0.0030 \text{ per unit}$$
$$Y_{comp}^{ca} = Y_{ca} + jB_{yca} = 0.1282 + j0.0770 \text{ per unit}$$

Let us assume $V_{ab} = 1$, $V_{bc} = a^2$ and $V_{ca} = a$. The phase currents are then $I_{ab} = 0.1555\angle -30.95°$ per unit, $I_{bc} = 0.2667 \angle -119.36°$ per unit and $I_{ca} = 0.1495\angle 150.98°$ per unit. We then get the following values of the line currents

$$I_a = 0.305\angle -30° \text{ per unit}$$
$$I_b = 0.305\angle -150° \text{ per unit}$$
$$I_c = 0.305\angle 90° \text{ per unit}$$

It can then be seen that even if the phase currents are unbalanced, the line currents are balanced. This implies that the source sees the compensated load consisting of three equal Y-connected resistors that have a value of $1/(\sqrt{3} \times 0.305) = 1.8929$ per unit. It is interesting to note that

$$\frac{1}{1.8929} = 0.5282 = G_{ab} + G_{bc} + G_{ca}$$

ΔΔΔ

The classical load balancing problem has been used in power systems using static var compensators. There are various ways in which loads can be balanced using SVCs. A direct implementation of the Steinmetz equations will allow a measured unbalanced load to be balanced by applying appropriate reactive elements between phases. We now discuss the balancing of the negative phase sequence (NPS) problem discussed in Chapter 2 (Section 2.2.5). Here the balancing compensator is on the secondary of a Y/Δ transformer and the Steinmetz equations required are modifird to allow for the phase shift of the sequences through the transformer.

Two Queensland Railway substations are shown in Figure 2.19, of which we consider the Grantleigh substation. The SVCs used in the load balancing in Queensland Rail have fixed capacitors and a thyristor controlled reactor (TCR) on the secondary of a star-delta transformer as shown in Figure 3.12. Note that the fixed capacitor bank is configured as a third and fifth harmonic filter with a combined unsymmetrical rating of 11, 5, 11 MVAr for phases RS, ST and TR respectively [18].

Traction loads that produce NPS currents which exceed the balancing capability of the SVC will cause the NPS voltage at the subject busbar to rise dependent only on the negative sequence impedance of the supply at the point of common coupling. Figure 3.13 shows the correlation between NPS voltage magnitude at the Grantleigh 132 kV busbar and railway loads in phase pair AB measured at Grantleigh on the 50 kV busbar. The point at which the SVC reaches its maximum capability and the NPS begins to rise occurs at approximately 220 A [18]. The residual NPS voltage of approximately 0.2% occurs because of the open-loop nature of the Steinmetz compensator. The correction for load imbalance is compensated but the background transmission line NPS is not seen and not corrected.

3.7.2 Closed-Loop balancing

An alternative is to directly control the busbar voltages in such a way as to minimize the NPS voltage component. The latter method takes into

3. Analysis and Conventional Mitigation Methods

account of any imperfectly transposed transmission lines and other sources of unbalance. For suppression of all NPS components, the load balancing is not based on impedance measurement but on feedback control of the reactive components. Consider the line currents contributed by the reactive elements. They are found from the currents in each element.

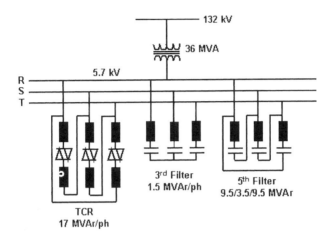

Figure 3.12. Structure of FC-TCR SVC used at Grantleigh

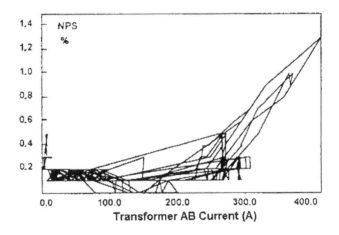

Figure 3.13. Correlation of NPS voltage at PCC with loads in the AB phase at Grantleigh

Provided the voltages at the compensator are close to positive sequence at angle θ the line currents can be found as

$$\begin{bmatrix} I_a \\ I_b \\ I_c \end{bmatrix} = j\angle(\theta + 30^o) \begin{bmatrix} 1 & 0 & -a \\ -1 & a^2 & 0 \\ 0 & -a^2 & a \end{bmatrix} \begin{bmatrix} B_{ab}|V_{ab}| \\ B_{bc}|V_{bc}| \\ B_{ca}|V_{ca}| \end{bmatrix} \qquad (3.79)$$

Let us choose a control law to make the reactive element current proportional to the magnitude of the voltage error from the balance

$$B_{ab}|V_{ab}| = k|\Delta V_{ab}|$$

When this current is flowing through a line of impedance jX, the resulting change in phasor voltage response is

$$\begin{bmatrix} \Delta V_a^r \\ \Delta V_b^r \\ \Delta V_c^r \end{bmatrix} = j^2 kX \angle(\theta + 30^o) \begin{bmatrix} 1 & 0 & -a \\ -1 & a^2 & 0 \\ 0 & -a^2 & a \end{bmatrix} \begin{bmatrix} |\Delta V_{ab}| \\ |\Delta V_{bc}| \\ |\Delta V_{ca}| \end{bmatrix} \qquad (3.80)$$

The line to line voltage response then satisfies

$$\begin{bmatrix} \Delta V_{ab}^r \\ \Delta V_{bc}^r \\ \Delta V_{ca}^r \end{bmatrix} = j^2 kX \angle(\theta + 30^o) \begin{bmatrix} 1 & -1 & 0 \\ 0 & 1 & -1 \\ -1 & 0 & 1 \end{bmatrix} \times \begin{bmatrix} 1 & 0 & -a \\ -1 & a^2 & 0 \\ 0 & -a^2 & a \end{bmatrix} \begin{bmatrix} |\Delta V_{ab}| \\ |\Delta V_{bc}| \\ |\Delta V_{ca}| \end{bmatrix}$$

(3.81)

For small changes in voltage the magnitude change is the component of the voltage aligned with the phase voltage. The phasor V_{ab} is at angle $\theta + 30^o$ and therefore we take the product of the above equation with a unit vector at $-\theta - 30^o$ to find the approximate magnitude change, i.e.,

$$\begin{bmatrix} |\Delta V_{ab}^r| \\ |\Delta V_{bc}^r| \\ |\Delta V_{ca}^r| \end{bmatrix} = j^2 kX \angle(\theta + 30^o) \begin{bmatrix} 1 & -1 & 0 \\ 0 & 1 & -1 \\ -1 & 0 & 1 \end{bmatrix} \begin{bmatrix} 1 & 0 & -a \\ -1 & a^2 & 0 \\ 0 & -a^2 & a \end{bmatrix}$$

$$\times \begin{bmatrix} 1 & 0 & 0 \\ 0 & a & 0 \\ 0 & 0 & a^2 \end{bmatrix} \begin{bmatrix} |\Delta V_{ab}| \\ |\Delta V_{bc}| \\ |\Delta V_{ca}| \end{bmatrix} \angle(-\theta - 30^o)$$

3. Analysis and Conventional Mitigation Methods

The above equation reduces to

$$\begin{bmatrix} |\Delta V_{ab}^r| \\ |\Delta V_{bc}^r| \\ |\Delta V_{ca}^r| \end{bmatrix} = -kX \begin{bmatrix} 2 & -1 & -1 \\ -1 & 2 & -1 \\ -1 & -1 & 2 \end{bmatrix} \begin{bmatrix} |\Delta V_{ab}| \\ |\Delta V_{bc}| \\ |\Delta V_{ca}| \end{bmatrix} \quad (3.82)$$

Note that the line-to-line voltage can have no zero sequence component, thus the equation reduces to a separate phase result

$$\begin{bmatrix} |\Delta V_{ab}^r| \\ |\Delta V_{bc}^r| \\ |\Delta V_{ca}^r| \end{bmatrix} = -3kX \begin{bmatrix} 1 & 0 & 0 \\ 0 & 1 & 0 \\ 0 & 0 & 1 \end{bmatrix} \begin{bmatrix} |\Delta V_{ab}| \\ |\Delta V_{bc}| \\ |\Delta V_{ca}| \end{bmatrix} \quad (3.83)$$

The correction in the compensator voltage from the line drop causes changes in the magnitudes of the line to line voltages which are in the opposite sense to the errors. Any zero sequence error is unable to be corrected because of the delta connection of the compensator. If the correction always has a finite gain, the voltage will converge towards positive sequence, but there would always be a residual offset due to the finite gain of the feedback law. From the analysis above we can conclude that integral controllers can provide for voltage balance and positive sequence correction, provided the unbalance of the voltage is limited.

Example 3.13: An experimental result is reported in [19]. The experimental setup contained an inductive transmission line, a switched Y-connected RL load and a transformer connected TSC. Three TSC banks are connected in parallel. The values of the capacitances are chosen such that the TSC can be switched in a binary fashion to give seven different levels. The TSCs banks are connected in delta and are rated 415 V (line-to-line) and 40 kVA. The reactance of inductor modeling the transmission line has been 6.75 Ω. The unbalanced load contains a 10 Ω resistor in each phase that is connected in series with an inductor. The reactances of the inductors are 116 Ω, 45 Ω and 233 Ω for the three phases. The schematic diagram of the TSC along with its control circuit is shown in Figure 3.14. The TSC circuit contains an inrush current limiting inductor, an RC snubber connected across the back to back thyristors and the main capacitor bank.

The capacitor bank in phase ab is incremented when the voltage V_{ab} is low. This independent control of the phases was shown to be satisfactory theoretically in this section and was confirmed in this experiment. Figure

3.15 shows the correction of voltage in one phase. Note that the use of discrete steps in the compensator means that there will be some residual imbalance in the system. The use of TSC is only suitable for correction of voltage drops and would require inductors to correct for undervoltages.

ΔΔΔ

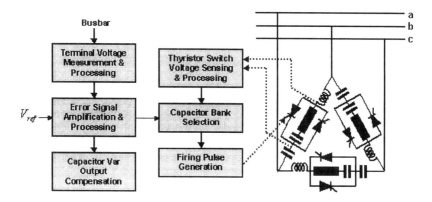

Figure 3.14. Schematic diagram of the TSC and its control scheme

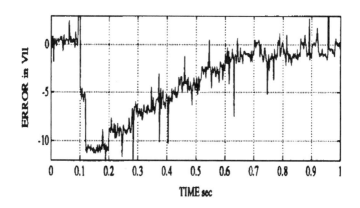

Figure 3.15. Error in line-to-line voltage with TSC control

3.7.3 Current Balancing

If the supply voltage is close to positive sequence, the current through the compensator can be made equal to the negative sequence current generated by an unbalanced load. Let us assume that the positive sequence line-to-neutral voltages are given by $V_a = 1$, $V_b = a^2$ and $V_c = a$. The line-to-line voltages then are

3. Analysis and Conventional Mitigation Methods

$$\begin{bmatrix} V_{ab} \\ V_{bc} \\ V_{ca} \end{bmatrix} = \begin{bmatrix} 1-a^2 \\ a^2-a \\ a-1 \end{bmatrix} \quad (3.84)$$

The current flowing in the delta are obtained from the above voltages as

$$\begin{bmatrix} I_{ab} \\ I_{bc} \\ I_{ca} \end{bmatrix} = j \begin{bmatrix} B_{ab}(1-a^2) \\ B_{bc}(a^2-a) \\ B_{ca}(a-1) \end{bmatrix} \quad (3.85)$$

The line currents are then given by

$$I_{abc} = \begin{bmatrix} I_a \\ I_b \\ I_c \end{bmatrix} = \begin{bmatrix} 1 & 0 & -1 \\ -1 & 1 & 0 \\ 0 & -1 & 1 \end{bmatrix} j \begin{bmatrix} B_{ab}(1-a^2) \\ B_{bc}(a^2-a) \\ B_{ca}(a-1) \end{bmatrix} \quad (3.86)$$

The sequence currents are found as

$$I_{012} = P I_{abc} = \frac{j}{\sqrt{3}} \begin{bmatrix} 1 & 1 & 1 \\ 1 & a & a^2 \\ 1 & a^2 & a \end{bmatrix} \begin{bmatrix} B_{ab}(1-a^2) - B_{ca}(a-1) \\ B_{bc}(a^2-a) - B_{ab}(1-a^2) \\ B_{ca}(a-1) - B_{bc}(a^2-a) \end{bmatrix}$$

Solving the above equation we get

$$I_{012} = j\sqrt{3} \begin{bmatrix} 0 \\ (B_{ab} + B_{bc} + B_{ca}) \\ -(a^2 B_{ab} + B_{bc} + a B_{ca}) \end{bmatrix} \quad (3.87)$$

Thus the compensator cannot deliver zero sequence current, the positive sequence contribution is at a fixed angle while the negative sequence component can be at an arbitrary angle. Let the desired negative sequence current be $c + jd$ and the desired positive sequence current is je. Then we have

$$\frac{e}{\sqrt{3}} = (B_{ab} + B_{bc} + B_{ca})$$

$$\frac{c+jd}{\sqrt{3}} = -j\left((-\frac{1}{2} - j\frac{\sqrt{3}}{2})B_{ab} + B_{bc} + (-\frac{1}{2} + j\frac{\sqrt{3}}{2})B_{ca}\right)$$

which can be expressed as

$$\begin{bmatrix} \dfrac{e}{\sqrt{3}} \\ \dfrac{c}{\sqrt{3}} \\ \dfrac{d}{\sqrt{3}} \end{bmatrix} = \begin{bmatrix} 1 & 1 & 1 \\ -\dfrac{\sqrt{3}}{2} & 0 & \dfrac{\sqrt{3}}{2} \\ \dfrac{1}{2} & -1 & \dfrac{1}{2} \end{bmatrix} \begin{bmatrix} B_{ab} \\ B_{bc} \\ B_{ca} \end{bmatrix} \tag{3.88}$$

From (3.88) we get

$$\begin{bmatrix} B_{ab} \\ B_{bc} \\ B_{ca} \end{bmatrix} = \frac{2}{3\sqrt{3}} \begin{bmatrix} \dfrac{1}{2} & -\dfrac{\sqrt{3}}{2} & \dfrac{1}{2} \\ \dfrac{1}{2} & 0 & -1 \\ \dfrac{1}{2} & \dfrac{\sqrt{3}}{2} & \dfrac{1}{2} \end{bmatrix} \begin{bmatrix} e \\ c \\ d \end{bmatrix} \tag{3.89}$$

Thus a solution can be found in terms of the required compensator current for the compensator admittances. Naturally when no zero sequence correction is required all reactances are equal. This result is only strictly true for positive sequence voltage at the compensator and load terminals.

3.8 Harmonic Reduction

The most common concentrated source of harmonics is the rectifier front end on motor drives and converters, fluorescent lamps and computer power supplies. When the load is a large rectifier then changing from six pulse rectifiers to twelve pulse or even up to 48 pulse can significantly reduce the strength of low order harmonics. Perfect cancellation of the low harmonics requires perfect symmetry of transformers and impedances and practical systems will always have some residual low order harmonics. The voltage at

3. Analysis and Conventional Mitigation Methods

the point of common coupling with other customers may have low harmonics if the source impedance at that point is low.

Because of the adverse effects that harmonics have on loads, standards have been developed to define the worst case environment that manufacturers should design equipment to operate within. The standards also set a basis for defining responsibility for correction if a new load is to be connected which could have adverse effects.

The Institute of Electrical and Electronics Engineers (IEEE) sets limits to the permitted voltage distortion at the point of common coupling (PCC) in IEEE 519 [20] which are largely adopted in North America. The PCC refers to the location in the network where other customers may be connected. This distinction can be important where a distorting load is fed from a dedicated line and the substation represents the lower distortion point of the network where other customers may be connected. Let us define V_n to be the per unit voltage (with respect to the fundamental) of the nth harmonic component. Then the individual harmonic components and the THD at the PCC are given in Table 3.6.

Table 3.6. IEEE 519: Voltage distortion limits [20]

Bus Voltage at PCC	Individual V_n (per unit)	Voltage THD (percent)
Less than 69 kV	3.0	5.0
Between 69 kV and 161 kV	1.5	2.5
Above 161 kV	1.0	1.5

IEC divides load into three classes. In this Class-1 refers to low voltage public supply and Classes 2 and 3 refer to industrial loads. The IEC recommended voltage distortion levels for the three classes are given in Tables 3.7 to 3.9 [21,22]. The total harmonic voltage distortion for Class-1 must be below 8 %, while those for Classes 2 and 3 must be below 10% for all harmonics up to 40. As we have mentioned before, the IEEE recommended practices are popular in the North America, while IEC standards are followed all over Europe. Therefore there is no uniform guidelines about the magnitude of the tolerable harmonic distortion.

For low power equipment, limits are also set on the permitted current distortion by IEC 61000-3-2 and are given in Table 3.10 [23]. IEEE 519 current harmonic standards are different for distribution, subtransmission and transmission systems [20]. The limits for distribution and transmission systems are given in Tables 3.11 and 3.12 respectively. The limits for subtransmission systems (69–161 kV) are half of the corresponding limits of the distribution systems.

Table 3.7. IEC 61000-2-2: Voltage distortion limits in public low-voltage networks (Class-1) [21]

Odd harmonics		Even harmonics		Triplen Harmonics	
n	V_n (pu)	n	V_n (pu)	n	V_n (pu)
5	6	2	2	3	5
7	5	4	1	9	1.5
11	3.5	6	0.5	15	0.3
13	3	8	0.5	≥ 21	0.2
17	2	10	0.5		
19	1.5	≥ 12	0.2		
23	1.				
25	1.5				
≥ 29	x				

$x = 0.2 + 12.5/n$

Table 3.8. IEC 61000-2-4: Voltage distortion limits in industrial plants (Class-2) [22]

Odd harmonics		Even harmonics		Triplen Harmonics	
n	V_n (pu)	n	V_n (pu)	n	V_n (pu)
5	6	2	2	3	5
7	5	4	1	9	1.5
11	3.5	6	0.5	15	0.3
13	3	8	0.5	≥ 21	0.2
17	2	10	0.5		
19	1.5	≥ 12	0.2		
23	1.5				
25	1.5				
≥ 29	x				

$x = 0.2 + 12.5/n$

Table 3.9. IEC 61000-2-4: Voltage distortion limits in industrial plants (Class-3) [22]

Odd harmonics		Even harmonics		Triplen Harmonics	
n	V_n (pu)	n	V_n (pu)	n	V_n (pu)
5	8	2	3	3	6
7	7	4	1.5	9	2.5
11	5	≥ 6	1	15	2
13	4.5			21	1.75
17	4			≥ 27	1
19	4				
23	3.5				
25	3.5				
≥ 29	y				

$y = 5\sqrt{(11//n)}$

The current limits in these tables are given for odd harmonics. The even harmonics are limited to 25% of the odd harmonics limits. For all power generation equipment, distortion limits are those with $I_{SC}/I_L < 20$, where I_{SC} is the maximum short circuit current at the PCC and I_L is the maximum

3. Analysis and Conventional Mitigation Methods

fundamental frequency of 15 or 30 minute load current at the PCC. Total demand distortion (TDD) is the total root-sum-square harmonic current distortion in percent of the maximum demand load current for 15 or 30 minute demand. This means that TDD is essentially the THD, but normalized by I_L.

Table 3.10. IEC 61000-3-2 maximum permissible harmonic currents for class D equipment (current limited to less than or equal to 16 A per phase) [23]

n	3	5	7	9	11	13	15 to 39
Max I_n (A)	2.3	1.14	0.77	0.40	0.33	0.21	0.15 – 0.15/n

Table 3.11. IEEE 519 current distortion limits for distribution systems (120 V – 69 kV) [20]

		I_n/I_L (%)				TDD (%)
I_{SC}/I_L	$n < 11$	$11 \leq n < 17$	$17 \leq n < 23$	$23 \leq n < 35$	$35 \leq n$	
< 20	4.0	2.0	1.5	0.6	0.3	5
20–50	7.0	2.5	2.5	1.0	0.5	8
50–100	12	5.5	5.0	2.0	1.0	15
100–1000	12	5.5	5.0	2.0	1.0	15
> 1000	15	7.0	6.0	2.5	1.4	20

Table 3.12. IEEE 519 current distortion limits for transmission systems (> 161 kV) [20]

		I_n/I_L (%)				TDD (%)
I_{SC}/I_L	$n < 11$	$11 \leq n < 17$	$17 \leq n < 23$	$23 \leq n < 35$	$35 \leq n$	
< 50	2.0	1.0	0.75	0.3	0.15	5
≥50	3.0	1.5	1.15	0.45	0.22	8

There are two approaches conventionally applied to reduce harmonic problems when they arise. These are

- Reduce the level of harmonics from the source.
- Correct at the PCC or deep into the network.

The source reduction may involve a higher pulse number in rectifiers to achieve some net cancellation of lower harmonics. It could include filtering components within the equipment. Network solutions could include changing the placement of capacitor banks to reduce resonance. However the more common requirement is to install harmonic filters to provide a preferential path for harmonic current. Typically this consists of a set of shunt LC filters tuned for each of the troublesome frequencies often combined with a general high pass filter to attenuate the higher frequency distortion.

The difficulties with the use of such filters are

- The power system Thevenin equivalent is variable and the design for the harmonic filter must make some worst case assumptions about the equivalent source impedance.
- The best filtering occurs for a sharply notched filter but the design of each filter needs to take into account the fact that line frequency is not rock steady and a variation of as much as 0.5 Hz is possible. Larger deviations can occur in small systems with low total inertia or in countries with inadequate generation resources. This deviation means that a 7th harmonic filter for a 50 Hz system may need to be designed for the required harmonic reduction between 346.5 Hz and 353.5 Hz.
- As components age their impedance value varies. This is particularly true for capacitors. The filter must thus be designed for some degree of tolerance of components as well as incorporation of tuning elements to restore the center frequency to nominal.

The design of tuned filters is reported in [15,24]. However the best option as per as harmonic reduction is active filtering. This will be discussed in detail in Chapters 7 and 8.

3.9 Voltage Sag or Dip Reduction

The duration of a voltage dip is limited by the ceasing of the transient causing the sag or by fault clearing through the operation of a protective device. For a motor start transient the duration is set by the run-up time of the motor. For a conductor clash the system may not experience a follow through arc and the fault will self clear at the next zero crossing. A branch could brush against a line and cause a temporary line fault. More commonly a circuit breaker or fuse must operate to break the fault current and provide sufficient time for the ionized air to dissipate and recombine. For many distribution circuit breakers and fuses there are characteristics which determine the clearing time setting. A very high fault current initiates a very rapid trip while a fault producing a mild overcurrent may take a minute or more before the protection operates.

The depth of the voltage dip and the number of customers affected to a given level depends on the fault model and the distribution and level of source impedance. Consider a high impedance transformer feeding two feeders as in Figure 3.16 (a). A severe fault on a feeder, such as on Bus 2 of the left feeder, will cause voltage dips for all customers on both feeders until the circuit breaker on the left feeder operates producing an outage for those customers. Those on the right feeder will experience a sag until the breaker

3. Analysis and Conventional Mitigation Methods

clears the faulted feeder. For the second case in Figure 3.16 (b) there will be only a mild sag on the right feeder for the fault on bus 2. All customers on the left feeder will experience an outage until the fault is cleared and the breaker reclosed or until the faulted section is isolated and supply is given through an alternate feeder.

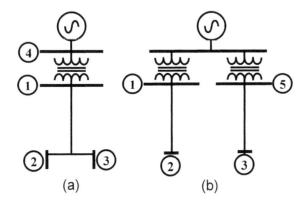

Figure 3.16. (a) Common source impedance and (b) separate source impedances

The vast majority of faults are single line to ground. This means that the fault current can be limited by inclusion of additional impedance in the zero sequence path. For example consider the distribution system of Figure 3.17. A standard adopted in many countries is the use of star delta transformers with the delta always on the high voltage side. For a line to ground fault on the line to the customer premises, the zero sequence current must flow through the neutral of the transformer at Bus-2. Inclusion of a neutral earthing Reactor or neutral resistor on the star side of transformers can significantly reduce fault currents. If phase-a of the line is faulted, the line voltages will show a zero sequence offset but this will not be seen by customers on the far side of the star delta transformers who will thus only experience a mild voltage dip. The disadvantage of a large impedance can be in the overvoltage transient created on the unfaulted phase leading to insulator breakdown.

If a feeder is of high impedance, then for faults at the far end even a high impedance fault can cause a severe depth of sag in the proximity of the fault. But the high impedance can mean that there is little sag for customers connected near the substation. For a low impedance line, a high impedance fault will cause only mild voltage dips. One of the severe difficulties for rural supplies using overhead lines is that there is a long length of exposed line with each of the branching feeders. In many cases there is a common source impedance so a line or medium voltage customer fault can cause a voltage dip over areas of hundreds of square kilometers.

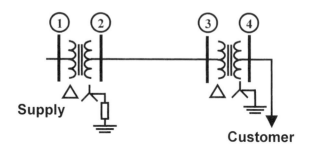

Figure 3.17. Fault current reduction through neutral earthing impedance

If there are some industrial customers very sensitive to voltage dips, the frequency of faults on that feeder can be reduced by setting switches such that only a short line section becomes part of the industrial feeder. This reduces the exposure to faults. This will not however greatly affect voltage dips unless the common source impedance is kept very low. For voltage drop under loads and for loss reduction it may be desirable to operate transformers as in Figure 3.16 (b) in parallel by joining buses 1 and 5. From the viewpoint of voltage dips it is essential to keep the buses split.

The expected sag level in a distribution system can be predicted by knowing the system structure, the distribution of faults per kilometer of line, and the distribution of fault impedances. For a radial distribution system the faults can be applied in sequence to each bus to determine the voltages for all other buses. The duration of the dip is determined by the protection characteristics. Thus knowing the fault location and impedance, the duration of the fault until it is cleared can be determined. Recording the depth and duration and based on the fault probability, a probability assignment table is formed. When all buses have been examined, the probability of occurrence and its duration are recorded, the probability of crossing the CBEMA curve is established. The probability contours can indicate sensitive system areas or where tuning of protection can aid in sag effect reduction.

3.10 Conclusions

The power quality aspects discussed in this chapter are: voltage dips, surges, harmonics, flickers, momentary outages and extended outages. There

3. Analysis and Conventional Mitigation Methods

are two main aspects of power quality that are reported frequently. These are momentary outages or dips sufficient to trip equipment and the extended outages. Harmonic levels do not usually result in high levels of customer complaint in most circumstances. One issue that does arise is where ripple control signals are used for switching street lights and hot water systems. High harmonic levels have been known to cause false switching of these ripple controlled hot water systems and the customer complaints are generated in response to customers experiencing cold showers.

In this chapter we have discussed the power quality standards and conventional mitigation techniques to the power quality problems. In addition to the mitigation techniques discussed above, the use of shunt capacitors to provide reactive power support is well known [17]. In addition, series capacitors are also used in distribution systems. The traditional method of voltage feeder mitigation involves reconfiguration of the feeders or substations. It is however reported that series capacitors can mitigate voltage flickers [25]. These capacitors however can lead to both ferroresonance problems involving transformers and subsynchronous resonance problems including motors. Usually such series capacitors have a bypass switch and a zinc-oxide arrester in parallel. The arrester restricts the voltage across the device and the bypass switch can remove the capacitor from the line in case of a fault.

It must however be emphasized here that none of the conventional mitigation techniques are flexible. For example a tuned filter, which is tuned to eliminate the 5th harmonic, can lead to resonance at some other frequency depending on the network configuration and harmonic component of the load. Therefore it is imperative that better and flexible mitigating devices are used for power quality problems. In the next chapter we introduce the concept of custom power and the custom power devices that provide us with a flexible choice to alleviate the problems.

3.11 References

[1] H. L. Lewis, *Power Distribution Planning Reference Book*, Marcel Dekker, New York, 1997.

[2] R. Billinotn and J. E. Billinoton, "Distribution system reliability indices," *IEEE Trans. Power Delivery*, Vol. 4, No. 1, pp. 561-568, 1989.

[3] J. J. Grainger and W. D. Stevenson, Jr., *Power System Analysis*, McGraw-Hill, New York, 1994.

[4] W. V. Lyon, *Transient Analysis of Alternating-Current Machinery*, John Wiley, New York, 1954.

[5] F. Z. Peng and J. S. Lai, "Generalized instantaneous reactive power theory for three-phase power systems," *IEEE Trans. Instrumentation & Measurements*, Vol. 45, No. 1, pp. 293-297, 1996.

[6] M. K. Mishra, A. Joshi and A. Ghosh, "A new compensation algorithm for balanced and unbalanced distribution systems using generalized instantaneous reactive power theory," *Electric Power systems Research*, Vol. 60, pp. 29-37, 2001.
[7] IEC 61000-2-1:1990, "Electromagnetic Compatibility", Part 2: Environment Sect. 1: Description of the Environment – Electromagnetic Environment for Low-Frequency Conducted Disturbances and Signalling in Public Power Supply Systems. First Edition, 1990-05.
[8] IEEE Interharmonic Task Force & CIGRE 36.05/CIRED 2 CC02 Voltage Quality Working Group, "Interharmonics in power systems," Draft, 1997.
[9] A. V. Oppenheim, A. S. Willsky and I. T. Young, *Signals and Systems*, Prentice-Hall, Englewood Cliffs, 1983.
[10] J. Arrillaga, N. R. Watson and S. Chen, *Power Quality Assessment*, John Wiley, New York, 2000.
[11] G. Ledwich and A. Ghosh, "A flexible DSTATCOM operating in voltage or current control mode," *Proc. IEE – Generation, Transmission & Distribution*, Vol. 149, No. 2, pp. 215-224, 2002.
[12] G. T. Heydt and W. T. Jewell, "Pitfalls of electric power quality indices," *IEEE Trans. Power Delivery*, Vol. 13, No. 2, pp. 570-578, 1998.
[13] R. S. Thallam and G. T. Heydt, "Power acceptability and voltage sag indices in the three phase sense," *Panel Session on Power Quality – Voltage Sag Indices, IEEE Power Engineering Society Summer Meeting*, Seattle, 2000.
[14] M. H. J. Bollen, "Voltage sag indices – Draft 1.1", with contribution from D. Sabin, *IEEE Panel on Voltage Sag Indices*, August 2, 2000.
[15] E. Acha and M. Madrigal, *Power System Harmonics*, John Wiley, Chichester, 2001.
[16] IEC Standard 61000-3-7 "Limitation of voltage fluctuation and flicker for equipment connected to medium and high voltage power supply system," 1995.
[17] T. J. E. Miller, *Reactive Power Control in Electric Systems*, John Wiley, New York, 1982.
[18] G. Ledwich and T. A. George, "Using phasors to analyze power system negative phase sequence voltages caused by unbalanced loads," *IEEE Trans. Power Systems*, Vol. 9, No. 3, pp. 1226-1232, 1994.
[19] G. Ledwich, S. H. Hosseini and G. F. Shannon, "Voltage balancing using switched capacitors," *Electric Power Systems Research*, Vol. 24, pp. 85-90, 1992.
[20] IEEE 519, Recommended Practices and Requirements for Harmonic Control in Electrical Power Systems, 1992.
[21] IEC 61000-2-2, Electromagnetic Compatibility (EMC) – Part 2: Environment – Section 2: Compatibility Levels for Low-Frequency Conducted Disturbances and Signalling in Public Low-Voltage Power Supply Systems, 1990.
[22] IEC 61000-2-4, Electromagnetic Compatibility (EMC) – Part 2: Environment – Section 4: Compatibility Levels in Industrial Plants for Low-Frequency Conducted Disturbances, 1994.
[23] IEC 61000-3-2, Electromagnetic Compatibility (EMC) – Part 3: Limits, Section 2 – Limits for Harmonic Current Emissions (Equipment Input Current $\leq$ 16A per phase).
[24] R. C. Dugan, M. F. McGranaghan and H. W. Beaty, *Electric Power Systems Quality*, McGraw-Hill, New York, 1996.
[25] L. Morgan, J. M. Barcus and S. Ihara, "Distribution series capacitor with high-energy varistor protection," *IEEE Trans. Power Delivery*, Vol. 8, No. 3, pp. 1413-1419, 1993.

Chapter 4

Custom Power Devices: An Introduction

The concept of custom power was introduced by N. G. Hingorani [1]. Like flexible ac transmission systems (FACTS) for transmission systems, the term custom power (CP) pertains to the use of power electronic controllers for distribution systems. Just as FACTS improves the reliability and quality of power transmission by simultaneously enhancing both power transfer volume and stability, the custom power enhances the quality and reliability of power that is delivered to customers. Under this scheme a customer receives a prespecified quality power. This prespecified quality may contain a combination of specifications of the following

- Frequency of rare power interruptions.
- Magnitude and duration of over and undervoltages within specified limits.
- Low harmonic distortion in the supply voltage.
- Low phase unbalance.
- Low flicker in the supply voltage.
- Frequency of the supply voltage within specified limits.

There are many custom power devices. The compensating power electronic devices are either connected in shunt or in series or a combination of both. In addition there are current breaking devices that are power electronic based. Any one or a combination of two or more of these devices are used to fulfill each one of the above mentioned objectives.

In this chapter we shall introduce the concept of all these custom power devices. We shall also discuss the concept of *Custom Power Park* that can serve customers who demand high quality power and are ready to pay a premium price for service provided to them. The high quality power supplied to the customers in a custom power park can also have different grades of supply quality. We shall start our discussion with utility-customer interface problem in which we shall discuss what a customer can expect from the utility and what the utility in return can expect from its customers.

It is however to be realized that the use of custom power is still in its infancy. Therefore, there is limited experience from custom power equipment installations. There is also an apparent reluctance in the community to disclose such experience even if it is positive. There is definitely no discussion of any negative experience. It is a common belief amongst the manufacturers and utilities alike that a substantial competitive advantage can be gained if the know-how of custom power equipment is not made available to others. We have therefore developed our models of the custom power equipment and tried to study these devices objectively. In this chapter we shall try to analyze the situation from a global standpoint rather than from the load perspective only.

4.1   Utility-Customer Interface

Consider the radial distribution system shown in Figure 4.1. It contains three load buses that are supplied by a source. As we have seen in Chapter 2 that if any of these three loads is unbalanced, it will cause unbalanced currents to flow through the feeder. This will also result in unbalanced bus voltages which will affect all the customers. Similarly, a load drawing harmonic current will also distort all the bus voltages. Further an arc furnace will cause the bus voltage to flicker thereby affecting all other loads connected to that bus. Therefore the question that needs to be asked is "Who has the responsibility to safeguard the quality of power being supplied?"

For example, let us assume that a load (or a group of loads) connected to Bus-3 is drawing unbalanced, harmonic current. In addition, the load power factor is poor. This will obviously lead to unbalance and distortion in other system quantities which will be undesirable to other customers. Also poor power factor will cause large resistive losses and voltage drop in the feeders that might be undesirable to the utility. The utility can then ask the customer of the polluting load to install corrective measures. It is well known that shunt capacitor is a good solution for correcting poor power factors. Similarly tuned filters are also used with power electronic loads to bypass harmonic currents. The problem of load balancing is also not new, as it dates back to Charles Steinmetz [2]. However, as we shall discuss later in this chapter, all these functions can be performed by a single shunt connected device that operates in the current control mode. It will be the responsibility of the customer of the polluting load to install the shunt connected device at his own premises such that he draws a pure sinusoidal current from the bus.

4. Custom Power Devices: An Introduction

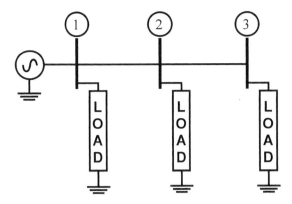

Figure 4.1. Single line diagram of a radial distribution system

Now suppose the customer of the polluting load refuses to install a shunt device. The utility can then impose a financial penalty on the customer. Since the shunt device will require heavy initial installation expenditure in addition to the day-to-day running expenditure, the customer may find it less expensive to pay the penalty. Then the distortion and unbalance in bus voltages will persist. Let us now assume that Bus-3 supplies an industrial area containing fairly large loads. Then their refusal to install a shunt device may create severe problem to customers at other buses. One option for the utility is to connect a shunt device at the bus itself such that the current at the left of this bus is sinusoidal and charge all the customers of Bus-3 for the installation and running cost.

Now due to the presence of large polluting loads at Bus-3, the customers of Bus-2 suffer the most. Since Bus-1 is situated near the source and may be separated from the source only by a small source impedance, the impact of the unbalance and distortion on its voltage will be minimal. This will not be true for the voltage of Bus-2. It will then be the responsibility of the utility to correct the bus voltage to within a specified limit. This can be done by installing a shunt device that operates in the voltage control mode. Suppose Bus-2 contains some sensitive loads that require nearly sinusoidal voltage, then a further level of compensation to meet the customer needs may be undertaken on negotiation between the sensitive customer and the utility.

As mentioned in Chapter 3, arc furnace can cause voltage flicker in the bus voltage. Suppose one of the loads in any of the buses is an arc furnace. Then it will cause the voltage of that bus to flicker. This problem can also be alleviated by installing a shunt device that operates in the voltage control mode. In fact as the custom power technology matures, it might be mandatory to install a shunt device with every arc furnace unless it is on a dedicated feeder.

Now consider a new situation and suppose Bus-3 is remotely located, i.e., the distance between Buses 2 and 3 is substantial. In addition, assume that Bus-3 has a process load that cannot tolerate voltage sag/swell or interruptions. In that event the utility can install a series device to protect the process industry from short duration voltage variations such that the rms value of the Bus-3 voltage is maintained constant. The series device can tightly regulate the Bus-3 voltage and can act as an active filter to prevent the harmonic component of the line current from distorting the Bus-3 voltage.

A unified power quality conditioner (UPQC) is a combination of a shunt and series device and can combine the functions of these two devices together. For example installation of this device at Bus-3 ensures that Bus-3 voltage is regulated and maintained sinusoidal while at the same time the total current drawn by the loads and compensator at Bus-3 is also maintained sinusoidal and at unity power factor. This can be achieved irrespective of unbalance or harmonics in the supply side or in the load side. It is however an expensive device and its use may be limited to particularly sensitive situations with a high value on power quality.

In addition to the shunt and series devices and their combination, there are solid state current limiting, breaking and transferring devices that use the advancements in the power semiconductor technology to provide very fast service. These devices are much faster than their mechanical counterparts. The custom power park concept envisages the use of these devices to provide near interruption-free power to the valued customers. Usually these devices are installed, maintained and operated in a coordinated fashion by the power distribution companies.

4.2 Introduction to Custom Power Devices

The power electronic controllers that are used in the custom power solution can be network reconfiguring type or compensating type. The network reconfiguration devices are usually called switchgear and they include current limiting, current breaking and current transferring devices. The solid state or static versions of the devices are called

- Solid state current limiter (SSCL)
- Solid state breaker (SSB)
- Solid state transfer switch (SSTS)

The compensating devices either compensate a load, i.e., correct its power factor, unbalance etc. or improve the quality of the supplied voltage.

4. Custom Power Devices: An Introduction

These devices are either connected in shunt or in series or a combination of both. The devices include

- Distribution STATCOM (DSTATCOM)
- Dynamic voltage restorer (DVR)
- Unified power quality conditioner (UPQC)

In this section we shall introduce these devices and will discuss only their idealized behavior.

4.2.1 Network Reconfiguring Devices

The schematic diagram of a solid state current limiter is shown in Figure 4.2. It consists of a pair of opposite poled switches in parallel with the current limiting inductor L_m. In addition, a series RC combination with a resistance of R_S and a capacitance of C_S is connected in parallel with the opposite poled switch. This RC combination constitutes the unpolarized snubber network [3]. The current limiter is connected in series with a feeder such that it can restrict the current in case of a fault downstream. In the healthy state the opposite poled switch remains closed. These switches are opened when a fault is detected such that the fault current now flows through the current limiting inductor. Let us illustrate its operating principle with the help of the following example.

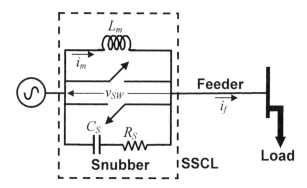

Figure 4.2. Schematic diagram of a solid state current limiter

Example 4.1: Let us consider the circuit of Figure 4.2 in which the source voltage is assumed to have a peak of 1.0 per unit and has a frequency of 50 Hz. The feeder has an impedance of $0.05 + j0.2$ per unit, while the load impedance is given by $2.0 + j1.5$ per unit. The SSCL parameters are

$$X_m = \omega L_m = 0.5 \text{ per unit}, \; R_S = 0.1 \text{ per unit and } X_S = \frac{1}{\omega C_S} = 1.0 \text{ per unit}$$

where ω is the system frequency in radians. It is assumed that the system is operating in the steady healthy state when a fault occurs at the load bus.

In the absence of the current limiter, the fault current will have a steady state value of 5.0 per unit but it can shoot up to about 20-25 per unit depending on the instant of the occurrence of the fault. To limit the fault current, the opposite poled switch opens once the occurrence of the fault is detected. In this case we have assumed that the fault is detected the moment it has occurred. This however is not a valid assumption and more realistic results will be discussed in Chapter 6.

The system response is shown in Figure 4.3, which depicts the fault current through feeder I_f, the voltage across the switch v_{SW} and the current through the limiting inductor I_m (see Figure 4.2). It can be seen that before the occurrence of the fault, both v_{SW} and I_m are zero. Once the fault occurs and the opposite poled switch opens, the capacitor C_S starts charging and the current through the limiting inductor rises linearly. The feeder current during this period rises sharply. However, the initial transient dies out quickly and the fault current is limited by the series combination of the SSCL and the feeder impedances. This condition persists if the fault is not cleared.

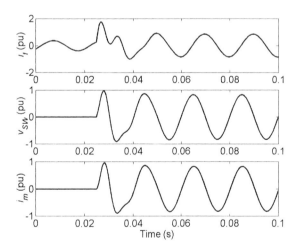

Figure 4.3. System behavior with SSCL during fault

Note that since the value of the snubber resistance R_S is negligible compared to the snubber reactance X_S the effective impedance of the SSCL during the faulted state will be the parallel combination of X_S and X_m. For the parameters chosen this will be inductive and will have a numerical value of 1.0 per unit. Therefore the total reactance in the circuit during the fault will be the sum of the feeder reactance and the effective SSCL reactance and this has a numerical value of 1.2 per unit. Therefore the peak of the feeder current assuming that the fault is not cleared is roughly 0.833 per unit as is evident from Figure 4.3.

Let us now investigate what happens when the snubber circuit is not present. As soon as the fault occurs, the inductance L_m is placed in series with the feeder inductance. However the current through the feeder has an initial value at that point while the initial value of the current through L_m is zero. But the same current must now flow through both these inductances. It is well known that the current through an inductance cannot change instantaneously. Therefore to bring the current through L_m to the same level as the current through the feeder, a large voltage equal to $L_m(di_m/dt)$ will be applied across the inductor. This is the same voltage v_{SW} that is applied across the opposite poled switch. This voltage is shown in Figure 4.4. It can be seen that the voltage has a spike with a maximum of about 27.0 per unit. It is needless to say that this will damage the semiconductor switch. Therefore the snubber circuit is essential for this configuration.

ΔΔΔ

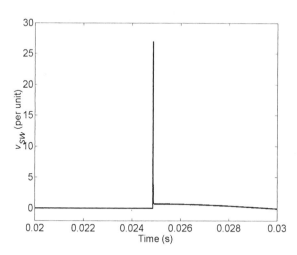

Figure 4.4. Voltage across the SSCL in the absence of the snubber

Note that the bidirectional switch of Figure 4.2 is usually made of GTOs (or may even be IGBTs) which have a fast current interrupting capability. A solid state circuit breaker (SSCB) has almost the same topology as that of an SSCL except that the limiting inductor is connected in series with an opposite poled thyristor pair. The thyristor pair is switched on simultaneously as the bidirectional switch is switched off once a fault is detected. This will force the fault current to flow through the limiting inductor in the same manner as discussed above. The thyristor pair is blocked after a few cycles if the fault still persists. The current through the thyristor pair will cease to flow at the next available zero crossing of the current. There still might be a small amount of current flow to the fault through the snubber circuit. However the magnitude of this current is small and this can be easily interrupted by a mechanical switch that is always placed in series with the SSCB. One example of the application of this structure is the earth current limiter where a SSCL is included in the neutral leg of transformers. Since 70% of faults are single line to ground faults this limiter will influence most faults and can often extinguish the arc of the faulted phase without actually breaking the curcuit. These circuits can often be implemented using SCR without the need of the more expensive current beak elements such as GTO.

The schematic diagram of a solid state transfer switch (SSTS) is shown in Figure 4.5. This device, which is also known as a static transfer switch (STS), is used to transfer power from the preferred feeder to the alternate feeder in case of voltage sag/swell or fault in the preferred feeder. The transfer switch would be used to protect sensitive loads. An SSTS contains two pairs of opposite poled switch. In this case the switch is made of thyristors. These switches are denoted by Sw_1 and Sw_2 in Figure 4.5. Suppose the preferred feeder supplies the power to the load. This is done through the switch Sw_1 while the switch Sw_2 remains open.

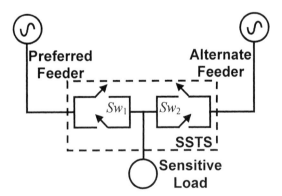

Figure 4.5. Schematic diagram of a static transfer switch

If a sudden voltage sag occurs in the preferred feeder, the SSTS then closes the switch Sw_2 such that current starts flowing through the alternate feeder to the load. The switch Sw_1 is then switched off. This switching scheme is known as make before break (MBB) switching in which the switch Sw_1 is disconnected only after the switch Sw_2 is connected. Note that if the local bus voltages appearing in both the preferred feeder side and the alternate feeder side have the same magnitude and phase there will be no transient in the load when the switch is operated in MBB. However this condition cannot be usually satisfied and in those cases a transient in the load current is unavoidable. Also note that it is not always possible to operate the device in MBB fashion when there is a fault in the preferred feeder. Depending on the direction of the current it may have to be operated in break before make (BBM) fashion otherwise the alternate feeder may start feeding the fault.

4.2.2 Load Compensation using DSTATCOM

The schematic diagram of a distribution system compensated by an ideal shunt compensator (DSTATCOM) is shown in Figure 4.6. In this it is assumed that the DSTATCOM is operating in current control mode. Therefore its ideal behavior is represented by the current source I_f. It is assumed that Load-2 is reactive, nonlinear and unbalanced. In the absence of the compensator, the current I_s flowing through the feeder will also be unbalanced and distorted and, as a consequence, so will be Bus-1 voltage.

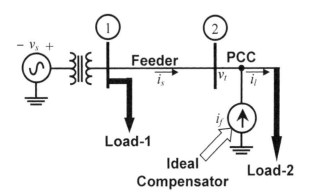

Figure 4.6. Schematic diagram of ideal load compensation

To alleviate this problem, the compensator must inject current such that the current I_s becomes fundamental and positive sequence. In addition, the compensator can also force the current I_s to be in phase with the Bus-2 voltage. This fashion of operating the DSTATCOM is also called load compensation since in this connection the DSTATCOM is compensating the load current. From the utility point of view, it will look as if the compensated load is drawing a unity power factor, fundamental and strictly positive sequence current.

The point at which the compensator is connected is called the utility-customer *point of common coupling* (PCC). Denoting the load current by I_l the KCL at the PCC yields

$$i_s + i_f = i_l \quad \Rightarrow \quad i_s = i_l - i_f \tag{4.1}$$

The desired performance from the compensator is that it generates a current I_f such that it cancels the reactive component, harmonic component and unbalance of the load current. Below we present two examples to illustrate the operating principle of the device.

Example 4.2: Let us first demonstrate the working principle of the DSTATCOM assuming that the source voltage is stiff. This implies that the source is connected at the point of common coupling and the feeder, Load-1 and the coupling transformer of Figure 4.6 is missing. The source voltage is assumed to be 50 Hz, balanced with the peak value of √2 per unit. It is supplying a three phase unbalanced RL load that are given by

$Z_a = 5 + j5$ per unit, $Z_b = 5 + j1$ per unit and $Z_c = 3 + j2$ per unit

where the subscripts *a*, *b* and *c* indicate the three phases. In addition to the RL, a three-phase controlled rectifier is also connected to the load. The rectifier draws a square wave current with a peak of 0.1 per unit and has a delay angle of 30°. The load currents are shown in Figure 4.7 (a). The distribution system is assumed to be a 3-phase, 4-wire system.

The ideal compensator now injects currents that cancel harmonics from load current and also balances the load. Furthermore, it also forces the current drawn from the source to be unity power factor. How the compensator achieves this will be discussed in detail in Chapter 7. Once these currents are injected, the source currents become balanced, harmonic free and unity power factor as shown in Figure 4.7 (b). In this figure a scaled version of the voltage of phase-a (denoted by v_{sa}) is also plotted to illustrate the source current of the corresponding phase is in phase with the voltage. The compensator currents are shown in Figure 4.7 (c), while the instantaneous powers are shown in Figure 4.7 (d). It can be seen that while the power supplied by the source (p_s) is constant, the power injected by the compensator has a zero mean. This implies that the compensator neither requires a real power from the source nor does it supply a real power to the load. The oscillating part of the load power is supplied by the compensator and the average part is supplied by the source such that the load power (p_l) is the sum of powers p_s and p_f.

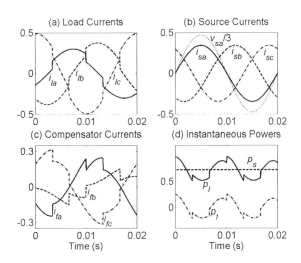

Figure 4.7. Load compensation with a stiff source

The structure of load compensation when the source is weak is shown in Figure 4.8. In addition to the ideal current source, this structure contains a capacitor. The reason for using this capacitor will be discussed in Chapter 8. It will be sufficient to mention at this point that it is used for filtering purposes. This provides a low impedance path to harmonics introduced by the compensator. Let us denote the voltage of PCC, which is the same as Bus-2 voltage, as v_t. The voltage across the capacitor is the same as the voltage at PCC or terminal voltage v_t. The current injected by the DSTATCOM is then given by (see Figure 4.8)

$$i_f = i_d - i_c \qquad (4.2)$$

Since $I_c = C(dv_t/dt)$, the current injected by the source I_d is known once I_f is known.

To explain how the compensator works, let us assume that the load current vector is given by

$$I_l = I_{lp} + jI_{lq} + I_h \qquad (4.3)$$

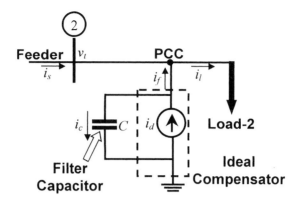

Figure 4.8. Compensator structure for non-stiff supply system

where the subscripts p, q and h denote the real, reactive and harmonic components respectively. Out of these three components, only the real component will be supplied from the source for unity power factor operation. The compensator must then inject the remaining part to the ac system. We than have

$$I_s = I_{lp}$$
$$I_f = jI_{lq} + I_h \qquad (4.4)$$

Since the current through the filter capacitor is given by $I_C = j\omega CV_t$, the current injected by the compensator is then given from (4.2) as

$$I_d = jI_{lq} + I_h + j\omega CV_t \qquad (4.5)$$

Note that this current has reactive and harmonic components only.

Example 4.3: Consider the distribution system shown in Figure 4.6 in which the source is taken to be the same as given in Example 4.2. It is assumed that it supplies two three-phase RL loads of which the load at Bus-1 is a balanced load of $0.5 + j0.2$ per unit per phase. Of and the load at Bus-2 is unbalanced. The balanced load at Bus-1 is, while the unbalanced load at Bus-2 is given by

$Z_a = 1.5 + j0.5$ per unit, $Z_b = 0.5 + j1$ per unit and $Z_c = 3 + j2$ per unit

4. Custom Power Devices: An Introduction

A 3-phase, 4-wire distribution system is assumed. The transformer of Figure 4.6 is neglected. The source reactance is assumed to be 0.1 per unit and the feeder impedance is $0.05 + j0.2$ per unit.

Let us first assume that the compensator is not connected to the system at the beginning, however the filter capacitor is. The steady state system response is then shown in Figure 4.9. It can be seen that all the system quantities are unbalanced. Also the load power oscillates with a frequency of 100 Hz and has a mean value of 0.9345.

Once the compensator is connected, the steady state system response is shown in Figure 4.10. It can be seen that the terminal voltage and currents are now balanced sinusoids and the terminal power is constant with a value of 1.0639 per unit. This is also the mean load power. There has been about 10% increase in the mean power because the terminal voltages are now balanced. Also since the power entering the terminal is the mean of the load power, the compensator only supplies the zero mean oscillating power required by the load.

ΔΔΔ

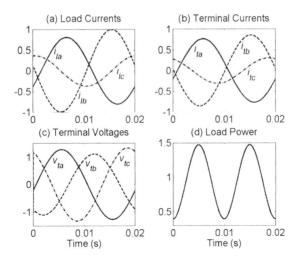

Figure 4.9. System response without compensator

The above two examples demonstrate the working principle of the shunt compensator. We must however note a few points.

- The compensators in these examples are realized by ideal current sources. In a practical circuit these are realized by an inverter, the output of the inverter being connected to the PCC by a transformer or an

interface inductor. The inverter then tracks the reference current required to achieve load compensation.
- An important aspect of the shunt load compensation is the generation of reference currents that achieve the desired performance. This will be discussed in Chapter 7.
- In a 3-phase, 4-wire distribution system, the compensator neutral is connected to the load neutral. This will force the zero-sequence current to circulate in the path such that the current drawn by the combined compensator-load combination becomes balanced.

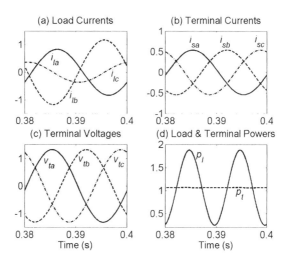

Figure 4.10. Steady state compensated waveforms in weak ac system

4.2.3 Voltage Regulation using DSTATCOM

The schematic diagram of an ideal shunt compensator acting as a voltage regulator is shown in Figure 4.11 (a). In this the ideal compensator is represented by a voltage source and it is connected to the PCC. However it is rather difficult to realize this circuit and the alternate structure is shown in Figure 4.11 (b). It can be seen that this is the same structure as used for load compensation in Figure 4.8. It has the advantage that the harmonics can be bypassed by the filter capacitor C.

The basic idea here is to inject the current i_d in such a way that the voltage v_t follows a specified reference. The compensator must be operated such that it does not inject or absorb any real power in the steady state. Therefore the relations given in (4.4) and (4.5) are also valid in this case. The magnitude of the voltage v_t can be arbitrarily chosen. However its phase angle must be chosen such that the relation $I_s = I_{lp}$ is satisfied.

4. Custom Power Devices: An Introduction

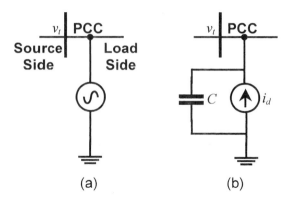

Figure 4.11. (a) Ideal voltage controller and (b) its practical realization

Example 4.4: Let us consider a system in which both the source voltages and load currents are unbalanced and distorted. Consider the distribution system shown in Figure 4.1. Assume that the loads at both Buses 1 and 3 are unbalanced and distorted. Then the Thevenin equivalent of the source side can be represented by an impedance that is connected to an unbalanced and distorted source. Similarly the Thevenin equivalent of the right side of the bus can be represented by an unbalanced and distorted load. The system parameters are not important for this discussion.

The source voltages are shown in Figure 4.12 (a). It can be seen that they are both unbalanced and distorted. The compensator is pressed into service at the end of the 1st cycle (at 0.02 s). The load currents, terminal voltages and the source currents prior to that instant are distorted both due to the load and source. Once the compensator is connected, the terminal voltages become balanced with two cycles. However the load currents are still unbalanced and distorted since the load is unbalanced and distorted. Similarly since the source is unbalanced and distorted, the currents entering the terminal (PCC) are still unbalanced and distorted. However, these distortions have now reduced significantly.

ΔΔΔ

4.2.4 Protecting Sensitive Loads using DVR

A dynamic voltage restorer (DVR) is used to protect sensitive loads from sag/swell or disturbances in the supply voltage. The implication of such a disturbance to food processing industry has been discussed in Section 2.2.6., The world's first DVR was installed in August 1996 at a 12.47 kV substation in Anderson, South Carolina. This was installed to provide protection to an automated rug manufacturing plant. Prior to this connection, the DVR was first installed at the Waltz Mill test facility near Pittsburgh for full power tests. The test results are discussed in [4]. The next commissioning of a DVR was in February 1997 at a 22 kV distribution system at Stanhope, Victoria, Australia. This was done to protect the diary milk processing plant mentioned in Section 2.2.6. The saving that may result from the installation of this DVR is estimated to be over $100,000 per year [5]. In the next phase of development, DVRs that can be mounted on an overhead platform supported by two poles were fabricated. The first platform mounted DVR is installed to protect Northern Lights Community College and several other smaller loads in Dawson Creek, British Columbia, Canada [6].

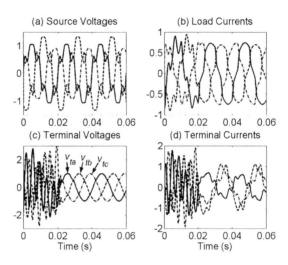

Figure 4.12. System quantities with DSTATCOM in voltage control mode

The schematic diagram of a sensitive load protected by an ideal series compensator (DVR) is shown in Figure 4.13. In this the DVR is represented by an ideal voltage source that injects a voltage v_f in the direction shown. There are two different ways of constructing this device. The DVR can be constructed such that it is either capable or not capable of supplying or absorbing real power. The DVR voltage control is simple if it is capable of supplying or absorbing real power. Note from Figure 4.13 that

$$v_l = v_t + v_f \qquad (4.6)$$

where v_l is the load bus voltage. The DVR then can regulate the bus voltage to any arbitrary value by measuring the terminal voltage v_t and supplying the balance through v_f.

4. Custom Power Devices: An Introduction

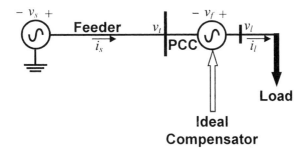

Figure 4.13. Schematic diagram of a sensitive load protected by a DVR

The solution to this problem however is not as straightforward when the DVR is not capable of supplying or absorbing any real power in the steady state. It may instaed have to supply or absorb real power during transients. Note from Figure 4.13 that the current through the line I_s is the same as the current through the load I_l. Also the phase angle difference between the load current I_l and the load voltage v_l is dictated by the power factor of the load. Also for no real power injection or absorption, the positive sequence fundamental frequency component of the voltage v_t must be in quadrature with the positive sequence fundamental frequency component with the load current I_l. The following example illustrates the DVR operation.

Example 4.5: Let us assume that an unbalanced and distorted source supplies the distribution system of Figure 4.13. The source voltages are shown in Figure 4.14 (a). The system frequency is 50 Hz. The feeder impedance is $0.05 + j0.3$ per unit. The load is assumed to be balanced RL with per phase impedance of $2 + j1.5$ per unit. The compensator is connected to the system at the end of the 1st cycle (0.02 s). It can be seen from Figure 4.14 that prior to this instant, the load voltages and currents are unbalanced and distorted and the power through the compensator is zero.

The compensator is represented by an ideal voltage source that is capable of generating fundamental as well as harmonic voltages. The compensator voltages are then obtained as per the following steps:

1. First extract the fundamental positive sequence of the voltages v_t and v_l and the current I_l. Then compute the power factor angle.
2. Generate the fundamental positive sequence component of the voltage v_f that is in quadrature with the line current from the measurements of step 1 such that the load terminal voltage is regulated to the desired value.

3. Since the load terminal voltage must be balanced sinusoidal, subtract the harmonic and unbalanced component of the terminal voltage from the fundamental positive sequence of the voltage v_f obtained from step 2 such that the injected voltage cancels out these components as per (4.6).

The system results are shown in Figure 4.14. It can be seen that both the load terminal voltage and, as a consequence, the load currents become balanced sinusoids within one cycle of the connection of the compensator at 0.02 s. The power through the compensator is oscillating with a mean of zero. Therefore the compensator neither absorbs nor injects any real power.

ΔΔΔ

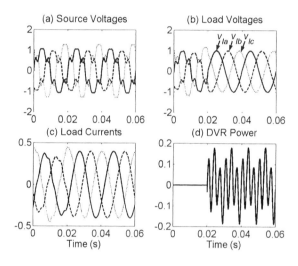

Figure 4.14. System quantities with an ideal series compensator

4.2.5 Unified Power Quality Conditioner (UPQC)

The schematic diagram of a unified power quality conditioner (UPQC) compensated distribution system is shown in Figure 4.15 (a). This is useful when both source and load are unbalanced and distorted. For example assume that the source voltage v_s is both unbalanced and distorted. Also the load current i_l is also unbalanced and distorted. As a consequence the terminal voltage v_t, the load voltage v_l and the source current i_s will also be unbalanced and distorted. Now suppose there are other customers connected to the load bus that draw purely balanced sinusoidal currents. Then both the source and load unbalance and distortion affect them. Again if there is a load bus upstream from the point of common coupling, the customers on that bus will equally get affected. A UPQC can alleviate this problem.

4. Custom Power Devices: An Introduction

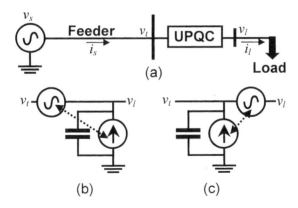

Figure 4.15. (a) Schematic diagram of a UPQC compensated system, (b) & (c) two alternate connections

A UPQC combines a series and a shunt compensator together. It can therefore yield the benefits of both these devices. For example it can tightly regulate the load bus voltage v_l shown in Figure 4.15 (a). Therefore all loads including the unbalanced and nonlinear load will have a supply voltage that is balanced and sinusoidal. The UPQC can also make the current drawn from the supply (i_s) balanced, sinusoidal and in phase with the terminal voltage (v_t). Therefore the voltage of any bus upstream from the PCC will not be affected due to a nonlinear and unbalanced load. However it will be impossible to correct the unbalance and distortion produced by the source voltage using this device. Therefore the upstream bus voltages will remain unbalanced and distorted.

There are two different ways of connecting a UPQC. These are shown in Figure 4.15 (b) and (c). In the connection of Figure 4.15 (b) the series device is placed before the shunt device while it is placed after the shunt device in Figure 4.15 (c). The operating principles of these two connections will be discussed in Chapter 10. The dotted line in these figures indicates any energy exchange path between the devices. Usually the inverter realizing the series device is supplied by a dc capacitor. Similarly the shunt inverter is also supplied by a dc capacitor. In a UPQC both these inverters are supplied by a common dc capacitor in the same way as a unified power flow controller used in bulk power transmission [7]. The energy exchange between the series and the shunt device takes place through this common dc capacitor.

4.3 Custom Power Park

In a custom power park all customers of the park should benefit from high quality power supply. Even the basic form of this supply is superior to the normal power supply from a utility. Electrical power to the park is supplied through two feeders from two independent substations as shown in Figure 4.16. Both these feeders are joined together via a solid state transfer switch. This can make a subcycle transfer from the preferred to the alternate feeder such that the duration of any voltage dip can be reduced to 4-8 ms. The custom power control center is fully equipped with a DSTATCOM and a DVR. The DSTATCOM eliminates harmonics and/or unbalance, while the DVR eliminates any voltage sag or distortion.

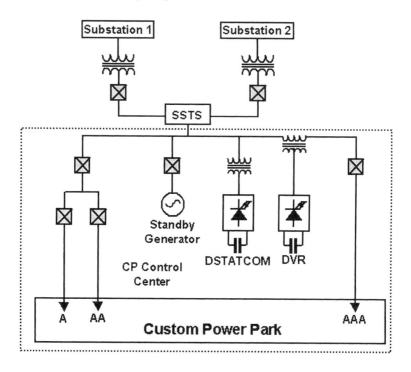

Figure 4.16. A custom power park

The incoming feeders to the park can be designed with improved grounding, insulation, arresters and reclosing. The SSTSs ensure that the feeder with higher voltage is selected in less than half a cycle in case of a voltage dip. Using similar switching principle, these SSTSs can also be used to protect the loads of the park from dynamic overvoltages as well. The DSTATCOM, when operated in the voltage control mode, can provide reactive power support to the park thereby maintaining the bus voltage. Under the configuration shown in Figure 4.16, there are three different grades of power that can be supplied to the park's customers. These are [8]

4. Custom Power Devices: An Introduction

- **Grade A**: This is the basic quality power. Since the transfer switches protect the incoming feeders, the quality of the power is usually superior to normal utility supply. In addition, this grade has the benefit of low harmonic power due to the DSTATCOM.
- **Grade AA**: This includes all the features of Grade A. In addition, it receives the benefit of a standby generator. The generator can be brought into service in about 10-20 seconds in case of a serious emergency such as power failure in both feeders.
- **Grade AAA**: This includes all the features of Grade AA. In addition, it has the benefit of receiving distortion or dip free voltage due to the DVR.

Under normal operating conditions, the standby generator stays off and is disconnected from the bus and customers of both Grade A and Grade AA receive the same service. The difference in their power quality becomes apparent when power to both feeders is lost simultaneously due to a fault in the power system. When both feeders are lost, the loads belonging to customers of both these grades are removed through circuit breakers. However, once the standby generator is switched on, the power to AA grade customers is restored by closing of the circuit breakers connected to the AA grade customers. This means that customers of AA grade do not lose power for more than 10-20 seconds under any situation. The customers of A grade however do not receive power until one of the feeders is back in service. The Grade AAA customers are not affected by the loss of both feeders as their voltage is maintained by the DVR till the standby generator comes into service. Since the standby generator has to be started instantaneously, brought to speed and synchronized within less than 20 seconds, it must be driven by a diesel or a gas turbine.

It is to be noted that the DVR must provide voltage support for the AAA grade supply in case of both feeder failures till the standby generator is synchronized with the bus. This implies that it may have to provide this support for about 20 seconds. It is therefore important to ascertain whether a DVR that is supplied by an ac storage capacitor is capable of holding the voltage for such a long time or not. Alternatively, the DSTATCOM and the DVR in the custom power control center can be replaced by a single UPQC that will be able to perform both the functions of harmonic neutralization and voltage balancing simultaneously.

Through the custom power park it is possible to supply power to different types of sensitive loads ranging from shopping malls and hospitals to semiconductor manufacturing. For example, a semiconductor manufacturing plant needs Grade AAA supply since a sudden voltage dip can cause the loss of a few hours of production. With modern and life support equipment, a hospital on the other hand, requires both AA and AAA grade supplies. The

AAA grade connection can be supplied to the operating theaters and life support systems, while the AA grade connection can be given to the rest of the building along with pathology and other testing facilities. Most shops in a shopping center (mall) or offices in an office building require grade A power. The grade of the quality of power a customer in the park receives depends on the nature of its load and the price he is ready to pay.

4.4   Status of Application of CP Devices

It was mentioned in the beginning of this chapter that the installations of the custom power (CP) devices are not very well documented. It is thus rather difficult to get the current installation status. The status given below is reported in [9] and by no means presents the complete picture.

- American Electric Power (AEP) has installed an indoor 15 kV, 600 A static transfer switch at an industrial park in Columbus, Ohio.
- Baltimore Gas and Electric (BGE) installed an indoor 15 kV, 600 A static transfer switch at an office building in downtown Baltimore in September 1995.
- In September 1996, BGE has also placed an outdoor 15 kV, 600 A static transfer switch in service at a chemical manufacturing plant in the Baltimore metropolitan area.
- In 1991, Chubu Electric Corporation of Japan installed three 7.2 kV, 300 A static transfer switches in a loop line configuration.
- Commonwealth Edison Company installed a 12.47 kV, 600 A static transfer switch on August 14, 1996 at a plastic film manufacturing plant. Between January 1 and October 15, 1997, there were 50 events with 40 successful transfers with no loss of production. Of the remaining 10 events that resulted in production interruptions, 5 events were due to voltage sags on both feeders.
- In November 10, 1996 Detroit Edison Company installed a static transfer switch at the Ford Motor Company Sheldon Road Plant, which provides components to all Ford's North American assembly plants. Its installation has saved a whole lot of expensive shutdown time.
- PG&E Energy Services installed two static transfer switches, both rated at 25kV, 300A, in September 1996.
- In October 1996 Texas Utilities placed in service an outdoor 15 kV, 600 A static transfer switch at an electric operations building in Fort Worth, Texas.
- The Tokyo Oil Industry Company of Japan installed a static transfer switch in 1997 for a generating unit transfer application.

4. Custom Power Devices: An Introduction

- Duke Power's installation in August 1996 has been reported in Chapter 9. The unit is rated at 2 MVA, can store 660 kJ, and operates at 12.47 kV. Normal load current is approximately 120 A. Its test results are given in [4].
- The DVR installed at Bonlac Foods Processing plant by Powercor Australia Ltd. Is rated at 2 MVA, can store 660 kJ, and operates at 22 kV.
- In April 1997 a DVR was installed at the Sappi Limited, Stanger Mill in South Africa that provides pulp to Sappi's paper making process. The DVR is supplied by a superconducting magnetic energy storage (SMES). The power to the mill is supplied by ESKOM. This DVR is rated at 750kVA and provides 2.4MJ of energy storage.
- Florida Power Corporation's 2 MVA inverter-based DVR provides protection to one of six 12.47 kV feeders at the Econ Substation (230/12.5 kV) of Orlando, Florida. This is placed in service in 1996 in a high density residential and commercial area.
- Two DVRs, each rated at 6 MVA, 12.47 kV with 1800 kJ energy storage, have been installed in July 1998 at a critical industrial site on the Salt River Project system at Phoenix metropolitan area in Arizona. Each of these two DVRs can boost a 20 MVA load as much as 30% and has a maximum capacity of 1200 A.
- In April 1998 a 4 MVA series compensator is installed by ScottishPower at the Caledonian Paper Mill at Irvine, Scotland.
- The installation of a platform mounted DVR by B.C. Hydro in Dawson Creek has been mentioned earlier in this chapter.
- Since February 1998 American Electric Power has been operating a DSTATCOM at a rock crushing facility. The rating of the device is ± 2 MVA at 12.47 kV and it utilizes two 1 MVAr capacitor banks that allow operation with output from 0 to 4 MVAr capacitive.
- British Columbia Hydro has demonstrated the operation of a trailer mounted DSTATCOM at the Adams Lake Lumber Company in Chase, British Columbia, Canada where it provided voltage regulation and voltage flicker mitigation caused by a large whole log chipping operation. This ± 2 MVA inverter-based trailer unit is now being readied by BC Hydro for relocation to another site.
- In 1993 an SVC of 60 MVAr capacity was installed at 77 kV line as a countermeasure against voltage fluctuations and unbalanced load due to an electric railroad by Central Japan Railway Company. Motivated by its success, the Central Japan Railway Company installed three more SVCs that are rated 60 MVAr at 154 kV, 34 MVAr at 77 kV and 48 MVAr at 154 kV.
- In December 1989 a DSTATCOM of 3.5 MVAr capacity using a bipolar transistor inverter for arc furnace flicker compensation was installed on

the Mitsubishi Steel Company's 33 kV feeder. In addition, Mitsubishi Steel Company is operating an SVC since 1984. The addition of the DSTATCOM enabled an increase in steel productivity of the arc furnace without any increase in the previous flicker level.
- A 1.2 MVA static series compensation device has been in service on the Public Service Electric and Gas system in New Jersey since September 1994.
- A static var compensator using an 8 MVAr GTO inverter for arc furnace flicker compensation was installed on the Sumitomo Steel Company's 22 kV feeder in July 1995.

4.5 Conclusions

In this chapter we have discussed the use of custom power devices for improving the distribution system quality. In this chapter we have only discussed the advantages of using these devices. However, as we have mentioned earlier, there are many important issues that need to be considered for obtaining an overall global picture. The most important issue is whether the installation of any of these devices adversely affects a customer in the vicinity that is not supported by such a device.

4.6 References

[1] N. G. Hingorani, "Introducing custom power," *IEEE Spectrum*, Vol. 32, No. 6, pp. 41-48, June 1995.
[2] T. J. E. Miller, ed., *Reactive Power Control in Electric Systems*, John Wiley, New York, 1982.
[3] M. H. Rashid, *Power Electronics: Circuits, Devices and Applications*, Prentice-Hall, Englewood Cliffs, 1993.
[4] N. H. Woodley, L. Morgan and A. Sundaram, "Experience with an inverter-based dynamic voltage restorer," *IEEE Trans. Power Delivery*, Vol. 14, No. 3, pp.1181-1185, 1999.
[5] N. H. Woodley, A. Sundaram, B. Coulter and D. Morris, "Dynamic voltage restorer demonstration project experience," 12th *Conf. Electric Power Supply Industry (CEPSI)*, Pattaya, Thailand, 1998.
[6] N. H. Woodley, K. S. Berton, C. W. Edwards, B. Coulter, B. Ward, T. Einarson and A. Sundaram, "Platform-mounted DVR demonstrated project experience," 5th *International Transmission & Distribution Conf, Distribution*, Brisbane, 2000.
[7] L. Gyugyi, "A unified power flow control concept for flexible ac transmission systems," *Proc. IEE*, Pt. C, Vol. 139, No. 4, pp. 323-331, 1992.
[8] N. G. Hingorani, "Custom Power and Custom Power Park," *Flexible Power HVDC Transmission and Custom Power, CIGRE Australian Panel 14*, Sydney, 1999.
[9] IEEE P1409 Distribution Custom Power Task Force 2, *Custom Power Technology Development*, 1999.

Chapter 5

Structure and Control of Power Converters

Apart from the breaking and transferring devices, all other power quality (PQ) enhancement devices like DSTATCOM, DVR, UPQC etc. are based on power converters. Furthermore modern FACTS devices like STATCOM, SSSC, UPFC etc. also employ power converters. However, FACTS devices have much higher power rating than PQ enhancement devices since they are used in bulk power transmission systems. Moreover, their operation philosophy is also different as they are assumed to work under balanced sinusoidal conditions. As a consequence, the control strategies of FACTS devices are different from the PQ enhancement or Custom Power devices. Since power converters have an important role to play in modern power systems, we discuss their topologies and control strategies in this chapter. For the background materials in the area of Power Electronics, there are numerous excellent textbooks, e.g., [1-3].

AC power converters are of two types – current source converter (CSC) and voltage source converter (VSC). The schematic diagrams of these converters are shown in Figure 5.1. A converter has a dc side, a power circuit made of power semiconductor switches and an ac side. The power circuit for both CSC and VSC is the same. The ac side of both converters is connected to loads or is interfaced to ac systems. It is the dc side that differentiates these two converters. The dc input to a CSC (also known as current source inverter or CSI) is a dc current source. This current source is usually realized by a controlled dc source that is connected to a large inductor in series. The dc input to a VSC (also known as voltage source inverter or VSI) is a dc voltage source that is usually realized by rectifying an ac voltage through a diode bridge rectifier. An LC filter with a small series inductor and a relatively large shunt capacitor is connected to the output of the rectifier bridge.

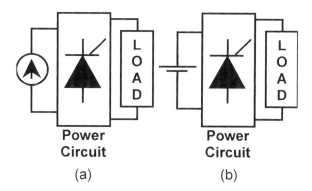

Figure 5.1. Schematic diagram of a (a) CSC and (b) VSC

A CSC is usually more reliable and fault tolerant than a VSC because the large series inductor limits the rate of rise of current in the event of a fault. However, CSCs have higher losses because of the need to store energy by circulating current in inductors which are more lossy than capacitive energy storage. In addition these inductors can limit the rate of response of the system. The thyristor based versions are mostly used for high power electric motor drives. In power line conditioning devices, the dc sources are not desired and pure energy storing devices are used – an inductor in a CSC and a capacitor in a VSC. Since capacitors are more efficient, smaller and less expensive than inductors, VSCs are most commonly used in Custom Power devices. Henceforth we shall restrict our discussions to VSCs only.

5.1 Inverter Topology

In this section we shall discuss simple single-phase and three-phase inverter topologies. We shall also present their dynamic models and illustrate the functioning of these inverters with the help of some simulation results. For the time being, we shall assume that the inverters are operated in the current control mode in which they track given reference currents. The inverter control will be discussed in detail later in the chapter.

5.1.1 Single-Phase H-Bridge Inverter

The schematic diagram of a single-phase H-bridge inverter is shown in Figure 5.2. It is called an H-bridge as it looks like the eighth letter of the English alphabet. The inverter contains four switches S_1–S_4, each comprising a power semiconductor device and an anti-parallel diode as also indicated in the figure. The power semiconductor device can be a power MOSFET for

5. Structure and Control of Power Converters

low power application or a gate turn-off (GTO) thyristor for high power application. However, for distribution system applications, the preferred device usually is the insulated gate bipolar transistor (IGBT) as it can carry fairly large current and has fast switching characteristics and low losses.

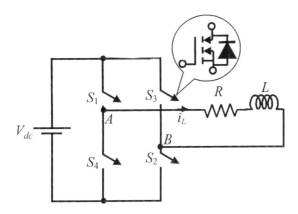

Figure 5.2. Schematic diagram of an H-bridge inverter

The load, which in this case is assumed to be a passive RL load, is connected between the two legs of the inverter. The inverter is supplied by a dc source with a voltage of V_{dc}. The switches of each leg usually have complementary values, e.g., when S_1 is on, S_4 is off and vice versa. Further the switches are operated in pairs. When the switches S_1 and S_2 are on, S_3 and S_4 are off. Similarly, when S_3 and S_4 are on, S_1 and S_2 are off. However, a small time delay is provided between the turning off a pair of switches and turning on the other pair. This period, called the blanking period, is provided to prevent the dc source from being short-circuited. Consider for example the transition when S_3 and S_4 are turned off and S_1 and S_2 are turned on. During this period if the switch S_1 gets turned on before the switch S_4 turns off completely, then it will connect the two leads of the dc source directly. It is therefore mandatory that switch S_4 turns off completely before the switch S_1 is turned on. To ensure this, the blanking period (deadtime) is used. The continuity of the current during the blanking period is maintained by the anti-parallel diodes. For deriving the system dynamic equations, we shall neglect the blanking period. Then the equivalent circuit diagrams for the two modes of operation are shown in Figure 5.3. The consequence of this blanking period means that the switches do not change state exactly when commanded, giving imperfections with the modulation of the inverters.

Let the current flowing through the load be denoted by i_L (see Figure 5.2). Then from the equivalent circuit shown, the expression for this current is given by

$$\frac{d}{dt}i_L = -\frac{R}{L}i_L + \frac{1}{L}V_{dc}u \qquad (5.1)$$

where the variable u is defined as

$$u = \begin{cases} 1 & \text{when } S_1 \text{ and } S_2 \text{ are on} \\ -1 & \text{when } S_1 \text{ and } S_2 \text{ are on} \end{cases} \qquad (5.2)$$

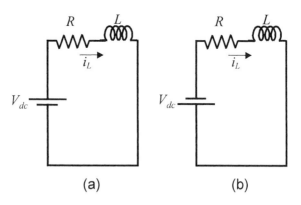

Figure 5.3. Equivalent circuit of the H-bridge inverter (a) when S_1 and S_2 are on and (b) when S_3 and S_4 are on

Example 5.1: In this example we shall demonstrate the operation of the inverter in a hysteresis band current control mode through digital computer simulation. The system parameters chosen for the study are

$V_{dc} = 200$ V and $R = 10\,\Omega$

The load inductance will be chosen later. Let us assume that the inverter has to track a reference current, the instantaneous value of which is given by

$$i_{Lref} = 10\left\{\sin\omega t + \frac{1}{5}\sin 5\omega t + \frac{1}{7}\sin 7\omega t\right\} \qquad (5.3)$$

where $\omega = 100\pi$.

The reference current is shown in Figure 5.4. Along with this current, we have also shown the *hysteresis band* which is the region between $I_{Lref} + h$ and

5. Structure and Control of Power Converters

$I_{Lref} - h$ for any scalar h. The inverter is then switched (i.e., the variable u is chosen) using the following logic

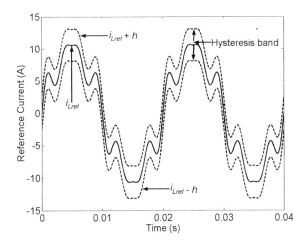

Figure 5.4. Reference current and hysteresis band

If $i_L \geq i_{Lref} + h$ then $u = -1$
elseif $i_L \leq i_{Lref} - h$ then $u = 1$ (5.4)

The rationale behind this logic is as follows: if the load current exceeds the upper limit of the band $i_{Lref} + h$, then bring the current down by applying a negative voltage across the load. This can be done by turning on the switches S_3 and S_4 (the equivalent circuit of which is shown in Figure 5.3 b). Similarly, when the current falls below the lower limit of $i_{Lref} - h$, it is then raised by turning on the switches S_1 and S_2 (the equivalent circuit of which is shown in Figure 5.3 a). Thus, theoretically speaking, if the load current waveform is monitored properly and the inverter switches are made to change their states the instant at which the current touches the limit, the load current will remain within the band. This however may not be achievable due to the finite time delays associated with the monitoring and gating circuits and also due to the blanking period. Moreover the load current is usually inductive and cannot change very rapidly. Therefore, it is likely that the load current may slightly overshoot the band.

The simulation results are shown in Figures 5.5 and 5.6. Figure 5.5 shows the load current while Figure 5.6 shows the tracking error. For the values of V_{dc} and R mentioned above, two different values of the inductor are chosen.

Also different values of the scalar constant h is chosen. The results are summarized in Table 5.1. From the results the following conclusions can be made:

- The tracking becomes better as the hysteresis band becomes narrower.
- As the band becomes narrower, the switching frequency becomes higher. This is obvious as the current touches and overshoots a narrower band more frequently resulting in the faster changing of switch states. Unfortunately, high switching frequency results in increased losses culminating in increased heating in the power semiconductor devices. Thus the choice of hysteresis band is a compromise between tracking error and inverter losses.
- For the same value of the hysteresis band, the switching frequency decreases as the value of the inductor increases. This is obvious as the rate of change in the inductor current (di_L/dt) decreases with the increase in the value of the inductor. It is needless to say that the tracking with bigger sized inductor becomes inferior compared to a lower sized inductor for the same value of h. This is evident from Figure 5.6 (b-c).
- Even by decreasing the hysteresis band, the tracking performance may not always be acceptable. This is evident from Figure 5.6 (d). The tracking performance can be improved by increasing di_L/dt. For a large inductor this can be achieved by increasing the dc voltage.

ΔΔΔ

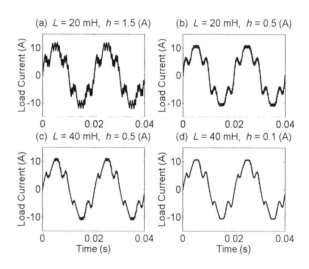

Figure 5.5. Load current for various load inductors and hysteresis bands

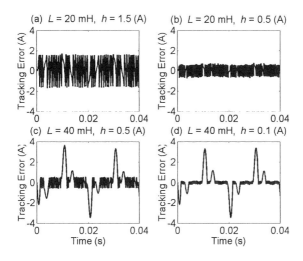

Figure 5.6. Tracking error for various load inductors and hysteresis bands

Table 5.1. Summarized results of Figures 5.5 and 5.6

L (mH)	H (A)	Results shown in Figures	Switching Frequency (Hz)	Comments
20	1.5	5.5 (a), 5.6 (a)	1150	Larger tracking error, but faster response
20	0.5	5.5 (b), 5.6 (b)	3175	Smaller tracking error and faster response
40	0.5	5.5 (c), 5.6 (c)	1075	Larger tracking error and slower response
40	0.1	5.5 (d), 5.6 (d)	4275	Smaller tracking error, but slower response

From the above example we can infer that the tracking performance is inversely proportional to the load inductor and the size of the hysteresis band and directly proportional to the magnitude of dc voltage. Furthermore, better tracking can be achieved at the expense of higher inverter losses.

5.1.2 Three-Phase Inverter

The schematic diagram of a three-phase inverter is shown in Figure 5.7. It contains six switches $S_1 - S_6$ each of which is made of a power semiconductor device and an anti-parallel diode. The switches of each leg are complementary. For example when S_1 is on, S_4 is off and vice versa. The inverter is supplied by two equal dc sources, the neutral point of which is denoted by N. A three-phase load is connected at the output of the inverter.

The load neutral is denoted by *n*. Note that whether the points *N* and *n* are joined depends on the system configuration required.

To demonstrate the derivation of the dynamic model of the system, we have assumed the load to be RL in this case also. Using the same framework, the system dynamics for other loads can also be derived. Because of the presence of the anti-parallel diodes, we can express the potential across the phase-a and the source neutral as

$$V_{aN} = \begin{cases} +\dfrac{V_{dc}}{2} & \text{if } S_1 \text{ is on} \\ -\dfrac{V_{dc}}{2} & \text{if } S_4 \text{ is on} \end{cases} \qquad (5.5)$$

Similar expressions can also be written for the other two phases.

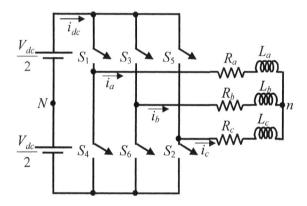

Figure 5.7. Schematic diagram of a three-phase inverter

Let us define the following switch states (or control variables)

$$u_a = \begin{cases} +1 & \text{when } S_1 \text{ is on} \\ -1 & \text{when } S_4 \text{ is on} \end{cases}$$

$$u_b = \begin{cases} +1 & \text{when } S_3 \text{ is on} \\ -1 & \text{when } S_6 \text{ is on} \end{cases} \qquad (5.6)$$

$$u_c = \begin{cases} +1 & \text{when } S_5 \text{ is on} \\ -1 & \text{when } S_2 \text{ is on} \end{cases}$$

5. Structure and Control of Power Converters

We also assume that the neutral path between the source and load (i.e., the path Nn) is open in a three-phase, three-wire configuration. Since the neutral point is floating, we can write the following equation

$$i_a + i_b + i_c = 0 \tag{5.7}$$

Also since the voltage V_{nN} between the points n and N is non-zero, the state equation for each phase can be written as

$$R_a i_a + L_a \frac{di_a}{dt} = \frac{V_{dc}}{2} u_a - V_{nN} \tag{5.8}$$

$$R_b i_b + L_b \frac{di_b}{dt} = \frac{V_{dc}}{2} u_b - V_{nN} \tag{5.9}$$

$$R_c i_c + L_c \frac{di_c}{dt} = \frac{V_{dc}}{2} u_c - V_{nN} \tag{5.10}$$

The dc side current is given by

$$i_{dc} = i_a u_a + i_b u_b + i_c u_c \tag{5.11}$$

Since (5.7) must always be satisfied, the three-phase, three-wire configuration is not suitable for tracking three independent currents.

Let us now assume that the two neutrals are connected together (i.e., the path nN is joined) in a three-phase, four-wire configuration. Then the constraint (5.7) can no longer be imposed on this circuit. Furthermore, since for this configuration $V_{nN} = 0$, each of the equations (5.8)-(5.10) becomes similar to (5.1). Therefore the three-phase inverter becomes equivalent to three single-phase inverters. Thus this configuration will allow a current to be tracked in the same manner as given in Example 5.1 irrespective of the currents in the other two phases. We can make the inverter to track currents, even unbalanced currents. In the general case there will be neutral current flowing in the path nN. Even when the load is balanced, high frequency current due to switching action will flow in this path. The following example illustrates the idea.

Example 5.2: Let us assume that the three-phase inverter is required to track a balanced load. The system parameters are

$$\frac{V_{dc}}{2} = 200\,\text{V},\ R = 10\,\Omega,\ L = 20\,\text{mH and } h = 0.1\,\text{A}$$

Let us assume that the inverter has to track the three-phase reference currents in hysteresis current control. These reference currents are

$$i_{aref} = I_{am}\left\{\sin \omega t + \frac{1}{5}\sin 5\omega t + \frac{1}{7}\sin 7\omega t\right\}$$

$$i_{bref} = I_{bm}\left\{\sin(\omega t - 120°) + \frac{1}{5}\sin 5(\omega t - 120°) + \frac{1}{7}\sin 7(\omega t - 120°)\right\}$$

$$i_{cref} = I_{cm}\left\{\sin(\omega t + 120°) + \frac{1}{5}\sin 5(\omega t + 120°) + \frac{1}{7}\sin 7(\omega t + 120°)\right\}$$

We shall demonstrate the functioning of the inverter first with balanced reference currents and then with unbalanced reference currents.

Let us first assume that the reference currents are balanced such that I_{am}, I_{bm} and I_{cm} are all equal to 10 A. The load currents are shown in Figure 5.8 (a-c). It can be seen that even though the load currents are balanced and do not contain any triplen harmonics, the neutral current (I_{nN}) is not zero in the steady state. The initial excursion neutral current is due to the delay in the inverter circuit in forcing the reference currents through the inductive load. However, the transient dies down very fast and the neutral current settles around zero. There is a high frequency ripple in the neutral current due to the switching action, but the average current has the desired value of zero.

Let us now choose an unbalanced reference current in which I_{am} = 10 A, I_{bm} = 7 A and I_{cm} = 12 A. The current tracking results are shown in Figure 5.9. It can be seen that there is a significant neutral current in this case. In addition, the high frequency switching ripples are also present in the neutral current. Note that this control directly aims to control the peak error in the phase currents but there is no direct control of the neutral current so its error can be the sum of the errors in the phase currents.

ΔΔΔ

5.2 Hard-Switched Versus Soft-Switched

The inverter circuits of Figures 5.2 and 5.7 are called hard-switched inverters since when they are switched, a full voltage is applied across the switches. This may result in stress in the power semiconductor switches reducing their life span but, more significantly, increased losses in the inverter. Soft-switched converters do not suffer from this problem as all the

5. Structure and Control of Power Converters   147

switches in a dc-to-ac converter are operated when the voltage across them is zero. This gives low loss in the switches. In the hard switched inverter however, when a switch is being turned off, first the voltage rises rapidly then the current falls depending on the recombination time of the carriers inside the semiconductor switch. It is the combination of high voltage with high current which generates a high level of switching loss in a converter.

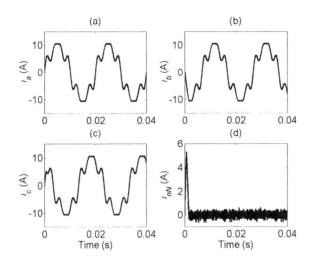

Figure 5.8. Balanced current tracking by the three-phase inverter

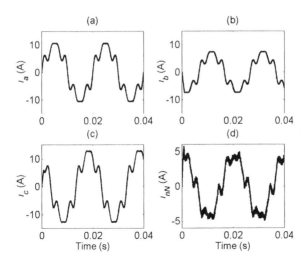

Figure 5.9. Unbalanced current tracking by the three-phase inverter

The schematic diagram of a soft-switched, single-phase inverter is shown in Figure 5.10. This is also called a resonant dc link inverter (RDCLI) as a resonant circuit is added in the dc side for soft switching [4]. The resonant circuit is represented by the inductance (L_r) and the capacitor (C_r). Ideally, the resistor R_r should be zero. However, this resistor has a small value owing to the finite Q-factor of the coil L_r. In this figure, the current $i_0 = I_L \times u$, where u is the switching function given by (5.2).

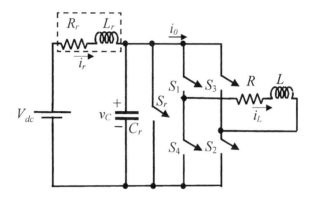

Figure 5.10. The schematic diagram of a resonant dc link inverter

The idea behind the RDCLI circuit is to generate a resonance using the pair (L_r, C_r). This forced resonance will enable the voltage v_C to periodically go to zero. The switch S_r is closed for a small interval of time when this voltage (v_C) is zero and, during this period, the switches $S_1 - S_4$ are switched on or off depending on the tracking requirements. The voltage across the switches is zero during this transition giving low loss. This condition is termed as zero voltage switching (ZVS). Assuming that the resonant cycle starts at an instant t_0, the link current (i_r) and capacitor voltage (v_C) waveforms are shown in Figure 5.11. The capacitor voltage must be zero at the beginning of every resonant cycle for successful ZVS. This can be done only when the inductor current is built up to certain value at the beginning of the cycle. For example, unless the inductor current is built up to the required value at instant t_2, the capacitor voltage will not be zero at instant t_3. Thus the initial inductor current $i_r(t_2)$ must be carefully chosen for successful ZVS [1]. Below we discuss a method of building up the initial current. In this approach it is assumed that the resonant period is fixed and is slightly less than the natural (undamped) resonant period of the pair (L_r, C_r). Then for building the initial current, the period for which the dc bus must be shorted is an important control parameter [5].

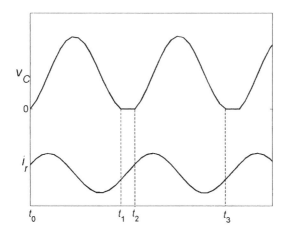

Figure 5.11. RDCLI capacitor voltage and link current waveforms with associated timings

Let us assume that the duration of the resonant cycle (i.e., $t_3 - t_2$) is fixed at ΔT microseconds. Since resonant cycle time is almost invariably much smaller than the time constant of the load circuit, the load current (i_0) is assumed to have a constant value equal to I_0 over a particular resonant cycle. This gives the equivalent circuit shown in Figure 5.12. In a practical circuit I_0 is estimated in the previous cycle, i.e., any time between t_0 and t_1 and is used for predicting $i_r(t_2)$. This is acceptable, as the resonant time is much smaller than the time constant of the load circuit.

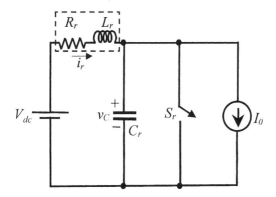

Figure 5.12. Equivalent circuit of an RDCLI

Referring to Figure 5.12, let us define a state vector and an input vector respectively as

$$x = \begin{bmatrix} v_C & i_r \end{bmatrix}^T \text{ and } v = \begin{bmatrix} I_0 & V_{dc} \end{bmatrix}^T$$

The state space equation of the circuit with S_r off is then given by

$$\dot{x} = Ax + Bv \tag{5.12}$$

where the matrices A and B are given by

$$A = \begin{bmatrix} 0 & 1/C_r \\ -1/L_r & -R_r/L_r \end{bmatrix}, \quad B = \begin{bmatrix} -1/C_r & 0 \\ 0 & 1/L_r \end{bmatrix}$$

Since $\Delta T = t_3 - t_2$, the solution of (5.12) at the instant t_3 based on the initial condition at the instant t_2 is

$$x(t_3) = e^{A\Delta T} x(t_2) + \int_0^{\Delta T} e^{A(\Delta T - \tau)} Bv(\tau) d\tau \tag{5.13}$$

Note that in the above equation V_{dc} is a known constant and I_0 is assumed to be known and therefore the input vector v is known. Further since both these quantities are assumed to be constant during the resonant cycle, (5.13) can be rewritten as

$$x(t_3) = \Phi x(t_2) + \Theta v(t_2) \tag{5.14}$$

where

$$\Phi = e^{A\Delta T} x(t_2) = \begin{bmatrix} \phi_{11} & \phi_{12} \\ \phi_{21} & \phi_{22} \end{bmatrix} \text{ and } \Theta = \int_0^{\Delta T} e^{A(\Delta T - \tau)} B d\tau = \begin{bmatrix} \theta_{11} & \theta_{12} \\ \theta_{21} & \theta_{22} \end{bmatrix}$$

The matrix Φ is known as the state transition matrix (STM).
From Figure 5.11 we define

$$x(t_3) = \begin{bmatrix} 0 & i_r(t_3) \end{bmatrix}^T$$

i.e., at the instant t_3, the capacitor voltage is zero and the link current is $i_r(t_3)$. Then premultiplying both sides of (5.14) by $C = \begin{bmatrix} 1 & 0 \end{bmatrix}$ we get

$$0 = C[\Phi x(t_2) \quad \Theta v(t_2)] \tag{5.15}$$

5. Structure and Control of Power Converters

Since A, B and ΔT are known a priori, the matrices ϕ and θ can be numerically evaluated beforehand. We can then expand (5.15) as

$$0 = [\phi_{11} \quad \phi_{12}]x(t_2) + [\theta_{11} \quad \theta_{12}]v(t_2) \tag{5.16}$$

Again from Figure 5.11 we get

$$x(t_2) = [0 \quad i_r(t_2)]^T \tag{5.17}$$

Substituting (5.17) in (5.16) and rearranging we get

$$i_r(t_2) = -\frac{1}{\phi_{12}}\{\theta_{11}I_0 + \theta_{12}V_{dc}\} \tag{5.18}$$

This gives the value of the desired current at the instant t_2 required to ensure successful zero crossing of the capacitor voltage at the instant t_3.

Once the desired computed value of $i_r(t_2)$ is obtained from (5.18), it is compared with the actual value of the link current by an analog comparator while the switch S_r is kept closed. The switch is opened when these two values are equal. This ensures that the link current is built up to the required level at the beginning of a resonant cycle such that the capacitor voltage goes to zero at the end of the resonant cycle.

Example 5.3: Let us consider an RDCLI with the following parameters

$$V_{dc} = 100\text{V}, L_r = 52\,\mu\text{H}, R_r = 0.128\Omega \text{ and } C_r = 0.89\mu\text{F}$$

The natural undamped frequency for this choice of inductor and capacitor is 23.5 kHz and the corresponding time for one cycle is 42.75 µs. We choose a resonant period (ΔT) of 37.5 µs, which is less than the cycle time of the natural frequency. Let us first investigate the no-load behavior (i.e., $I_0 = 0$) of the circuit by removing all loads. This is shown in Figure 5.13. It can be seen from this figure that the link voltage (v_C) periodically goes to zero after about every 37.5 µs. Also, the shorting switch is closed for about 5 µs for the link current (i_r) to built up, to ensure correct ZVS. Note that the link voltage has a peak that is above 200 V even when the dc source voltage is 100 V.

We now consider a load that is given by

$$R = 10\Omega \text{ and } L = 20\text{mH}$$

The reference current that has to be tracked is the same as that given in (5.3). The result of the current tracking is shown in Figure 5.14 (a) and the tracking error is shown in Figure 5.14 (b) when the dc source voltage is 100 V. From Figure 5.13 we know that the peak of the link voltage exceeds 200 V for a supply voltage of 100 V. However, the tracking performance is still unsatisfactory. The maximum current that can be tracked by this circuit, irrespective of the resonant circuit, is nearly equal to V_{dc}/R. Thus for better tracking, the dc source voltage must be raised. A perfect tracking is obtained when $V_{dc} = 200$ V. This is shown in Figure 5.14 (c) and (d). However the peak of the link voltage in this case rises above 400 V.

ΔΔΔ

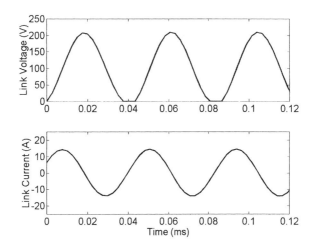

Figure 5.13. No-load behavior of RDCLI

Comparing the results of Examples 5.1 and 5.3 it can be concluded that for the same load both hard and soft-switched inverters have similar tracking performance for the same dc supply voltage. However the link voltage in a resonant inverter peaks around twice the dc supply voltage, thereby requiring a larger rating capacitor and switch voltage rating. Furthermore, to ensure ZVS, the control circuit becomes more complex. Even then the circuit is prone to failures unless proper safety margins are provided. The major advantage of this circuit is that the switching losses are minimized. However the limitations, especially the complex control logic, far outweigh the advantage gained in terms of losses. Advances have been made in voltage limiting such that the resonance may require less than 20% overvoltage rating. This approach makes soft switching more attractive but the advances in the switching speed of IGBTs and other power level devices have made a

5. Structure and Control of Power Converters

more significant impact on the applicability of inverters to power delivery applications. In the remainder of the book we shall only consider hard-switched inverters.

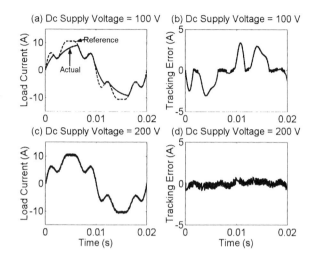

Figure 5.14. Tracking performance of RDCLI

5.3 High Voltage Inverters

The inverter applications that have been examined thus far assume that the dc bus voltage is higher than the line to neutral voltage of the main supply connection point. These inverter structures require switches that can turn on and off and this restricts attention to switch devices such as

- metal oxide field effect transistor (MOSFET)
- insulated gate bipolar transistor (IGBT)
- gate turn-off thyristor (GTO)
- field controlled thyristor (FCT)

Table 5.2 lists the rating and the switching speed of the above devices.

Table 5.2. Rating and switching speed of the power semiconductor devices

Devices	Upper voltage rating	Switching speed
MOSFET	1000 V	50 ns
IGBT	3300 V	0.5 µs
GTO	8000 V	5 µs
FCT	5000 V	0.1 µs

For direct connection to ac systems, the GTO is the preferred device for the construction of the highest voltage rated inverters. However GTOs have lower switching speed, implying that the switching losses are high. MOSFETs have very low switching losses but are not suitable for very high voltage ratings.

Using series connections of IGBTs, a new range of inverter applications have appeared. Using an inverter structure as in Figure 5.7, dc link systems have been developed commercially carrying 60 MW at 80 kV. The difference is that each switch shown may be the series connection of a large number of IGBTs. The main difficulty for series connection of switches is timing differences during turn off. If two 1000 V rated devices were in series, they would be expected to operate safely at 2000 V. However if one of the devices turns off while the other is still conducting, it will experience the full 2000 V which can damage the device or cause it to turn on again. The current systems use resistor and capacitor strings to force the voltage sharing of the devices during turn off. This process can require some compromise in the achieved switching speed with a corresponding increase in switching loss. The advantage however is that we can now construct inverters that can be connected to 11 kV distribution lines without the additional cost of interposing transformers. Other solutions such as multilevel converters are discussed in the next section.

5.4 Combining Inverters for Increased Power and Voltage

In this section we shall discuss three ways of combining inverters which can improve

- effective switch frequency
- system voltage rating
- power handling capacity

The basic inverter has a switch frequency, voltage rating and power capacity limited by the individual switches. Combining many switches can address one or more of the inverter limitations. A common approach is to combine inverter modules. One connection is to combine six step inverters such as in Figure 5.7, but where each switch turns on and off not more than once per fundamental cycle. This constraint permits the efficient use of GTOs, which have high switching losses and cannot successfully operate at high frequency.

5.4.1 Multi-Step Inverter

A multi-step inverter is constructed using many 6-step inverters of Figure 5.7 and a magnetic circuit to provide phase shift between the inverters. This principle of construction has been around for quite some time [6]. However with the advent of GTOs, they have gained considerable attention. The details of the construction of these inverters are fairly complicated. We shall only outline the principle of construction below.

Consider the basic 6-step inverter shown in Figure 5.15, which basically is the inverter of Figure 5.7 with the loads removed. Each switch leg makes a voltage step up and down every mains cycle giving a total of six steps. A $6n$-step inverter is built using n 6-step inverters. For example two such inverters are used to construct a 12-step inverter and four inverters are required to construct a 24-step inverter. The inverter of Figure 5.15 produces a set of three quasi-squarewave voltage waveforms of a given frequency. Let us connect the dc voltage source sequentially to the three output terminals via appropriate inverter switches. The output voltage waveforms, when each switch is conducting for a period of 180°, are then shown in Figure 5.16.

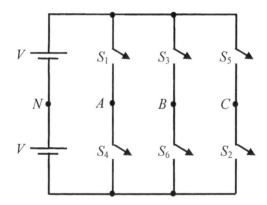

Figure 5.15. A basic 6-step inverter

The Fourier series of the output voltage waveforms are given by

$$f_A(t) = \sum_{n=1,3,5,\ldots} \frac{4V}{n\pi} \sin(n\omega t) \tag{5.19}$$

$$f_B(t) = \sum_{n=1,3,5,\ldots} \frac{4V}{n\pi} \sin(n\omega t - 2n\pi/3) \tag{5.20}$$

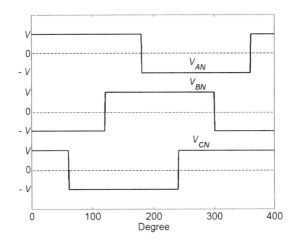

Figure 5.16. Output voltage waveforms of a 6-step inverter

$$f_C(t) = \sum_{n=1,3,5,\ldots} \frac{4V}{n\pi} \sin(n\omega t + 2n\pi/3) \qquad (5.21)$$

where ω is the fundamental frequency in rad/s.

From the above three equations it is clear that there are no even harmonics on the ac side. It can also be seen that for a particular value of n, the three functions are either in phase, or appear in the sequence *A-B-C* or appear in the sequence *A-C-B*. Let us denote them as zero-sequence, positive-sequence and negative-sequence respectively. Note that these sequences are not the symmetrical component sequences used for the representation of unbalanced circuits. These sequences for various values of n are given in Table 5.3. From this table it is evident that the triplen harmonics (3, 9, 15, ...) are zero-sequence and thus can circulate in the windings of a Δ-connected circuit. If we now connect the output of the inverter to a Δ-connected primary of a transformer, the triplens will not appear in the secondary winding of the transformer. Alternatively, if the inverter output is connected to an ungrounded Y-connected transformer primary winding, the zero-sequence components of the output voltage will be forced to zero.

Thus the output (ac) side voltage harmonics that appear in the transformer secondary for either of the above two transformer connections will be $6q \pm 1$ where $q = 1, 2, 3, \ldots$ including both positive and negative-sequences. The dc side current harmonics generated are given by $6q$ where

5. Structure and Control of Power Converters

q = 1, 2, 3, ... [7]. Hence we can state that the ac side harmonics are complementary of the dc side harmonics.

Table 5.3. Harmonic sequences of a 6-step inverter

Sequences	Harmonic number n				
Zero-sequence	3	9	15	21	...
Positive-sequence	1	7	13	19	...
Negative-sequence	5	11	17	23	...

Lower order voltage harmonics are of concern to any power system. Since the power system is mostly inductive, the currents produced by higher order voltage harmonics get attenuated. Since the effective impedance of an inductive circuit increases with the increase in frequency, the magnitude of the nth harmonic current reduces with the increase in the value of n. This implies that the reduction in the current magnitude associated with lower order harmonics is not as significant as the reduction achieved for higher order harmonics. We must therefore try to eliminate the lower order voltage harmonics. Let us consider two 6-step converters that are connected as shown in Figure 5.17. In this diagram only the dc side is shown. It can be seen that both these inverters are supplied by a common dc source that has a potential of $V_{dc} = 2V$.

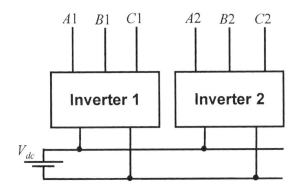

Figure 5.17. DC side connection diagram of two 6-step inverters

Let us consider the 6th harmonic component on the dc side of the 1st inverter. It can be eliminated if another 6th harmonic component that is shifted by 180° at that frequency is added to it. Their sum will then combine to zero. To effect a phase shift of 180° from the 2nd inverter, firing instants of all the switches of the 2nd inverter must lag the corresponding switches of the 1st inverter by 30° at the fundamental frequency. This will ensure that the 6th

harmonic component of the 2nd inverter lags that of the 1st inverter by 180°. Furthermore, such a phase shift will ensure the cancellation of all the odd multiples of the 6th harmonic components. For example, the 18th harmonic component of the 2nd inverter will lag that of the 1st inverter by 540° making them in phase opposition. Thus only the even multiples of the 6th harmonic component will remain in the dc side.

Let us now investigate what happens in the ac side if the firing angles of all the switches of the 2nd inverter are delayed by 30° on the fundamental frequency from the corresponding switches of the 1st inverter. The respective phase-A output voltage waveform of the two inverters are shown in Figure 5.18 when the fundamental voltage of the 2nd inverter lags the fundamental voltage of the 1st inverter by an angle θ. The Fourier series of the phase-A voltage of these two inverters are given by

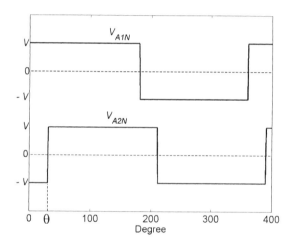

Figure 5.18. Output waveforms of two phase shifted inverters

$$f_{A1} = \frac{4V}{\pi}\sin \omega t + \sum_{n=5,7,...} \frac{4V}{n\pi}\sin n\omega t \qquad (5.22)$$

$$f_{A2} = \frac{4V}{\pi}\sin(\omega t - \theta) + \sum_{n=5,7,...} \frac{4V}{n\pi}\sin(n\omega t - n\theta) \qquad (5.23)$$

From (5.22) and (5.23) it is obvious that for $\theta = 30°$, the 5th and 7th harmonic component of the 2nd inverter are phase shifted by $-150°$ and $-$

5. Structure and Control of Power Converters

210° respectively with respect to the 1st inverter. In order to eliminate these two harmonic components, the 5th harmonic component needs to be further phase shifted by − 30° and the 7th harmonic component need a phase shift of + 30°. This means that negative sequence is phase shifted by − 30°, while the positive sequence is phase shifted by + 30°.

We can connect any three-phase transformer in Δ-Y to provide + 30° phase shift between their primary and secondary windings [6]. In that case if the positive sequence is phase shifted by + 30°, the negative sequence gets phase shifted by − 30° and vice versa. To explain this, consider the Δ-Y connected transformer that is shown in Figure 5.19. In this the uppercase letters denote the primary side and the lowercase letters denote the secondary side. Also the addition of a suffix 1 to winding names differentiates them from the terminal names.

The vector diagram when the terminals ABC are connected to a balanced supply with phase sequence ABC is shown in Figure 5.20. It is evident that the induced secondary voltages (e.g., $V_{a1a1'}$) lead the corresponding primary line to neutral voltage by 30°. This phase shift same for all frequencies, i.e., for all positive sequence harmonics (7th, 13th etc.). For negative sequence harmonics (5th, 11th etc.), the phases B and C exchange positions and the phase angle is − 30°.

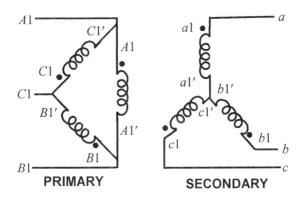

Figure 5.19. Schematic diagram of three single-phase transformers

The transformer connection for the addition of the output of two inverters is shown in Figure 5.21. The Δ-connected primary windings are connected to outputs of two inverters at points $A1$, $B1$, $C1$ (Inverter 1) and $A2$, $B2$, $C2$ (Inverter 2). The inverter outputs $A2$, $B2$, $C2$ are arranged to lag the outputs $A1$, $B1$, $C1$ by 30° through a phase shift in their gating signals. The transformer connected to Inverter 1 (Δ-open star) has a turns ratio of 1:1, while that connected to Inverter 2 (Δ-Δ) has a turns ratio of 1:√3.

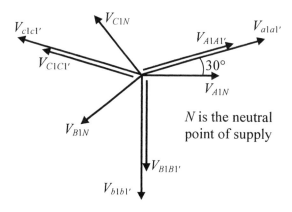

Figure 5.20. Positive-sequence phasor diagram

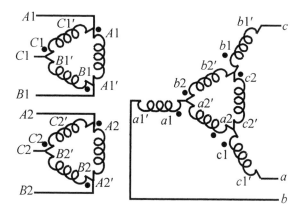

Figure 5.21. Transformer connection to add the outputs of two inverters

The secondary windings $a2a2'$, $b2b2'$ and $c2c2'$ of the transformer connected to Inverter 2 are connected in Δ. Thus there is no phase shift introduced by this inverter transformer. The secondary windings of the transformer at the output of Inverter 1 are connected in an open star with the terminals $a1$, $b1$, $c1$ forming the open star. Note the dotted terminals in Figure 5.21. Then from this figure we get

$$V_{ab} = V_{a1a1'} + V_{a2a2'} - V_{c1c1'}$$

If we consider only the fundamental component of the inverter outputs, then the vector diagram corresponding to the above equation is shown in Figure 5.22 (a). It shows that V_{ab} is in phase with the primary reference voltage

5. Structure and Control of Power Converters

V_{A1N}, which is the fundamental output of Inverter 1. The other voltages can be written by symmetry from Figure 5.21

$$V_{bc} = V_{b1b1'} + V_{b2b2'} - V_{a1a1'}$$
$$V_{ca} = V_{c1c1'} + V_{c2c2'} - V_{b1b1'}$$

A vector diagram corresponding to the 5th harmonic is shown in Figure 5.22 (b). The phase angles of the three terms in the equation for V_{ab} are such that they cancel out completely. A similar diagram can also be drawn for the 7th harmonic and it can be shown that this component also becomes equal to zero.

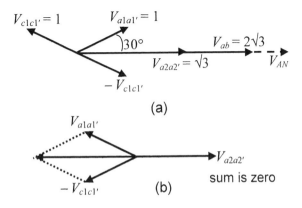

Figure 5.22. Phasor diagram (a) fundamental and (b) 5th harmonic

A time domain construction corresponding to the expression V_{ab} is shown in Figure 5.23 (a). The levels produced are 1, 1+√3 and 2+√3 times the dc bus voltage of the inverter supplying the primary, which is 1/√3 per unit for this figure. The circuit discussed above is a 12-step inverter. The output line-to-line voltage waveform of this inverter is shown in Figure 5.23 along with its harmonic spectrum. The dc side voltage V_{dc} is chosen to be 1.0 per unit to obtain the output voltage. It is evident from Figure 5.23 (b) that the most dominant ac side harmonic components are 11th and 13th.

It is seen that by providing a phase shift of 30° between two inverter-transformer combinations, it is possible to generate a 12-step waveform with a spectrum that contains 11th and higher order harmonics only. If we now combine two such inverters, we shall be able to get a 24-step inverter. In this case however the phase difference between the successive inverter-transformers must be 15°. In a similar way we can construct a 6q-step inverter by providing a phase shift of 360°/6q between the successive inverter-transformers.

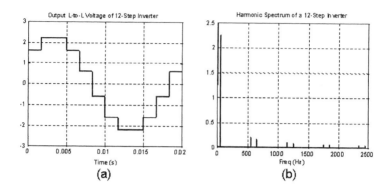

Figure 5.23. (a) Output voltage waveform of a 12-step inverter and (b) its harmonic spectrum

The construction of multi-step inverter requires a number of single-phase transformers that can be connected to provide any feasible degree of phase shift. The development of a 48-step inverter using eighteen single-phase three-winding transformers and six single-phase two winding transformers is reported in [8] and its use in bulk power transmission is given in [9-11]. The basic idea of construction here is to provide phase shift using these transformers. In the 48-step inverter the phase shift is 7.5° between the successive inverters.

5.4.2 Multilevel Inverter

The inverter configuration shown in Figure 5.15 is basically that of a 2-level inverter as the output voltage can take on the values $+V$ and $-V$. However there is a class of inverters that can take on more than two values. They are called multilevel inverters [12-18]. Multilevel inverters are usually named after the level of voltages that can be obtained from them. For example a 3-level inverter can produce voltage levels of $+V$, 0 and $-V$ where V is the voltage of the dc supply. Similarly four different voltage levels can be obtained from a 4-level inverter, five different voltage levels can be obtained from a 5-level inverter and so on. Below we shall discuss the construction and operating characteristics of some of these inverters.

The schematic diagram of one leg of a 3-level inverter is given in Figure 5.24. It contains four switches (S_{1a}, S_{1b}, S_{2a} and S_{2b}) and two diodes (D_1 and D_2). Each of these switches consists of a power semiconductor device and an anti-parallel diode. Each of the two dc sources supplying the inverter has a magnitude of V. The neutral point of these two sources is denoted by N. A three-phase inverter can be constructed by duplicating the leg shown in Figure 5.24. The output voltage across A and N is given in Table 5.4 for various switch combinations. It is evident from this table that the output voltage has three levels.

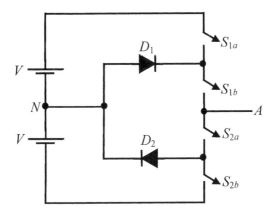

Figure 5.24. Schematic diagram of a 3-level inverter

Table 5.4. Output voltage of a 3-level inverter

Switch Status				V_{AN}
S_{1a}	S_{1b}	S_{2a}	S_{2b}	
Off	Off	On	On	$-V$
Off	On	On	Off	0
On	On	Off	Off	$+V$

The schematic diagram of one leg of a 4-level inverter is shown in Figure 5.25. This inverter requires the levels $+ V/3$ and $- V/3$ in addition to the levels of $+ V$ and $- V$. Since the number of levels is even, the level 0 is not achievable by this inverter. To achieve the above-mentioned dc levels, three capacitors of equal size are connected as shown in the figure. Assuming that the capacitors are fairly large, the voltage levels in the various parts of the circuit are also shown in Figure 5.25. Note that all these voltages are measured with respect to the neutral point (N). Table 5.5 lists the four voltage levels across A and N for various switch combinations. Note that for each of these four voltage levels, any three of the six switches must be closed. This is done to facilitate current flow in either direction. For example consider the second row of Table 5.5 in which the desired level is $+ V/3$. When a current flows out of terminal A, it flows through the switches S_{1a} and S_{1b}. However when the current is flowing into terminal A, a path is provided through the switch S_{2a}. The voltage level between A and N remains constant at $+ V/3$ in either case.

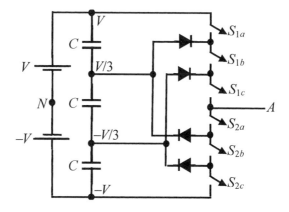

Figure 5.25. Schematic diagram of a 4-level inverter

Table 5.5. Output voltage of a 4-level inverter

| \multicolumn{6}{c}{Switch Status} | V_{AN} |
S_{1a}	S_{1b}	S_{1c}	S_{2a}	S_{2b}	S_{2c}	
On	On	On	Off	Off	Off	$+V$
Off	On	On	On	Off	Off	$+V/3$
Off	Off	On	On	On	Off	$-V/3$
Off	Off	Off	On	On	On	$-V$

The schematic diagram of one leg of a 5-level inverter is shown in Figure 5.26 in which only the voltage levels are shown. These voltage levels can be obtained using four equal capacitors in the same way as shown in Figure 5.25. In the 5-level inverter any four switches are closed at any given time. Table 5.6 lists the inverter output voltage for various switch combinations.

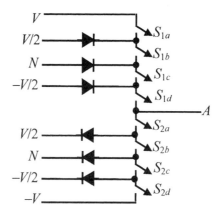

Figure 5.26. Schematic diagram of a 5-level inverter

5. Structure and Control of Power Converters

Table 5.6. Output voltage of a 5-level inverter

| \multicolumn{8}{c|}{Switch Status} | V_{AN} |
S_{1a}	S_{1b}	S_{1c}	S_{1d}	S_{2a}	S_{2b}	S_{2c}	S_{2d}	
On	On	On	On	Off	Off	Off	Off	$+V$
Off	On	On	On	On	Off	Off	Off	$+V/2$
Off	Off	On	On	On	On	Off	Off	0
Off	Off	Off	On	On	On	On	Off	$-V/2$
Off	Off	Off	Off	On	On	On	On	$-V$

In a similar fashion we can construct six or higher level inverters. The structure of a generalized m-level inverter is given in [18]. We shall now discuss voltage waveform synthesis techniques using multilevel inverters. Selective voltage harmonic elimination has long been proposed in [19] for 2-level inverters. This well-established technique has been extended to a 5-level inverter in [20]. Consider the voltage waveform of a 5-level inverter shown in Figure 5.27 (a) in which only the positive half cycle is shown. The negative half cycle is the mirror image of the positive half cycle. In this figure the magnitude of the dc voltage (V) is assumed to be 1.0 per unit.

The output voltage waveform is reconstructed by firing appropriate switches at prespecified instants. For example, at instant C_1, the switches S_{1b}, S_{1c}, S_{1d} and S_{2a} are turned on (see Table 5.6). In a similar way the other switches are turned on and off to reconstruct the voltage of Figure 5.27 (a). As we mentioned earlier that the voltage waveform has a halfwave symmetry. This implies that all the even harmonics and dc component are zero. Now notice that the voltage waveform has an odd quarterwave symmetry. Thus all the a_n terms in the Fourier series (c.f. equation 3.40) expansion will be zero. Then the magnitude of the nth harmonic component of the voltage is given by

$$b_n = \frac{2}{\pi} \int_0^\pi V_{AN} \sin(n\theta) d\theta \qquad (5.24)$$

Evaluating the above equation for odd n we get

$$b_n = \frac{2}{n\pi}\left[-\cos nC_1 - \cos nC_2 + \cos nC_3 - \cos nC_4\right] \qquad (5.25)$$

To eliminate 3rd, 5th, 7th and 9th harmonics we obtain four simultaneous nonlinear equations substituting n by 3, 5, 7 and 9 in (5.25). The solutions of these equations are given in [20] as

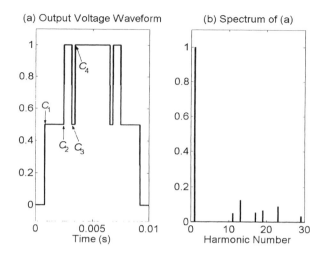

Figure 5.27. (a) Output voltage of a 5-level inverter and (b) its harmonic spectrum

$$C_1 = 14.57°, C_2 = 45.43°, C_3 = 57.43° \text{ and } C_4 = 62.57°$$

The voltage waveform of Figure 5.27 (a) is drawn with these values and the corresponding harmonic spectrum is shown in Figure 5.27 (b). The fundamental frequency is assumed to be 50 Hz. The voltage spectrum is normalized with respect to the fundamental. It can be seen that the highest order harmonic in this case is 11th.

Multilevel inverters can also be operated in PWM control mode [16,21]. One major drawback of this form of multilevel inverters is the requirement of a large number of capacitors. The switching of these inverters can cause unequal charging of the capacitors resulting in voltage imbalance [21-23]. Usually the following factors are responsible for capacitor voltage imbalance

- Unequal capacitor leakage currents.
- Unequal delay in the semiconductor devices.
- Asymmetrical charging of capacitors during transients.
- Asymmetrical circuit configuration.

A solution to capacitor imbalance problem has been given in [21], which uses voltage feedback to cancel out imbalance. The other possible solution to this problem is to use chopper circuit that can transfer charge from one capacitor to another thereby maintaining and equalizing the voltages [24].

5.4.3 Chain Converter

The basic chain converter consists of a string of H-bridge inverters around capacitors. Each capacitor may be inserted in a positive or negative sense or bypassed. The use of GTOs to reduce losses often implies that the capacitors are switched in or out once per half cycle. The maximum voltage is achieved by switching all capacitors inserted with the same direction. The schematic diagram of a chain converter is shown in Figure 5.28.

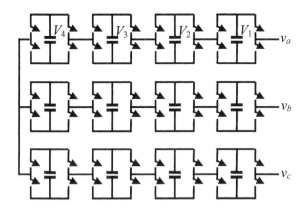

Figure 5.28. Schematic diagram of a chain converter

Conceptually the modulation would then be built up as in Figure 5.29. The turn on and turn off of each of the levels determine the voltage to be synthesized. The control is made more difficult since the capacitor size is finite and there is a change in voltage of each of the capacitors particularly those on the longest period. In the chain converter, the energy is not shared between the levels or phases and thus large energy storage capacity is required. The result shown in Figure 5.29 uses a very large capacitor so the voltage changes for the levels are not visible.

The control does not anticipate droops but does measure capacitor voltages at the beginning of the switch period. Each switch period is assumed to be equal. The idea is to choose the time to turn each level on so that the average voltage over the switch interval matches the desired voltage. For N levels of capacitor, there will be N switchings to a new level in each quarter of a cycle. For analysis here equal N switch intervals are chosen. More sophisticated analysis would shape the intervals to produce the minimum current ripple. The main factor to be considered here is that the finite capacitor size will cause the voltage of each link in the chain to change as the on time proceeds.

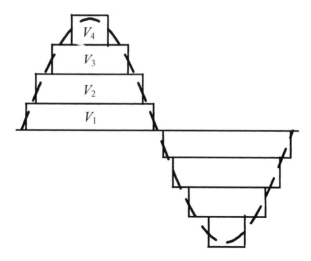

Figure 5.29. Basic addition of levels for one phase of a chain converter

Alstom has developed the chain converter for use by the national grid in the United Kingdom. The aim was to develop controllable reactive elements with low losses. The control scheme used by Alstom is based on off-line optimization for a range of line currents. The controller interpolates between the computed firing points when the current is in between the computed values. In common with all such optimizations, there will be no compensatory changes in the firing when the system changes due to voltage harmonics or variations in the capacitor parametric values.

The chain converter can provide reactive power and harmonic compensation. It can combine switch modules for higher voltage higher power operation with an effective increase in switch frequency. The converter is not usually suitable to connect between a dc source and supply mains because the energy is not drawn from a single source. Each dc stage needs to draw energy from electrically separate sources. This may however be a strength for photovoltaic or fuel cell applications.

So far in this chapter we have discussed the topology of the basic inverters and how their operation will influence their use in distribution compensators. While the soft switch version is not receiving strong attention for compensators, the multi-step or multi-inverter topologies offer a very attractive path for the application in which they are connected to ac buses, particularly at high voltages. We shall now discuss the inverter control techniques.

5.5 Open-Loop Voltage Control

There are many forms of modulation used for communicating information. When a high frequency carrier has an amplitude varied in response to a lower frequency signal we have what is called an amplitude modulation (AM). When the carrier frequency is varied in response to the modulating signal we get frequency modulation (FM). These signals are used for radio modulation because the high frequency carrier signal is needed for efficient radiation of the signal. When communication by pulses was introduced, the amplitude, frequency and pulse width become possible modulation options. In many power electronic converters pulse width modulation (PWM) is the most reliable way of reconstructing a desired output voltage waveform. For a reliable signal representation it is necessary that the frequency of the switching be significantly higher than that of the desired signal.

Modulation in power electronics is a process of forming a switched representation of a waveform. The switched representation is more efficient compared with linear amplifiers. The linear amplifier has a voltage across the switch while it is carrying current and thus dissipates as much energy as is delivered to the load. Ideally the switched signal can handle high power with an efficiency approaching 100%.

There are two approaches to forming a switched waveform – open and closed loop. The most well-known modulation scheme performs well in open loop. This modulation is based on having the average value of the switched waveform to match the average value of the modulating signal over an interval. The concept is most easily applied for fixed switch frequency applications.

5.5.1 Sinusoidal PWM for H-Bridge Inverter

The schematic diagram for the generation of PWM control signal is shown in Figure 5.30. It contains a carrier signal and a modulating signal. The magnitudes of these two signals are compared through an analog comparator. The PWM control signal is set high when the modulating signals has a higher numerical value than the carrier signal and is set low when the carrier signal has a higher numerical value. This implies that in the H-bridge, switches S_1 and S_2 (see Figure 5.2) are closed when the modulating signal is higher than the carrier and switches S_3 and S_4 are closed when carrier is higher than the modulating signal. Since the output voltage of the inverter for such an arrangement is $\pm V_{dc}$, this is called bipolar voltage switching.

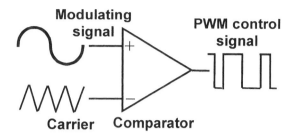

Figure 5.30. PWM control signal generation scheme

The most popular form of pulse width modulation synthesis is the sinusoidal PWM (SPWM). In an SPWM scheme the modulating signal is sinusoidal and carrier signal is a triangular wave. The frequency of the modulating signal is chosen to be the fundamental frequency of the output waveform to be synthesized. The SPWM output waveforms are defined in terms of modulation index (m_a) and frequency ratio (m_f). The modulation index is the ratio of the peak of modulating waveform to the peak of the carrier wave, while the frequency ratio is the ratio of the frequency of the carrier wave to that of the modulating wave. Typical bipolar sinusoidal PWM waveforms are shown in Figure 5.31. In this figure the frequency ratio (m_f) is chosen to be 9 and the dc source voltage (V_{dc}) is assume to be 1.0 per unit. Two different PWM output voltages are shown – one when m_a is less than 1 and the other when m_a is 0. The former is called modulated output and later is called unmodulated as the modulating waveform is zero in this case.

Let us define the frequency of the carrier wave as f_c and that of the modulating wave as f_m. It is evident from Figure 5.31 that the SPWM output waveform is not sinusoidal. In fact it will contain fundamental and harmonics of the frequency f_m. The peak amplitude of the fundamental is equal to m_a times the dc source voltage [1]. The harmonic spectrum of a bipolar SPWM output voltage with $m_f = 9$, $m_a = 0.4$ and $V_{dc} = 1.0$ per unit is shown in Figure 5.32. Since the SPWM output waveforms have mirror image or halfwave symmetry, the even harmonics are absent. The most dominant harmonics present in the output voltage of an SPWM is m_f, which in this case happens to be the 9th. The other possible harmonics are $m_f \pm 2$, $m_f \pm 4$, $2 m_f \pm 1$, $2 m_f \pm 3$, $3 m_f$, $3 m_f \pm 2$ etc.

The peak amplitude of the harmonic components given in Figure 5.32 is listed in Table 5.7 [1]. Note that the harmonic numbers and their numerical values vary with both m_f and m_a. It is thus desirable to obtain a general expression for computing the numerical values of harmonics. One such method, given in [25], uses Bessel function to form closed form expressions of the harmonics. However with modern day high-speed computers, the numerical values can be computed with ease.

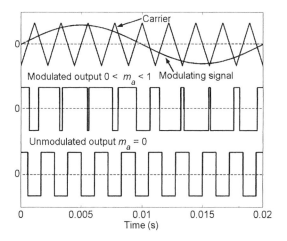

Figure 5.31. Bipolar SPWM waveforms

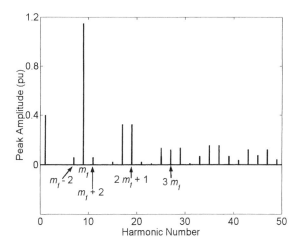

Figure 5.32. Harmonic spectrum of bipolar SPWM output voltage for $m_f = 9$ and $m_a = 0.4$

From Table 5.7 it is evident that the harmonics in the output voltage appear at a multiple of the frequency ratio and its sidebands. Note that for

SPWM m_f is always chosen as an odd integer. This choice results in both quarterwave and halfwave symmetry eliminating the even harmonics and the dc component. Another important point to be noted is that the modulation index is always chosen to be in the range $0 < m_a < 1$. A value of m_a greater than 1 results in over modulation. An over-modulated output voltage for m_a of 1.2 shown in Figure 5.33. The harmonic spectrum of the output voltage, shown in Figure 5.34, contains all the odd harmonics.

Table 5.7. Numerical values of harmonics for $m_f = 9$ and $m_a = 0.4$

Harmonic Number		Peak Amplitude (per unit)
In Terms of m_f	Actual Values	
	1	0.4
m_f	9	1.15
$m_f \pm 2$	7, 11	0.061
$2m_f \pm 1$	17, 19	0.326
$2m_f \pm 3$	15, 21	0.024
$3m_f$	27	0.123
$3m_f \pm 2$	25, 29	0.139
$3m_f \pm 4$	23, 31	0.012

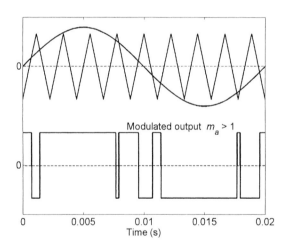

Figure 5.33. Over-modulated SPWM output waveform for $m_f = 9$ and $m_a = 1.2$

So far we have discussed bipolar SPWM technique. It is also possible to run the H-bridge inverter of Figure 5.2 in an unipolar fashion. In this the inverter output voltage varies between $+ V_{dc}$ and 0 in the positive half cycle of the modulating signal and between $- V_{dc}$ and 0 in the negative half cycle. Let us denote the modulating voltage by v_{mod}. Then the unipolar switching scheme is based on the modulating waveforms v_{mod} and $-v_{mod}$. This is shown in Figure 5.35.

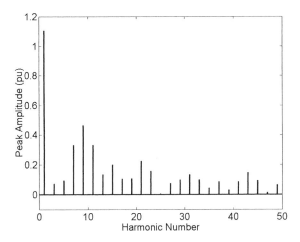

Figure 5.34. Harmonic spectrum of the over-modulated output of Figure 5.33

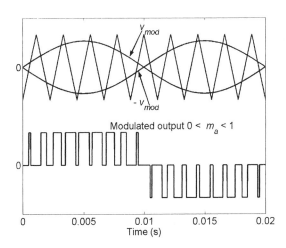

Figure 5.35. Unipolar SPWM output waveform for $m_f = 9$ and $m_a = 0.8$

The uniploar PWM control law is as follows. Unlike the operation of the H-bridge inverter discussed so far, the switches of each leg do not take complementary values. Table 5.8 lists the four switch states that the inverter of Figure 5.2 can have. When both the two upper switches are on, the output voltage is zero. The current will then circulate, depending on its direction, either through the power semiconductor device of S_1 and anti-parallel diode of S_3 or through the power semiconductor device of S_3 and anti-parallel diode of S_1. Similarly when both the lower switches are on, the power semiconductor devices and anti-parallel diodes of S_2 and S_4 maintain the continuity of current.

Table 5.8. Four switch states of the H-bridge inverter of Figure 5.2

Switches On	Output Voltage
S_1 and S_2	$v_{AB} = + V_{dc}$
S_1 and S_3	$v_{AB} = 0$
S_2 and S_4	$v_{AB} = 0$
S_3 and S_4	$v_{AB} = - V_{dc}$

Let us denote the carrier waveform as v_{carr}. Then the uniploar SPWM switching waveforms are generated as follows.

If $v_{mod} > v_{carr}$ then switch S_1 on
elseif $v_{mod} < v_{carr}$ then switch S_4 on
If $-v_{mod} > v_{carr}$ then switch S_3 on
elseif $-v_{mod} < v_{carr}$ then switch S_2 on

This implies that in the positive half cycle of v_{mod}, when the value of v_{mod} is more than the value of v_{carr} and the value of $-v_{mod}$ is less than that of v_{carr}, the inverter output is V_{dc}; it is zero otherwise. Similarly in the negative half cycle of v_{mod}, the inverter output voltage is $-V_{dc}$ only when the value of v_{mod} is less than the value of v_{carr} and the value of $-v_{mod}$ is more than that of v_{carr}. Otherwise it is zero. The advantage of the unipolar switching is that the harmonics gets shifted and appear as the side bands of the even multiples of m_f. For the output voltage of Figure 5.35, the harmonic spectrum is shown in Figure 5.36.

5.5.2 Sinusoidal PWM for three-phase Inverter

Let us consider the three-phase inverter shown in Figure 5.15. In a three-phase PWM three sinusoidal modulating signals are generated, each phase shifted from the other by 120°. These are denoted by $v_{mod,a}$, $v_{mod,b}$ and $v_{mod,c}$ in Figure 5.37. The switches of the leg-A of the inverter in Figure 5.15 are switched by the intersection of $v_{mod,a}$ in the same manner as discussed above, i.e., when $v_{mod,a}$ is higher than v_{carr}, then switch on S_1 such that $v_{AN} = + V$. Again when $v_{mod,a}$ is lower than v_{carr}, then switch on S_4 such that $v_{AN} = - V$.

5. Structure and Control of Power Converters

Therefore the output voltage v_{AN} varies between $\pm V$. In a similar way, the switches of the inverter leg-B and C are turned on or off by comparing the carrier wave with $v_{mod,b}$ and $v_{mod,c}$ respectively.

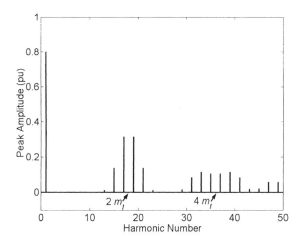

Figure 5.36. Harmonic spectrum of unipolar switching for $m_f = 9$ and $m_a = 0.8$

The line-to-line voltage between any two legs of the inverter then looks like the output waveform of a unipolar inverter. Note that these voltage vary between $\pm V$. This is also shown in Figure 5.37 in which the fundamental components of the voltages v_{AB}, v_{BC} and v_{CA} are phase shifted from each other by 120°. Also note that since the line-to-line voltage waveforms look like the unipolar SPWM output, their harmonic components also appear as sidebands of the even multiple of m_f similar to the ones shown in Figure 5.36.

5.5.3 SPWM in Multilevel Inverter

The sinusoidal PWM method can also be applied to a multilevel inverter. One such technique is outlined for odd-level inverters in [16]. In this there are $(N-1)$ carrier waveforms where N is the number of levels. These carrier waveforms can be placed in various fashions. We shall however consider the case in which the carriers are in phase but have different dc offsets. The waveforms for a 5-level inverter are shown in Figure 5.38. In this figure Ca_1 varies between 0.5 to 1, Ca_2 varies between 0 to 0.5, Ca_3 varies between -0.5 to 0 and Ca_4 varies between -1 to -0.5. To restrict the inverter from over modulating, we assume that the sinusoidal modulating wave varies between -1 and 1.

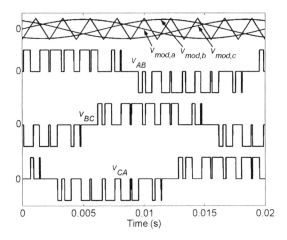

Figure 5.37. Three-phase SPWM inverter output waveforms

Let us now consider the 5-level inverter shown in Figure 5.26 and its switching operations listed in Table 5.6. We assume that the dc voltage V is 1.0 per unit. Then the inverter operates in the following fashion. The inverter output voltage is made + 0.5 per unit when the instantaneous value of the modulating waveform becomes more than that of the carrier wave Ca_2. From Table 5.6 it can be seen that this is achieved by closing switches S_{1b}, S_{1c}, S_{1d}, and S_{2a}. The inverter output voltage is made 1.0 per unit through appropriate switches shown in Table 5.6 when the modulating waveform similarly crosses Ca_1. In the negative half cycle the inverter output is made $-$ 0.5 per unit when the instantaneous value of Ca_3 becomes more than that of the modulating wave and it is made $-$ 1.0 per unit when Ca_4 similarly crosses the modulating wave. The switches to be utilized for achieving these voltage levels are given in Table 5.6. It is also assumed that the capacitor voltages of the 5-level inverter are maintained such that regulated $\pm$ 0.5 per unit and $\pm$ 1.0 per unit voltage levels are available.

The modulated output waveform of Figure 5.38 is obtained for $m_f = 9$ and $m_a = 0.8$ where m_a is defined with respect to the outer level of carrier wave. The harmonic spectrum of this waveform is shown in Figure 5.39. It can be seen from this figure that the magnitude of the fundamental is almost 0.8. However, it contains all odd harmonics with the most prominent being at m_f. Other dominant harmonics also appear as sidebands of the multiple of m_f.

Compared to the magnitudes of the unipolar SPWM output harmonic components of Figure 5.36 the magnitudes of harmonic components of Figure 5.39 are smaller. In fact the total harmonic distortion (THD) of the output of the multilevel inverter output voltage is 31.61% while that of the output voltage of the unipolar SPWM is 69.36%. Therefore the increased voltage level has caused a reduction in the THD by more than half. Note that an increase in the frequency ratio m_f will not result in the reduction of THD any further. However, the dominant harmonics will now be higher order that can be easily filtered by inductive loads.

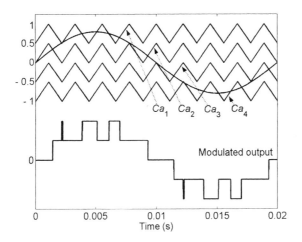

Figure 5.38. Output waveform of a 5-level inverter

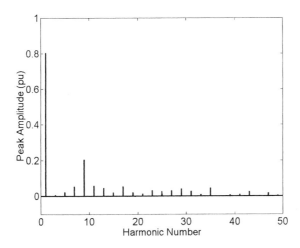

Figure 5.39. Harmonic spectrum of 5-level inverter voltage output for $m_f = 9$ and $m_a = 0.8$

5.5.4 Space Vector Modulation

This is essentially an averaging technique that takes into consideration that a three-phase inverter has only eight switch states. Consider the inverter shown in Figure 5.15. There are two states for each leg. Therefore, for the three independent legs of the inverter, a total combination of 2^3 states can be obtained. Let us consider the positive instantaneous symmetrical component given by

$$v_1 = \frac{1}{\sqrt{3}}\left(v_{AN} + a v_{BN} + a^2 v_{CN}\right) \tag{5.26}$$

where $a = e^{j120°}$ and v_{AN}, v_{BN} and v_{CN} are the voltages of the three output terminals of the inverter (see Figure 5.15). Note that each of the voltages v_{AN}, v_{BN} and v_{CN} can have a value of either $+V$ or $-V$. The eight states are then listed in Table 5.9 along with the switch positions for $V = 1.0$ per unit. Out of these eight states two states correspond to the voltage vector v_1 being zero. These two states are obtained by either closing all the three top switches or closing all the bottom switches.

Table 5.9. Eight switch states of a three-phase inverter

State No.	S_1	S_2	S_3	S_4	S_5	S_6	v_{AN}	v_{BN}	v_{CN}	v_1
I	On	Off	Off	Off	On	On	+1	−1	+1	$0.5774 - j1.0$
II	On	On	Off	Off	Off	On	+1	−1	−1	$1.1547 + j0.0$
III	On	On	On	Off	Off	Off	+1	+1	−1	$0.5774 + j1.0$
IV	Off	On	On	On	Off	Off	−1	+1	−1	$-0.5774 + j1.0$
V	Off	Off	On	On	On	Off	−1	+1	+1	$-1.1547 + j0.0$
VI	Off	Off	Off	On	On	On	−1	−1	+1	$-0.5774 - j1.0$
VII	On	Off	On	Off	On	Off	+1	+1	+1	0
VIII	Off	On	Off	On	Off	On	−1	−1	−1	0

These eight switch states are pictorially depicted in Figure 5.40. Each vector is displaced from the contiguous vector by 60°. The entire voltage vector space is divided into six regions. Each region is the triangular area between two contiguous vectors, e.g., Region 1 is the space between vector I and II. Now suppose at any given instant of time we want to recreate the vector v_p shown in Figure 5.40. This can be done by time averaging of the nearest inverter state vectors I and II and the zero vector (VII or VIII). The averaging is done over a suitably chosen time interval t_a. This technique is known as space vector modulation [26].

5. Structure and Control of Power Converters

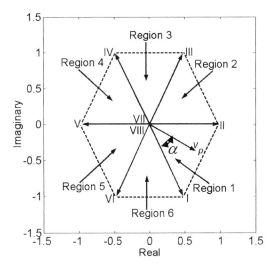

Figure 5.40. Inverter space vectors

Let us denote the inverter voltage for states I, II and VII respectively as v_I, v_{II} and v_{VII}. Then assuming that the inverter spends time t_1 in state I and time t_2 in state II we get the following equation

$$v_p t_a = v_I t_1 + v_{II} t_2 + v_{VII}(t_a - t_1 - t_2) \tag{5.27}$$

Let us assume that v_p has a magnitude of V_{pm} and an angle α with vector v_I. Note from Table 5.9 that v_I, v_{II} and v_{VII} have numerical values of $0.5774 - j1.0$, $1.1547 + j0.0$ and 0, respectively. Then (5.27) can be resolved into real and imaginary parts as

$$V_{pm} t_a \cos(60° - \alpha) = 0.5774 t_1 + 1.1546 t_2 = \frac{t_1}{\sqrt{3}} + \frac{2 t_2}{\sqrt{3}} \tag{5.28}$$

$$V_{pm} t_a \sin(60° - \alpha) = t_1 \tag{5.29}$$

Note that Table 5.9 is generated for $V = 1.0$ per unit. A general expression for any V then can be written from (5.28) and (5.29) as

$$\frac{t_1}{t_a} = \frac{V_{pm}}{V} \sin(60° - \alpha) \tag{5.30}$$

$$\frac{t_2}{t_a} = \frac{V_{pm}}{V} \sin \alpha \qquad (5.31)$$

The inverter then can be operated with t_1 time in state I, t_2 time in state II and the remaining $t_a - t_1 - t_2$ time in zero voltage state (VII or VIII). In order to reduce commutations per cycle, it is necessary to calculate the state transitions carefully. For example assume that the inverter is required to reconstruct two consecutive values of v_p that are in the Region 1. Then the inverter switching sequence will be I – II – VIII – II – I – VII. This means that for the reconstruction of the first value of v_p, the inverter first spends t_1 time in state I, then spends t_2 time in state II and finally spends $t_a - t_1 - t_2$ time in state VIII. Subsequently for reconstructing the second value of v_p the inverter first spends t_2 time in state II, then spends t_1 time in state I and finally spends $t_a - t_1 - t_2$ time in state VII. This pattern ensures that two switches remain in their old state during any transition. It is thus important that the entire switching pattern is mapped before operating the inverter.

An example of space vector output voltage is given in Figure 5.41 in which V is taken to be 1.0 per unit. This is accomplished by placing four equidistant vectors in each region and then maintaining the switching cycle mentioned above. The harmonic spectrum of modulated voltage is shown in Figure 5.42. It can be seen that it contains all odd harmonics including the triplens. The total harmonic distortion of 64.1% is comparable to that obtained for unipolar switching.

5.5.5 Other Modulation Techniques

In a trapezoidal modulation scheme the gating signals are obtained by comparing the triangular carrier waveform with a modulating trapezoidal wave [27]. This type of modulation can increase the peak of the fundamental wave. However, the output contains lower order harmonics. In a staircase modulation the modulating wave is a staircase [28] while in stepped modulation the modulating wave is a stepped waveform [29]. The carrier wave in both these schemes is triangular. Note that neither the staircase nor the stepped waveform is a sampled and zero-order hold approximation of a modulating sinewave. While both these methods produce high quality output waveforms, the amplitude of the fundamental is lower for the staircase modulation and higher for stepped modulation. However, to obtain high quality output the number of pulse per cycle has to be computed beforehand for both these methods.

PWM techniques often produce high acoustic noise at the PWM switching frequency and at its multiples. Randomizing the pulse position in a

PWM inverter can reduce this objectionable noise [30,31]. Such strategy, when employed in ac motor drive systems, can not only reduce the acoustic noise but also can reduce EMI. However, its scope of application to power systems is very limited.

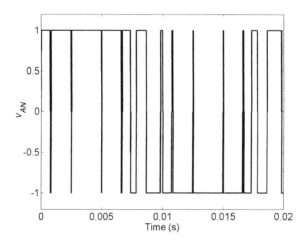

Figure 5.41. Space vector modulated output

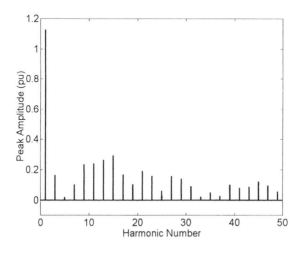

Figure 5.42. Harmonic spectrum of the space vector output voltage of Figure 5.41

5.6 Closed-Loop Switching Control

In the open-loop control, the objective is to produce a set of voltages with a specified fundamental magnitude and phase. Such synthesis has application in power transmission systems, especially with the FACTS controller that employ synchronous voltage source (SVS). However, most custom power devices are operated in closed-loop fashion in which they either track a specified current reference or a voltage reference or both. We have already discussed one such controller while discussing current tracking in hysteresis band. In this section we shall discuss closed-loop switching control in detail.

5.6.1 Closed-Loop Modulation

The open-loop PWM modulation makes the following assumptions

- The voltage being switched is always exactly constant.
- The switching of the output occurs exactly at the instant defined by the logic signal.
- There are no voltage drops associated with the switches.

These imperfections, which are present in all switching systems, mean that the exact modulation will not be produced. The synthesis of the imperfect switched voltage waveform will mean that the resulting current, power or harmonic correction will not be perfectly produced. One of the earliest applications of feedback was to overcome the nonlinearities of the early valve amplifiers. High gain feedback reduces susceptibility of the amplification to amplifier nonlinearities. The standard feedback control systems have an amplifier driven by the error between the reference signal and the output. This is shown in Figure 5.43. The PWM modulator can be modeled as an amplifier of the low frequency terms provided the system attenuates the additional switch frequency terms. In practice the gain of the modulator must be limited because there will always be some switch frequency terms present which can dominate over the error signal creating an imperfect modulation.

Another form of modulation relies on the system to ramp up and down as the switching amplifier changes state. When there are measurable changes in the system output at the desired switch frequency then the switching decision can be made based on the error signal. The simplest form of this control is in very familiar room heater. The heater is turned on when the room temperature falls too low and is turned off again when the room temperature is too high. If the temperature sensing system has a measurement noise equivalent to 1 degree deciding to switch the heater when the temperature falls by 0.1 degree will result in the switching being dominated by the noise and creating a very rapid wear for the switch. At a more reasonable temperature band of say 3 degrees both the band and the room time constant will determine the switch frequency.

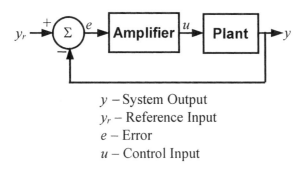

y – System Output
y_r – Reference Input
e – Error
u – Control Input

Figure 5.43. Closed-loop control structure

In power electronic applications, excessively high frequency switching of the electronic switches will result in unacceptable losses in the switch since typically there is a finite energy dissipated in the switch at each switching event.

5.6.2 Stability of Switching Control

Consider the first order system given by

$$\dot{x} = ax + u \quad (5.32)$$

where x is the system state variable and u is the control input. Let us constrain the control input u such that it takes values between +1 and −1. Then we choose the following Lyapunov function candidate

$$V = \frac{1}{2}x^2 \quad (5.33)$$

The Lyapunov function has a derivative of

$$\dot{V} = x(ax + u) = ax^2 + ux \quad (5.34)$$

When the system parameter a is negative, the system is stable for $u = 0$. The fastest convergence occurs when the control input is a signum function, i.e., $u = -\text{sgn}(x)$. The signum function $\text{sgn}(x)$ is defined as

$$\text{sgn}(x) = \begin{cases} +1 & \text{for } x > 0 \\ -1 & \text{for } x < 0 \\ 0 & \text{for } x = 0 \end{cases} \tag{5.35}$$

The minimum time problem, treated in [32], shows that the optimal minimum time problem will be bang-bang, meaning that the control will only select the largest positive or largest negative signals.

Let us now consider the case when a is positive. Let the value of the state variable x be positive. If we use the same control as discussed before we then get $u = -\text{sgn}(x) = -1$. The derivative of the Lyapunov candidate function is

$$\dot{V} = x(ax + u) = x(ax - 1) \tag{5.36}$$

The derivative will be negative and the control will converge if $|ax| < 1$. If, on the other hand, the value of the state variable x is negative, the control signal is $u = +1$. We then get

$$\dot{V} = x(ax + u) = x(ax + 1) \tag{5.37}$$

Since x is negative, the derivative of the Lyapunov function will be negative and the control will again converge for $|ax| < 1$. Thus we can conclude that the control for a positive a converges for $|ax| < 1$.

Let us now consider the tracking problem in which the output is required to follow a reference input x_{ref}. For this problem we can use the Lyapunov function candidate

$$V = \frac{1}{2}(x - x_{ref})^2 \tag{5.38}$$

The derivative of the above function is

$$\dot{V} = (ax + u - \dot{x}_{ref})(x - x_{ref}) \tag{5.39}$$

The control law for fastest convergence at a particular instant of time is

5. Structure and Control of Power Converters

$$u = -K \operatorname{sgn}(x - x_{ref}) \tag{5.40}$$

where K is a positive number. Let us assume that $x - x_{ref}$ is negative such that $u = K$. Then from (5.39) we get the following condition for stable control

$$ax + K - \dot{x}_{ref} > 0 \implies K > \dot{x}_{ref} - ax \tag{5.41}$$

Similarly when $x - x_{ref}$ is positive, we get the following condition from (5.39) for closed-loop stability

$$ax - K - \dot{x}_{ref} < 0 \implies K > ax - \dot{x}_{ref} \tag{5.42}$$

Therefore the system is stable when

$$\left| ax - \dot{x}_{ref} \right| < K \tag{5.43}$$

The control law given in (5.40) is very intuitive in that when the error is positive the control should be negative. In practice such a control would quickly enter a chatter mode, changing as rapidly as the implementation would permit.

Example 5.4: To illustrate the tracking performance, let us consider a reference waveform of $x_{ref} = 15\sin(100\pi t)$ that has to be tracked by the switching controller of the form given by (5.40). Let us assume that the parameter $a = -50$. Note that derivative of x_{ref} has maximum of 1500π. Therefore to satisfy the constraint (5.43) we choose K to be 5000. The system response is shown in Figure 5.44 in which it is assumed that the initial value of x is 20. It can be seen that x rapidly converges to the desired state. However, the switching frequency is extremely high.

ΔΔΔ

Usually, the heating in the semiconductor devices or wear on mechanical contacts limit the switching frequency and thus such a control action may not be desirable in practice at all. We must therefore investigate alternate forms of control to limit the switching frequency.

5.6.3 Sampled Error Control

In this method the switching pulses are generated at equidistant discrete intervals of time (regular samples). If the error of the tracking were sampled at a set rate and the control chosen based on the sign of the error, the chatter frequency is limited to below the sample rate. This will mean that the system can converge to a region of the target but will not be guaranteed to reach the target and remain there. The size of the chatter region depends on the sample time and the system time constants.

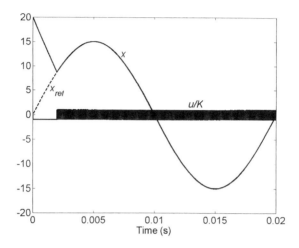

Figure 5.44. System output and control signal with switching controller

Consider the error waveform shown in Figure 5.45 (a). The error is compared with zero and a switching pulse is generated every time it crosses this level between the sample instants. The error is positive at the 0th sample. It is therefore necessary to bring the error closer to zero. The error however remains positive for 1st, 2nd and 3rd samples and becomes negative in the 4th sample. It remains negative in the 5th sample before becoming positive again in the 6th sample. The switches during these samples are arranged such that the error is forced towards zero.

The switching waveforms are shown in Figure 5.45 (b). In this figure, the desired output, indicated by the dashed line, is $x_{ref} = 15 \sin(100 \pi t)$ and the parameters a and K are chosen respectively as -50 and 5000. The actual and reference waveforms are compared at each sampling instant. These two waveforms are shown in figure. A switching decision is taken based on the comparison. For example, a negative going pulse is generated every time the actual output becomes greater than the desired output. Similarly, a positive going pulse is generated once the actual output falls below the desired one.

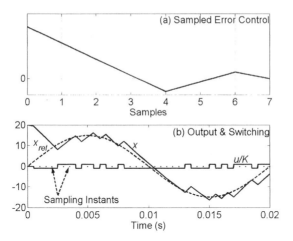

Figure 5.45. Sampled error control switching strategy

5.6.4 Hysteresis Control

We have already discussed hysteresis controller for current tracking earlier in the chapter. This is the most common form of tracking control. It adds a hysteresis band $\pm h$ to the reference signal. The control is generated when the system state crosses the band, i.e.,

$$\begin{aligned}&\text{If}\quad (x - x_{ref}) > h \quad \text{then}\quad u = -K\\&\text{elseif}\quad (x - x_{ref}) \leq -h \quad \text{then}\quad u = +K\end{aligned} \quad (5.44)$$

where K is a scalar constant. For this case we can assume that the control is in the range $-K \leq u \leq K$. The chatter frequency will then depend on the hysteresis band and the speed of transition. This form of control is always stable for first order systems. However the tracking performance may deteriorate when the system contains a source or the state value is large. For example, consider the system

$$\dot{x} = ax + u + V_s \sin(\omega t) \quad (5.45)$$

where $V_s \sin(\omega t)$ is a sinusoidal exogenous signal, which can be a voltage source. The state velocity will depend on the system state value as well as the source value. When the trajectories are curved, using hysteresis to force the peak values of x to be equidistant from x_{ref}, the average will not necessarily track x_{ref} perfectly.

An example of hysteresis band control is shown in Figure 5.46. The reference waveform is given by $x_{ref} = 15 \sin(100\pi t)$ and the parameters a and K are chosen respectively as -50 and 5000. For the waveforms shown in Figure 5.46, the value of h is chosen as 0.95. It can be seen that once the system output enters the band, it stays inside the band. The switching pulses are also shown in Figure 5.46. It can be seen that the switching frequency in this case is higher than that sampled error control case shown in Figure 5.45 (b). However, as the peaks are well controlled, the mean squared error in this case is much better compared to that shown in Figure 5.45. Consequently, the actual waveform is closer to the desired waveform with the hysteresis control. The effect for narrowing the hysteresis band is shown in Fig. 5.44. The hysteresis band for this case can be assumed to be zero resulting in a perfect tracking but very high switching frequency.

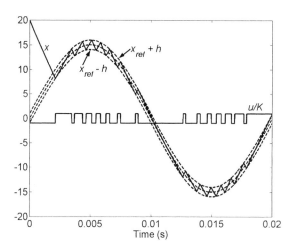

Figure 5.46. Hysteresis band control with $h = \pm 0.95$

5.7 Second and Higher Order Systems

The simple tracking error based switching controllers discussed in the previous section can be viewed as high gain proportional feedback. The switching control $u = -\text{sgn}(x - x_{ref})$ will give a unit control even with very small errors and thus can be seen as a gain nearing infinity. First order systems are readily stabilized by proportional controllers, even when the gain approaches infinity, while many resonant systems become unstable under high proportional control. To illustrate the idea, let us consider the following example.

Example 5.5: Consider the circuit shown in Figure 5.47. In this the source generates a voltage equal to $\pm K$. In the direction of the currents i_1 and i_2 and the polarity of the voltage v_C are as shown. We shall now investigate the closed-loop system stability when we use a hysteretic controller to track the currents i_1 and i_2.

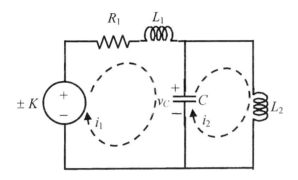

Figure 5.47. A simple two-loop circuit driven by a switched source

Let us define a state vector as

$$x = [i_1 \quad i_2 \quad v_C]^T$$

The system state space equation is then given by

$$\dot{x} = Ax + Bu \qquad (5.46)$$

where

$$A = \begin{bmatrix} -\dfrac{R_1}{L_1} & 0 & -\dfrac{1}{L_1} \\ 0 & 0 & \dfrac{1}{L_2} \\ \dfrac{1}{C} & -\dfrac{1}{C} & 0 \end{bmatrix}, \quad B = \begin{bmatrix} \dfrac{1}{L_1} \\ 0 \\ 0 \end{bmatrix}, \quad u = \pm K$$

Let us assume that we are required to follow a reference current given by $y_{ref} = 15 \sin(100\,\pi t)$. We use this reference to track i_1 and i_2 separately in the hysteresis current control mode. The system parameters are chosen as $R_1 = 0.02\ \Omega$, $L_1 = 20$ mH, $L_2 = 40$ mH and $C = 100\ \mu$F. The value of K is chosen as 700 and h is chosen as 0.95. The tracking results are shown in Figure 5.48. It can be seen that when the current i_1 is chosen as the output variable, the controller can track the reference current in the hysteresis band (shown in Figure 5.48 a). However, the system becomes unstable when the current i_2 is chosen as the output variable and the controller is required to track the reference in the hysteresis band (shown in Figure 5.48 b). A scaled version of the switching controller is also shown in Figure 5.48 (b). Every time the controller switches, the current changes in the opposite direction without being able to stabilize the system.

△△△

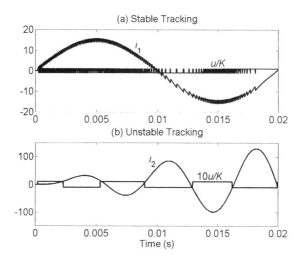

Figure 5.48. Hysteresis control of a third order system

To investigate the closed-loop stability of the hysteresis tracking controller of Example 5.5, let us first consider that I_1 is the output variable. Then defining $y = I_1 = Cx$, where $C = [1\ 0\ 0]$, we get the open-loop transfer function of the system as

$$\frac{Y(s)}{U(s)} = C(sI - A)^{-1} B = \frac{\pm \left(K/L_1\right)\left(s^2 + 1/CL_2\right)}{s^3 + s^2\left(R_1/L_1\right) + s\left(1/CL_2 + 1/CL_1\right) + R_1/CL_1L_2}$$

Let us further assume that K is positive and the switching controller is approximated by a proportional control with a gain of K_P. Then defining $K' = KK_P$, the characteristic equation for the closed-loop system is

5. Structure and Control of Power Converters

$$s^3 + s^2\left(\frac{R_1}{L_1} + \frac{K'}{L_1}\right) + s\left(\frac{1}{CL_2} + \frac{1}{CL_1}\right) + \left(\frac{R_1}{CL_1L_2} + \frac{K'}{CL_1L_2}\right) = 0 \quad (5.47)$$

Note that R_1, L_1, L_2, C and K' are positive. Then from Routh-Hurwitz criterion it can be shown that the following condition must be satisfied for the system to be stable

$$R_1 + K' > 0 \quad (5.48)$$

Since the condition (5.48) is always true, the system is stable irrespective of the values of the system resistor, inductor, capacitor or the control gain. This result is reflected in Figure 5.48 (a).

Let us now consider that I_2 is the output variable, i.e., $y = I_2$. Then the output matrix becomes $C = [0\ 1\ 0]$. The open-loop transfer function of the system is

$$\frac{Y(s)}{U(s)} = C(sI - A)^{-1}B = \frac{\pm K/CL_1L_2}{s^3 + s^2\left(R_1/L_1\right) + s\left(1/CL_2 + 1/CL_1\right) + R_1/CL_1L_2}$$

The characteristic equation for the closed-loop system is then

$$s^3 + s^2\left(\frac{R_1}{L_1}\right) + s\left(\frac{1}{CL_2} + \frac{1}{CL_1}\right) + \left(\frac{R_1}{CL_1L_2} + \frac{K'}{CL_1L_2}\right) = 0 \quad (5.49)$$

Then Routh-Hurwitz criterion puts the following condition on stability

$$K' < \frac{R_1 L_2}{L_1} \quad (5.50)$$

The condition (5.50) can never be satisfied for reasonable values of the components and for the fact that K' is large. Therefore the system will always be unstable as shown in Figure 5.48 (b).

Using analysis in the similar lines as presented above, it is possible to show that a hysteresis controller may be able to function properly for tracking current in a second order RLC circuit. However, a hysteresis controller will fail for an LC circuit. The closed-loop characteristic equation for an LC circuit will be of the form $s^2 + (1/LC + K')$. This means that the circuit will go through a sustained oscillation. It is therefore imperative that more suitable controllers are used for stabilizing 2nd and higher order systems.

5.7.1 Sliding Mode Controller

Sliding mode switching design is one useful approach for higher order systems. In this approach a sliding line is defined and the system states slides along this line to a stable equilibrium point. Consider the system

$$\ddot{x} = -\alpha x + u \tag{5.51}$$

Let us consider the time varying surface

$$S = x + \beta \dot{x} \tag{5.52}$$

Note that $S = 0$ is called the sliding surface, which is a linear differential equation whose unique solution is $x = 0$. Thus the regulating problem reduces to choosing a control such that S is kept at zero for $t > 0$. The system motion on the sliding surface can be given as an average of the system dynamics on both sides of the surface. This has been termed as the Fillippov's equivalent dynamics [33].

Let us now examine the positive function $V = S^2$. Its derivative is

$$\dot{V} = 2S\dot{S} = 2(x + \beta\dot{x})\{\dot{x} + \beta\{-\alpha x + u\}\} \tag{5.53}$$

For u sufficiently large compared with the states and for α positive, the control

$$u = -K \, \text{sgn}(x + \beta\dot{x}) \tag{5.54}$$

will ensure that the derivative of V is always negative which, for the positive hysteresis function, implies that S will converge to zero.

This $S = 0$ condition implies

$$\dot{x} = -\frac{1}{\beta}x \tag{5.55}$$

which defines the sliding line. The switching control will chatter about the switching line and the dynamics of the second order system reduce to the first order dynamics of the switching line. This is a two-stage process. In the

5. Structure and Control of Power Converters

first stage the main focus is in converging to the sliding line. The effort in the second stage is on staying on the line. This implies exponential convergence with time constant β. The convergence requires

$$|u| > \left|\frac{\dot{x} - \alpha\beta x}{\beta}\right| \tag{5.56}$$

Thus choosing a small α implies a greater difficulty in ensuring convergence in the first place. Sliding mode controllers are discussed in [33-35].

5.7.2 Linear Quadratic Regulator (LQR)

Let us consider the state space equation of the form given in (5.46). For the single-input case, the following linear quadratic cost function is chosen

$$J = \int_0^\infty (x^T Q x + r u^2) dt \tag{5.57}$$

where Q is a symmetric positive semi-definite matrix and $r > 0$ is a scalar constant which puts a penalty on the maximum control action. The minimization of the cost function results in a feedback control of the form

$$u = -Kx \tag{5.58}$$

where the gain matrix K is obtained by the steady state solution of the Riccati equation

$$0 = A^T P + PA - \left(PBB^T P\right)/r + Q \tag{5.59}$$

$$K = \left(B^T P\right)/r \tag{5.60}$$

A Linear Quadratic Regulator is shown to produce an infinite gain margin and a phase margin of at least 60° [36]. Another important aspect of the LQR is that it is tolerant of input nonlinearities, as shown in Figure 5.49. The LQR design is stable provided that the effective gain of the input nonlinearity is constrained in the sector between ½ and 2 [36]. When the errors are large, and the control is bounded between + 1 and – 1 the elements of the gain matrix K must be small. For a set of decreasing values of r, we get a corresponding set of increasing values of K. Thus there always exists a value of r such that Kx is bounded appropriately.

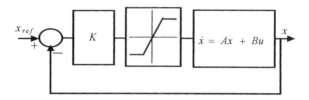

Figure 5.49. State feedback in the presence of nonlinear element in the forward path

Note that for switching controller, the control input u in (5.46) can only take on values $\pm V$, where V is a large number. Let us assume for the time being that the control input is a continuous-time signal equal to u_c. The LQR based full state feedback tracking control law is then given by

$$u_c = -K(x - x_{ref}) \qquad (5.61)$$

where K is the feedback gain matrix calculated from (5.59) and (5.60). The switching variable u is then obtained from this continuous-time signal u_c by a hysteresis action around zero, i.e.,

$$\begin{aligned}&\text{If } u_c > h \text{ then } u = V \\ &\text{elseif } u_c < -h \text{ then } u = -V\end{aligned} \qquad (5.62)$$

Example 5.6: Let us consider the same system given in Example 5.5 and shown in Figure 5.47. We choose a current reference of $y_{ref} = 15 \sin(100\pi t)$ and then generate a feedback signal which will force the current i_2 to track this reference. In a full state feedback control system of the form (5.61), the reference values for all the state variables must be provided. Let us define the fundamental frequency as $\omega = 100\pi$. Assuming that $i_{2ref} = 15 \sin(\omega t)$, we get the following two equations from Figure 5.47

$$v_{Cref} = L_2 \frac{di_{2ref}}{dt} = 15\omega L_2 \cos(\omega t)$$

$$i_{1ref} = C\frac{dv_{Cref}}{dt} + i_{2ref} = (1 - \omega^2 L_2 C)i_{2ref}$$

We then get

$$x_{ref}^T = \begin{bmatrix} i_{1ref} & i_{2ref} & v_{Cref} \end{bmatrix} \tag{5.63}$$

We now use the state reference computed in (5.63) in the LQR control law (5.61). The state feedback matrix is generated using

$$Q = \begin{bmatrix} 0 & & \\ & 10 & \\ & & 1 \end{bmatrix} \text{ and } r = 10^{-6}$$

From the choice of the weighting matrix Q it can be seen that the maximum weight is given to i_2, followed by v_C. No weighting is given to i_1. Also, the penalty on control r determines the control effort. The smaller this number, the more the control effort, which in this case means increase in the switching frequency. The variable V of (5.62) is chosen to be 300 V. The value of h in (5.62) is chosen to be 0.001.

Figure 5.50 shows the tracking results in which all the three state variables and their respective references are shown. The tracking errors are shown in Figure 5.51. Since we have used a full state feedback in which the references of all the state variables are computed accurately, it can be seen from these figures that all the state variables converge to their respective reference values.

ΔΔΔ

5.7.3 Tracking Controller Convergence

When designing a switching control system to track a reference, the stability and robustness of the control is of concern. In addition to the stability of control system to force the system error to zero, we must also evaluate if the tracking process will work well for a given system. To facilitate this, let us assume that system state space description is given as

$$\dot{x} = Ax + Bu \tag{5.64}$$

Let us form a reference that is defined by

$$\dot{x}_{ref} = Ax_{ref} + Bu^* \tag{5.65}$$

Implicit in this formulation is the assumption that the reference signal is attainable for some control input. Many reference formulations can create arbitrary output signals without addressing the issue of whether the other states are compatible with the reference.

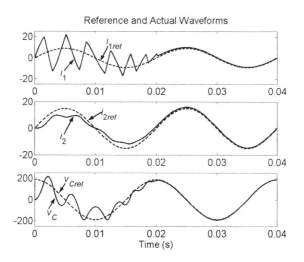

Figure 5.50. State variable tracking using LQR

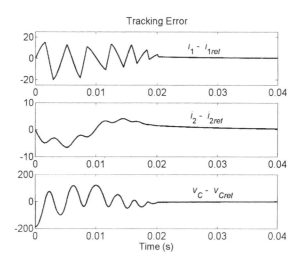

Figure 5.51. Tracking error using LQR

Defining an error vector as $x_e = x - x_{ref}$, we can write by subtracting (5.65) from (5.64)

$$\dot{x}_e = Ax_e + B(u - u^*)$$ (5.66)

From the control law given in (5.61), if we remain on the switching surface, then we get [36]

$$K(x - x_{ref}) = 0 \Rightarrow Kx_e = 0$$

We can then write

$$K\dot{x}_e = 0 \Rightarrow KAx_e + KB(u - u^*) = 0$$

From the above equation we get

$$(u - u^*) = -\frac{KA}{KB}x_e$$ (5.67)

Substituting (5.67) into (5.66) the error equation becomes

$$\dot{x}_e = \left(I - \frac{BK}{KB}\right)Ax_e = Mx_e$$ (5.68)

Hence the tracking error will converge to zero for any initial condition provided that the following three conditions satisfy.

- **Condition-1**: All the eigenvalues of the matrix M of (5.68) have negative real parts,
- **Condition-2**: The reference signal can be generated by an equation in the form of (5.65) and
- **Condition-3**: Since u is bounded between $+V$ and $-V$, (5.67) requires that

$$-V \leq u^* - \frac{KA}{KB}x_e \leq V.$$ (5.69)

It is a property of switching law based on LQR design that Condition-1 above is satisfied. Most tracking problems are posed as the output of the plant following some reference y_{ref}. Our task is to define a state reference satisfying $y_{ref} = Hx_{ref}$. In frequency domain the output equation can be given as

$$Y_{ref}(s) = H(sI - A)^{-1} BU^*(s) \tag{5.70}$$

Assuming that we are dealing with a single-input, single-output (SISO) system we can write from (5.70)

$$U^*(s) = \frac{Y_{ref}(s)}{H(sI - A)^{-1} B} \tag{5.71}$$

Thus the corresponding reference state can be formed from

$$x_{ref}(s) = (sI - A)^{-1} BU^*(s) \tag{5.72}$$

Hence any single output-tracking problem can be translated into a state tracking problem and expressed in the form of (5.66). In practice, the circuit analysis would be easiest method to form x_{ref} from y_{ref}, particularly for sinusoidal tracking. This is illustrated in Example 5.6.

Condition-3 given above cannot be truly satisfied for arbitrary references. From (5.71) we can see that a bounded y_{ref} implies a bounded u^*. The system states are not necessarily bounded. But if y_{ref} and the system states are sufficiently small, then the Condition-3 can be satisfied and perfect tracking can occur. In practice however unmodeled disturbances such as faults or large step load changes can be sufficiently large that the tracking conditions are lost and a finite tracking error will occur.

5.7.4 Condition for Tracking Reference Convergence

The change from output tracking to state tracking is not a problem if the system model is known perfectly. The approach analyzed here is where the reference for the switching controller is fixed for a period. The system reaches steady state then the voltages and currents of the circuit used to define the new reference values. In the case of an LC system with a voltage reference, a consistent reference value for the current may perhaps be set from the fundamental component of current found flowing in the circuit when the particular reference voltage is used.

If we satisfy the stability and magnitude conditions, the system will stay in the tracking band and the system performance will approximate

5. Structure and Control of Power Converters

$$K(x - x_{ref}) = 0$$

This implies

$$K(\dot{x} - \dot{x}_{ref}) = 0$$

For the system model given in (5.64) the above equation yields

$$K(Ax + Bu - \dot{x}_{ref}) = 0 \tag{5.73}$$

The effective control in this mode can be found as

$$u_{eff} = -(KB)^{-1}(KAx - K\dot{x}_{ref}) = 0 \tag{5.74}$$

Applying the control to the system (5.64) we get

$$\dot{x} = Ax - B(KB)^{-1}KAx + B(KB)^{-1}K\dot{x}_{ref} \tag{5.75}$$

which is stable if the eigenvalues of $[I - B(KB)^{-1}K]A$ are in the left half plane. This is the same as the Condition-1 of Section 5.7.3. For an LQR designed system this stability is guaranteed, provided that controller saturation is avoided.

Expressing this state space model in frequency domain then when steady state is reached

$$X(s) = \left[sI - \left(I - B(KB)^{-1}K\right)A\right]^{-1} sB(KB)^{-1}KX_{ref}(s) \tag{5.76}$$

Expressing the above equation for the nth harmonic we get

$$X(j n\omega) = \left[j n\omega I - \left(I - B(KB)^{-1}K\right)A\right]^{-1} j n\omega B(KB)^{-1}KX_{ref}(j n\omega) \tag{5.77}$$

If the reference value for the $k + 1$ iteration is based on a linear combination of output reference (y_{ref}) and the steady system solution for other system states, e.g.,

$$X_{ref}(j n\omega)_{k+1} = PX(j n\omega)_k + TY_{ref}(j n\omega)_k$$

we get

$$X_{ref}(j\omega)_{k+1} = P\{[j\omega I - (I - B(KB)^{-1}K)A]^{-1} \\ j\omega B(KB)^{-1}K\}X_{ref}(j\omega)_k + TY_{ref}(j\omega)_k \qquad (5.78)$$

Then the stability of the reference formulation depends on the eigenvalues of

$$P\{[j\omega I - (I - B(KB)^{-1}K)A]^{-1} j\omega B(KB)^{-1}K\}$$

being inside the unit circle.

5.7.5 Deadbeat Controller

This is a discrete-time output feedback controller in which the penalty on control is assumed to be zero. Let the output of the system is given by the difference equation

$$A(z^{-1})y(k) = B(z^{-1})u_c(k) \qquad (5.79)$$

where A and B are polynomials given by

$$A(z^{-1}) = 1 + a_1 z^{-1} + a_2 z^{-2} + \cdots + a_n z^{-n} \\ B(z^{-1}) = b_0 + b_1 z^{-1} + b_2 z^{-2} + \cdots + b_m z^{-m} \qquad (5.80)$$

with $n \geq m$ and z^{-1} being a unit delay operator satisfying the following relation

$$z^{-1}y(k) = y(k-1)$$

In a deadbeat controller the following cost function is chosen

$$J = [y(k) - y_{ref}(k)]^2 \qquad (5.81)$$

It is then minimized by taking its derivative with respect to $u_c(k)$ and equating the result to zero, i.e.,

$$\frac{\partial J}{\partial u_c(k)} = 2[y(k) - y_{ref}(k)]\frac{\partial y(k)}{\partial u_c(k)} = 0 \qquad (5.82)$$

5. Structure and Control of Power Converters

Note from (5.79) and (5.80) that the system difference equation can be written as

$$y(k) = -a_1 y(k-1) - \cdots - a_n y(k-n) + b_0 u_c(k) + \cdots + b_m u_c(k-m)$$

such that

$$\frac{\partial y(k)}{\partial u_c(k)} = b_0$$

Therefore the control law from (5.82) is given as

$$u_c(k) = \frac{1}{b_0} [y_{ref}(k) + a_1 y(k-1) + \cdots + a_n y(k-n) \\ - b_1 u_c(k-1) - \cdots - b_m u_c(k-m)] \quad (5.83)$$

Once $u_c(k)$ is known, the switching variable $u(k)$ can be obtained in the same manner as (5.62). Deadbeat PWM control is reported in [37].

Example 5.7: Let us consider the same system as given in Examples 5.5 and 5.6. The continuous state equation (5.46) is converted into a discrete-time form as

$$x(k+1) = \phi x(k) + \theta u_c(k) \quad (5.84)$$

where k is the kth sample, ϕ is the state transition matrix and θ is the input matrix. These matrices are given by

$$\phi = e^{A t_d} \text{ and } \theta = \int_0^{t_d} e^{At} B \, dt$$

where t_d is the sampling period. Let the output be defined by $y = Hx$. Then the deadbeat control action can be computed from (5.83) as

$$u_c = \frac{y_{ref}(k+1) - H\phi x(k)}{H\theta} \quad (5.85)$$

A sample time of 100 μs is chosen and the values of V and h are the same as those given in Example 5.6. The system response when the capacitor voltage v_C is taken as reference (i.e., $H = [0 \ 0 \ 1]$) is shown in Figure 5.52 (a). It can be seen that the tracking is reasonably good in this case. However the tracking fails when the current i_2 is taken as output (i.e., $H = [0 \ 1 \ 0]$). This is shown in Figure 5.52 (b). The reference in either case is the same as those given in Example 5.6.

The main problem with a deadbeat controller is that it is very sensitive to system parameters. Substituting (5.83) into (5.79) we obtain the system closed-loop equation of $y(k) = y_{ref}(k)$, which implies that the controller poles cancels all the open-loop zeros. It will cause problem when the system is nonminimum-phase, i.e., when the system zeros are lying outside the region of stability (unit circle) [38]. The current tracking failed, as one of the poles in this case is located at -3.73. Therefore the deadbeat controller must be judiciously used even if it has a very fast convergence property.

△△△

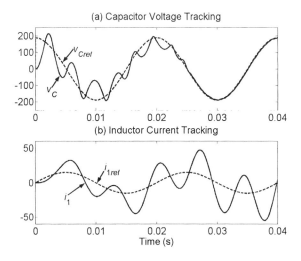

Figure 5.52. Tracking with deadbeat controller

5.7.6 Pole Shift Controller

This is a discrete-time control technique in which the open-loop system poles are radially shifted towards the origin (i.e., more stable locations) to form the closed-loop poles [39,40]. Consider the discrete-time state space equation given in (5.84). The eigenvalues of the open-loop system are given by the solution of the characteristic equation

$$|zI - \phi| = z^3 + a_1 z^2 + a_2 z + a_3 = 0 \tag{5.86}$$

5. Structure and Control of Power Converters

In the pole shift controller, the eigenvalues are shifted by a factor λ, which is called the pole shift factor and is constrained by $0 < \lambda < 1$. The desired closed-loop poles are then obtained by multiplying the open-loop eigenvalues by λ. This shifting will results in the desired closed-loop system characteristic equation of the form

$$\Delta_{cl}(z) = z^3 + \lambda a_1 z^2 + \lambda^2 a_2 z + \lambda^3 a_3 = 0 \qquad (5.87)$$

The closer λ is to unity, the more is the penalty on control as this will prohibit a large amount of shifting of the open-loop poles.

If the control law is given by $u_c(k) = -K(x - x_{ref})$, then the characteristic equation of the closed-loop system will become

$$|zI - \phi + \theta K| = 0 \qquad (5.88)$$

The elements of the gain matrix K are then obtained by equating (5.87) to (5.88). The switching signal u is then obtained in the same manner as given in (5.62). The pole shift design is a suboptimal design that results in smooth control action. Also since the penalty on control action can be easily adjusted using a single parameter λ, the closed-loop system will unlikely to be unstable for small pole shifts. In this regard this scores over the deadbeat controller which forces all the closed-loop poles to the origin thereby requiring excessive control effort.

Example 5.8: Let us consider the same system as given in Example 5.6 in which the references are the same and they are generated in the same manner as given in Example 5.6. The values of V and h are the same as given before, i.e., 300 V and 0.001 respectively. The value of the pole shift factor λ is chosen to be 0.8. The system response is shown in Figure 5.53. It can be seen that the tracking of the current i_1 is not smooth as it ripples around the reference. However the tracking of the other two state variables is smooth and tracking convergence is very fast.

ΔΔΔ

5.7.7 Sequential Linear Quadratic Regulator (SLQR)

We have discussed the linear quadratic controller in Section 5.7.2 where it was mentioned that provided that the effective gain of the input nonlinearity is constrained in the sector between ½ and 2 the LQR design is stable. When the errors are large, and the control is bounded to $\pm V$, the gain K must be small. For a set of decreasing values of r we get a corresponding set of increasing values of K. Thus there will always exist a value of r such that Kx is bounded appropriately. Finite time convergence of the regulator problem can be shown, provided Q is chosen such that as $r \to 0$, $K \to 0$. This gives a better performance than the exponential convergence of sliding line solutions.

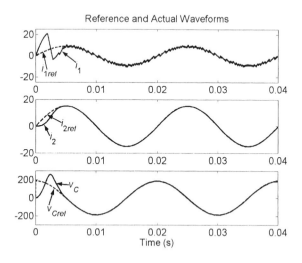

Figure 5.53. Pole shift controller tracking performance

In many cases the switching is based on

$$u = -proj\{K(x - x_{ref})\} \tag{5.89}$$

where *proj* refers to the projection of the LQR control onto the permissible set of switch states. This result helps in designing the control that will create the final portion of the trajectory along a switching line. The design also provides for good convergence in reaching the switching line for large system errors provided that they are stable. For control of an oscillatory system with small control limits the projection design will provide good damping until the control can force the system onto the final sliding line.

There are strict conditions discussed on this design [41] for when r should be changed as the system converges. But for most cases the simplistic solution of increasing K if the control value has been less than ½ for extended periods usually provides a satisfactory solution.

For a well-known environment, the range of errors is known beforehand and one value of gain K can be found which, on the average, would keep the errors in the desired range. Since the LQR parameters Q and r are chosen by a user, the design is not unique. For SISO systems a good set of parameters

5. Structure and Control of Power Converters

is found from the consideration of the mirror root locus [42]. The tracking problem can be expressed in terms of the output $y = Hx$ in which the following cost function is minimized.

$$J = \int_0^\infty \left(y^2 + ru^2\right) dt \qquad (5.90)$$

The locus of the roots of the closed loop system as r varies can be found from the root locus of the open loop system. The poles and zeros of the open-loop system and their values mirrored on the imaginary axis are used to define the nominal system and the standard root locus rules are applied. Those trajectories in the left half plane define the closed loop poles that can be obtained by adjustment of the LQR parameter r. The poles will move infinitely fast as the value of r approaches zero, provided there are no finite zeros. The choice of H helps in defining the zeros of the system. If the system is transformed into controllable canonical form and H is chosen to measure only the final state, then there will be no zeros. The design process would be to find the transformation matrix T to the controllable canonical form and to use $\Phi = HT$ in the LQR design. This choice translates to the standard Riccati equation with $Q = \Phi^T \Phi$.

For multi input systems no such unique canonical form is available to ensure the lack of finite zeros. Experience to date has shown that Q can be easily chosen keeping the design requirement in mind in most cases.

Example 5.9: Let us consider a series RLC circuit that is driven by a voltage source v_S. Let us denote the inductor current as i_L and the capacitor voltage as v_C. The system state space equation of the form (5.46) is obtained, where

$$x = \begin{bmatrix} i_L \\ v_C \end{bmatrix}, \quad A = \begin{bmatrix} -R/L & -1/L \\ 1/C & 0 \end{bmatrix}, \quad B = \begin{bmatrix} 1/L \\ 0 \end{bmatrix}, \quad u = v_S \qquad (5.91)$$

Also let the output be the capacitor voltage such that $y = v_C = [0 \ 1]x$. Let us choose $R = 1.5\ \Omega$, $L = 9$ mH and $C = 100\ \mu$F. Then the system matrices are given by

$$A = \begin{bmatrix} -166.67 & -111.11 \\ 10^4 & 0 \end{bmatrix}, \quad B = \begin{bmatrix} 111.11 \\ 0 \end{bmatrix}, \quad H = [0 \ 1]$$

Let us transform the system into phase variable canonical form so the relation between the state space form and the pole-zero form is easily seen. This can be done by choosing a transformation matrix T which will transform to a new state space description with relationship $x = Tz$. The required transformation matrix, for a general n^{th} order system is [43]

$$T = \begin{bmatrix} t_1 \\ t_1 A \\ \vdots \\ t_1 A^{n-1} \end{bmatrix} \tag{5.92}$$

where

$$t_1 = \begin{bmatrix} 0 & 0 & \cdots & 0 & 1 \end{bmatrix} S^{-1}, \quad S = \begin{bmatrix} B & AB & A^2 B & \cdots & A^{n-1} B \end{bmatrix}$$

The matrix S is the controllability matrix. Note that the transformation is possible only when the pair $\{A,B\}$ is controllable such that the matrix S is full rank. The transformed state space system is then given by

$$\begin{aligned} \dot{z} &= T^{-1} A T z + T^{-1} B u = \Lambda z + \Gamma u \\ y &= H T z = \Phi z \end{aligned} \tag{5.93}$$

For the system given in (5.91) the transformed matrices are

$$\Lambda = \begin{bmatrix} 0 & 1 \\ -1.11 \times 10^6 & -166.67 \end{bmatrix}, \quad \Gamma = \begin{bmatrix} 0 \\ 1.23 \times 10^6 \end{bmatrix}, \quad \Phi = \begin{bmatrix} 0.009 & 0 \end{bmatrix}$$

In this canonical form, the entries in the last row of Λ are the negative of the coefficients of the system transfer function denominator polynomial. The elements of Φ define the coefficients of the numerator polynomial of the transfer function. To illustrate the LQR solutions as the control penalty changes, we can use the mirror root locus. To construct the mirror root locus we note that

$$\Phi (sI - \Lambda)^{-1} \Gamma = \frac{\beta(s)}{\alpha(s)} \tag{5.94}$$

where the polynomials β and α define the zeros and poles of the system respectively and are given by

$$\alpha(s) = (s + p_1)(s + p_2)\cdots(s + p_n)$$
$$\beta(s) = (s + z_1)(s + z_2)\cdots(s + z_m)$$
(5.95)

We now define a $2n$-degree polynomial as

$$\Delta(s) = \alpha(s)\alpha(-s) + r^{-1}\beta(s)\beta(-s) \tag{5.96}$$

As is shown in [42] the closed loop poles of the LQR controlled system satisfy $\Delta(s)=0$ as the control penalty r changes. It is obvious that if λ is a root of $\Delta(s)$, then $-\lambda$ is also a root of $\Delta(s)$. Therefore $\Delta(s)$ will have n left half s-plane roots with n roots in the right half plane that are mirror images. We would like draw the root locus of $\Delta(s) = 0$ to examine the closed loop poles. The points $r \to 0$ and $r \to \infty$ are the extreme points of the mirror root locus. For $r \to \infty$ the roots of $\Delta(s)$ are the roots of $\alpha(s)$ and their mirror image. For $r \to 0$ the roots of $\Delta(s)$ are the roots of $\beta(s)$ and their mirror image. Thus if the zeros of the original system were not finite then the roots of $\Delta(s)$ will not be finite. Given that the LQR solution is the stable set of roots the LQR solution will become infinitely fast as $r \to 0$. If the original system has finite zeros the LQR solution will still give a finite convergence rate even as the control penalty approaches zero.

The equation $\Delta(s) = 0$ can be written in a standard form as [42]

$$(-1)^{n-m} r^{-1} \frac{\prod_{i=1}^{m}(s + z_i)(s - z_i)}{\prod_{i=1}^{n}(s + p_i)(s - p_i)} = -1 \tag{5.97}$$

The mirror root locus of the open-loop system of (5.91) is shown in Figure 5.54. Note that the system has eigenvalues at $-83 \pm j1051$ and no zeros. There are four branches of the loci. The left half plane branches start from the points $-83 \pm j1051$ when $r \to \infty$. Similarly the right half plane branches start at $83 \pm j1051$. Crosses in the figure mark the starting points. Since there are no zeros, all the branches move towards infinity as $r \to 0$.

Let us assume that the system is now controlled by an LQR designed linear controller. For this we have chosen $Q = \Phi^T \Phi$ and varied r. For each pair of Q and r, the gain matrix K is computed. The eigenvalues of the closed-loop system are then given by the solution of the equation $|sI - \Lambda + \Gamma K| = 0$. A portion of the root locus as the gain varies is shown in Figure 5.55. It can be seen that the system poles move further away from the imaginary axis as the control penalty r tends to zero.

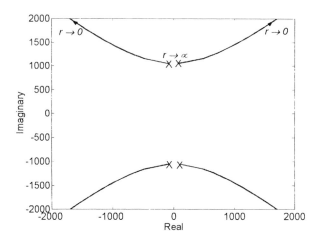

Figure 5.54. Mirror root locus of the system of (5.91)

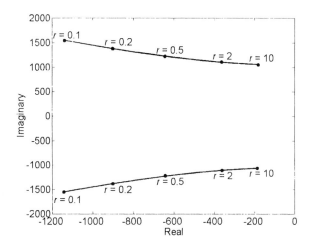

Figure 5.55. Closed-loop eigenvalues with LQR

ΔΔΔ

For practical systems, the control value is limited so when the error is large the gains must be kept small to avoid saturation. As the errors decrease, the gains can be increased and if the errors could approach zero the convergence could be achieved infinitely fast as in finite time to reach the target. For a switching control problem the projection of the desired LQR

5. Structure and Control of Power Converters

control value onto the possible switch states will determine the effective gain variation. Provided that the LQR control value and the switch value do not differ by more than a ratio of 2, over the system trajectory, the system will be stable. For tracking problems however, the errors in a finite switch rate controller cannot be kept at zero. For this reason control gains cannot increase indefinitely without causing saturation, and the control penalty cannot be made zero, particularly for switching control.

Consider the RLC circuit problem discussed before. Since it is a single input case, the projection onto ± 1 can be expressed as

$$u = \text{sgn}\{K_1(i_L - i_{ref}) + K_2(v_C - v_{ref})\} \quad (5.98)$$

We can rewrite the above equation as

$$u = \text{sgn}\left\{(v_C - v_{ref}) + \frac{K_1}{K_2}(i_L - i_{ref})\right\} \quad (5.99)$$

If only the first term were present this would be the same as a standard hysteresis controller with zero hysteresis band. The presence of the second term acts as the stabilizing term. Note that the actual gains themselves are not important but only the ratio of gains. When the tracking error is large, a high r must be used to get the low gain required to satisfy the project rule. For this system a high r corresponds to a high gain ratio; as the error decreases a low r is possible and the LQR solution for this problem shows that the gain ratio would be decreased.

A set of LQR problems can be presolved with an index to these levels of r. Initially the index is high selecting a high value of r and thus the system starts with a low gain. As the length of the position vector decreases, the projected value of Kx is checked and when it is smaller than 2 for a significant segment of trajectory, the level is decreased. The results are shown in Figures 5.56 to 5.58. The system states are shown in Figure 5.56. In this figure the inductor current is multiplied by a constant defined by $z_0 = \sqrt{(L/C)}$. The LQR control output, switching pulses and the control gain are shown in Figure 5.57. The control gain is computed by taking the inner product of the gain matrix K. The log-magnitude of this inner product is then plotted to show the changes in levels with time. The phase-plane trajectory is shown in Figure 5.58. It can be seen that this trajectory starts as a damping controller and ends as a sliding line with the slope increasing.

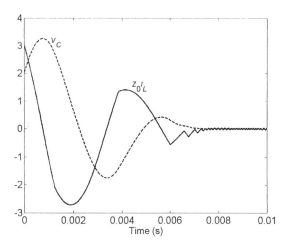

Figure 5.56. System states with SLQR control

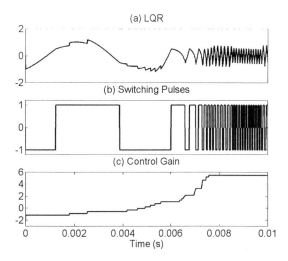

Figure 5.57. Response of the control system

5.8 Conclusions

In this chapter we have presented the structures of various power converter circuits and converter control techniques. Out of the large number of converter circuits available, the most suitable topology depends on various factors. The issues that need to be addressed are whether these converters will be required to operate at a very high switching speed. In that

5. Structure and Control of Power Converters

event soft-switched converters can be used such that the switching losses are kept low. The other consideration is the power level. For high power applications, the converters discussed in Section 5.4 can be used. Since this book discusses distribution system applications for which neither high power nor switching frequency is of great concern, we shall consider mostly hard-switched converters in the remainder of the book.

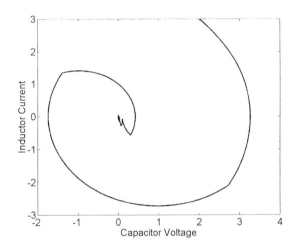

Figure 5.58. Phase-plane trajectory

The other topic that is of importance is the converter control. In this chapter we have presented some of the converter control techniques and their closed-loop stability aspects. We have mostly concentrated on hysteresis band control in which the converter switching actions are generated from the feedback control action through a hysteresis band around zero error. This is because we want the controller to force system to operate with a zero tracking error, i.e., $x - x_{ref} = 0$. The use of hysteresis control may result in high switching frequency and may be undesirable for some applications. Therefore the usability of some other form of zero tracking controller (e.g., sampled error, adaptive hysteresis etc.) can also be investigated. Since the basic aim of this book is to discuss the improvements in power quality of distribution system, we shall only use the hysteresis band control of (5.62) when we discuss practical power quality enhancing devices in Chapters 8 to 10. However the tracking performance of other controllers, such as regular sampled deadbeat, can approach that of hysteresis controller.

5.9 References

[1] N. Mohan, T. M. Undeland and W. P. Robbins, *Power Electronics: Converters, Applications and Design*, John Wiley, New York, 1989.
[2] M. H. Rashid, *Power Electronics: Circuits, Devices and Applications*, Prentice-Hall, Englewood Cliffs, 1993.
[3] G. K. Dubey, S. R. Doradla, A. Joshi and R. M. K. Sinha, *Thyristorised Power Controllers*, Wiley Eastern, New Delhi, 1986.
[4] D. M. Divan, "The resonant dc link inverter – a new concept in static power conversion," *IEEE Trans. Industry Applications*, Vol. IA-25, No. 2, pp. 317-325, 1989.
[5] K. K. Mahapatra, A. Ghosh, A. Joshi and S. R. Doradla, "A novel current initialization scheme for parallel resonant dc link inverter," *International Journal of Electronics*, Vol. 87, No. 9, pp. 1125-1137, 2000.
[6] B. D. Bedford and R. G. Hoft, *Principles of Inverter Circuits*, John Wiley, New York, 1964.
[7] L. Gyugyi, N. G. Hingorani, P. R. Nannery and N. Tai, "Advanced static var compensator using gate turn-off thyristors for utility applications," *CIGRE*, Paper No. 23-203, 1990.
[8] L. Sunil Kumar, *Design, Modeling and Control of a 48-step inverter base S^3C*, M. Tech. Thesis, IIT Kanpur, 1998.
[9] L. Sunil Kumar and A. Ghosh, "Modeling and control design of a static synchronous series compensator," *IEEE Trans. on Power Delivery*, vol. 14, no. 4, pp. 1448-1453, 1999.
[10] L. Sunil Kumar and A. Ghosh, "Static synchronous series compensator – design, control and applications," *Electric Power Systems Research*, vol. 49, pp. 139-148, 1999.
[11] G. N. Pillai, A. Ghosh and A. Joshi, "Torsional Oscillation Studies in an SSSC Compensated Power System," *Electric Power Systems Research*, Vol. 55, pp. 57-64, 2000.
[12] A. Nabae, I. Takahashi and H. Akagi, "A new neutral-point-clamped PWM inverter," *IEEE Trans. Industry Applications*, Vol. IA-17, No. 5, pp. 518-523, 1981.
[13] P. M. Bhagwat and V. R. Stefanovic, "Generalized structure of a multilevel PWM inverter," *IEEE Trans. Industry Applications*, Vol. IA-19, No. 6, pp. 1057-1069, 1983.
[14] N. S. Choi, J. G. Cho and G. H. Cho, "A general circuit topology of multilevel inverter," *Proc. IEEE Power Electronics Specialist Conference (PESC)*, pp. 96-103, 1991.
[15] M. Marchesoni, "High-performance current control techniques for applications to multilevel high-power voltage source inverters," *IEEE Trans. Power Electronics*, Vol. 7, No. 1, pp. 189-204, 1992.
[16] G. Carrara, S. Gardella, M. Marchesoni, R. Salutari and G. Sciutto, "A new multilevel PWM method: A theoretical analysis," *IEEE Trans. Power Electronics*, Vol. 7, No. 3, pp. 497-505, 1992.
[17] J. S. Lai and F. Z. Peng, "Multilevel converters – a new breed of power converters," *IEEE Trans. Industry Applications*, Vol. 32, No. 3, pp. 509-517, 1996.
[18] R. W. Menzies and Y. Zhuang, "Advanced static compensation using a multilevel GTO thyristor inverter," *IEEE Trans. Power Delivery*, Vol. 10, No. 2, pp. 732-738, 1995.

5. Structure and Control of Power Converters 213

[19] H. S. Patel and R. G. Hoft, "Generalized techniques of harmonic elimination and voltage control in thyristor inverters: Part I – Harmonic Elimination," *IEEE Trans. Industry Applications*, Vol. IA-9, No. 3, pp. 310-317, 1973.

[20] J. B. Ekanayake, N. Jenkins and C. B. Cooper, "Experimental investigation of an advanced static var compensator," *Proc. IEE – Generation, Transmission & Distribution*, Vol. 142, No. 2, pp. 202- 210, 1995.

[21] Y. Chen, B. Mwinyiwiwa, Z. Wolanski and B. T. Ooi, "Regulating and equalizing dc capacitance voltages in multilevel STATCOM," *IEEE Trans. Power Delivery*, Vol. 12, No. 2, pp. 901-907, 1997.

[22] M. D. Manjrekar, P. K. Steiner and T. A. Lipo, "Hybrid multilevel power conversion system: A competitive solution for high-power applications," *IEEE Trans. Industry Applications*, Vol. 36, No. 3, pp. 834-841, 2000.

[23] X. Yuan and I. Barbi, "Fundamentals of a new diode clamping multilevel inverter," *IEEE Trans. Power Electronics*, Vol. 15, No. 4, pp. 711-718, 2000.

[24] M.K. Mishra, A. Ghosh and A. Joshi, "A new STATCOM topology to compensate loads containing ac and dc components," *IEEE Power Engineering Society Winter Meeting*, Singapore, 2000.

[25] T. H. Barton, "Pulse width modulation waveforms – the Bessel Approximation," *Proc. IEEE Industry Applications Society Annual Conference*, pp. 1125-1130, 1978.

[26] H. W. Van der Broeck, H. C. Skudelny and G. V. Stanke, "Analysis and realization of a pulsewidth modulator based on space vector," *IEEE Trans. Industry Applications*, Vol. 24, No. 1, pp. 142-150, 1988.

[27] K. Taniguchi and H. Irie, "Trapezoidal modulating signal for three-phase PWM inverter," *IEEE Trans. Industrial Electronics*, Vol. IE-3, No. 2, pp. 193-200, 1986.

[28] K. Thorborg and A. Nystrom, "Staircase PWM: an uncomplicated and efficient modulation technique for ac motor drives," *IEEE Trans. Power Electronics*, Vol. 3, No. 4, pp. 391-398, 1988.

[29] J. C. Salmon, S. Olsen and N. Durdle, "A three-phase PWM strategy using a stepped reference waveform," *IEEE Trans. Industry Applications*, Vol. 27, No. 5, pp. 914-920, 1991.

[30] J. T. Boys, "Theoretical spectra for narrow-band random PWM waveforms," *Proc. IEE*, Pt. B, Vol. 140, No. 6, pp. 393-400, 1993.

[31] R. L. Kirlin, S. Kwok, S. Legowski and A. M. Trzynadlowski, "Power spectra of a PWM inverter with randomized pulse position," *IEEE Trans. Power Electronics*, Vol. 9, No. 5, pp. 463-472, 1994.

[32] S. Barnett, *Introduction to Mathematical Control Theory*, Clarendon Press, Oxford, 1975.

[33] J. J. E. Slotine and W. Li, *Applied Nonlinear Control*, Prentice-hall, Englewood Cliffs, 1991.

[34] R. A. DeCarlo, S. H. Zak and G. P. Mathews, "Variable structure control of nonlinear multivariable systems: A tutorial," *Proc. IEEE*, Vol. 76, No. 3, pp. 212-232, 1988.

[35] J. Y. Huang, W. Gao and J. C. Hung, "Variable structure control: A survey," *IEEE Trans. Industrial Electronics*, Vol. 40, No. 1, pp. 2-22, 1993.

[36] B. D. O. Anderson and J. B. Moore, *Linear Optimal Control*, Prentice-Hall, Englewood Cliffs, 1971.

[37] A. Kawamura, T. Haneyoshi and R. G. Hoft, "Deadbeat controlled PWM inverter with parameter estimation using only voltage sensor," *IEEE Trans. Power Electronics*, Vol. 3, No. 2, pp. 118-125, 1988.

[38] K. J. Astrom and B. Wittenmark, *Computer Controlled Systems*, Prentice-Hall, Englewood Cliffs, 1990.
[39] A. Ghosh, G. Ledwich, O. P. Malik and G. S. Hope, "Power system stabilizers based on adaptive control techniques," *IEEE Trans. Power App. & Systems*, Vol. PAS-103, pp. 1983-1989, 1984.
[40] A. Ghosh, G. Ledwich, G. S. Hope and O. P. Malik, "Power systems stabilizers for large disturbances," *Proceedings of IEE*, Vol. 132, Pt. C, No. 1, pp. 14-25, 1985.
[41] G. Ledwich, "Linear switching controller convergence," *Proc. IEE – Control Theory & Applications*, Vol. 142, No. 4, 1995.
[42] T. Kailath, *Linear Systems*, Prentice-Hall, Englewood Cliffs, 1980.
[43] B. C. Kuo, *Digital Control Systems*, Holt, Rinehart & Winston, New York, 1980.

Chapter 6

Solid State Limiting, Breaking and Transferring Devices

The benefits of a solid state transfer switch in a Custom Power Park have been outlined in Chapter 4. In addition to the transfer switch a combination of power semiconductor switches and passive elements is used to construct solid state current limiting, breaking and transferring devices. The devices that are discussed in this chapter are

- Solid state current limiter
- Solid state breaker
- Solid state transfer switch

By inserting a reactance in series with a faulted circuit a fault current limiting device limits the damaging fault current. To prevent the fault current from rising to a dangerous level, the reactance must be inserted as soon as the fault is detected. The solid state breaker must be able to disconnect any faulted circuit within less than one cycle of the fundamental supply frequency. In addition, the solid state breaker can also have a built in current limiting device. The load transfer switch is usually used for sensitive loads that are connected to two incoming feeders. At any given time the load is supplied by one of the two feeders. The transfer switch quickly transfers the load from one feeder to another in a *make-before-break* fashion in case of a severe voltage sag in the supplying feeder.

For a sub-cycle make-before-break operation, a rapid detection of the occurrence of voltage sag/swell must be performed. An instantaneous detection is possible only in the case of three-phase sags (or swells). However algorithms that can perform sub-cycle detection of less severe sags (e.g., sag in one phase only) are more appropriate for transfer switch applications. In this chapter we shall also discuss some detection algorithms.

There are two important issues that must be considered before connecting a solid state current limiting or breaking device in a distribution network. The first and foremost issue is the identification of locations in the network in which such devices can be placed. The second issue to be considered is the coordination of the protective devices. Once a limiter or breaker is placed in a network, it must not adversely affect downstream protective devices. These issues are also discussed in this chapter.

6.1 Solid State Current Limiter

Distribution systems current limiting devices are classified into two categories – single-shot devices and multi operation devices [1]. A typical single-shot device is a current limiting fuse. In addition to limiting fault currents, these devices can also break a faulted circuit. However, such a device needs human intervention for replacement once the fault is cleared. Multi-operation devices can be vacuum, semiconductor or superconductor type [1]. The superconductor device operates through the quenching of the superconducting property at high magnetic fields. The superconducting cylinder shields the conductor coil from the magnetic core giving a low inductance. At high currents the field is sufficient to cause the material to leave the superconducting state and have the flux link the core creating a high reactance device. In this chapter we shall focus the discussion on the topology of the semiconductor type limiter.

6.1.1 Current Limiter Topology

The topology of a typical current limiter is shown in Figure 6.1. It contains an anti-parallel (back-to-back) gate turn-off thyristor (GTO) switch, a current limting inductor and a zinc oxide (ZnO) arrester, all connected in parallel [2,3]. The combination is connected in series with the distribution feeder that needs to be protected. The schematic diagram of the back-to-back GTO switch is shown in Figure 6.2. It includes a series of opposite poled GTO pairs, each of which has an RC snubber connected in parallel. The number of GTO switches depends on the rated peak voltage level across the current limiter. The ZnO arrester can limit this voltage level.

It is to be noted that a GTO can be switched off at any time by applying a negative gate pulse. Therefore it has the capability to interrupt a current at any time. A thyristor, on the other hand, switches off only when the current through it changes polarity. The magnitude of the fault current will depend on the instant of the fault occurrence – the closer the fault to the zero

6. Solid State Limiting, Breaking and Transferring Devices

crossing of the current, the more will be the magnitude of the current. If an anti-parallel thyristor switch is in a current limiter, it will keep on conducting till the next zero crossing irrespective of the occurrence of the fault. Since this will defeat the purpose for which a current limiter is installed, thyristor switches are not considered favorably for current limiter application.

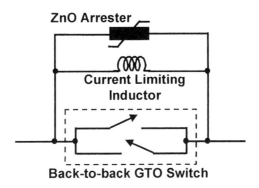

Figure 6.1. A GTO based fault current limiter

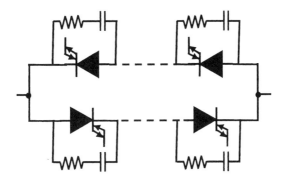

Figure 6.2. Schematic diagram of anti-parallel GTO switch

6.1.2 Current Limiter Operating Principle

Under normal (unfaulted) operating conditions, the GTOs are gated for full conduction. This implies that the GTOs in the forward path are gated at the positive going zero-crossing of the current through them, while the GTOs in the reverse path are gated at the negative going zero-crossing. Once a fault occurs, the GTOs are turned off as soon as the fault is detected. A

GTO can respond within a few microseconds of the gate signal being applied. Furthermore, note that the prospective peak magnitude of the turn off current can be very high. Therefore, the GTOs help by blocking the fault current flowing through them before it can reach damaging level. As soon as the GTOs are turned off, the fault current is diverted to the snubber capacitor that limits the rate of rise in voltage across the GTOs. The voltage across the anti-parallel GTO switch rises until it reaches the clamping level established by the ZnO arrester [2]. Note that the same voltage also appears across the current limiting reactor and once the clamping level of the voltage is reached, the current across the reactor will rise linearly. This linear rise will continue till it becomes equal to the instantaneous level of current flowing in the line. Thereafter, the current will be limited by total effective series impedance, i.e., by a combination of the impedance of the limiting reactor and feeder impedance till the fault point.

The current limiting device must resume normal operation once the fault is cleared. To restore normal operation, the line current is sensed and a turn on command is issued once the current drops to the normal level. The anti-parallel GTO switch is then turned on at the nearest zero crossing of the periodic voltage across the switch. Otherwise, the snubber capacitor will discharge a high magnitude current through the line. Let us now consider the following example.

Example 6.1: Consider a simple radial distribution system that is supplying a series RL load. The line-to-neutral source voltage is 6.35 kV (rms) and the system frequency is 50 Hz. The feeder has a resistance of 3.025 Ω and an inductance of 38.5 mH while the load resistance and inductance are given by 60.5 Ω and 770.3 mH respectively. This implies that for a base voltage of 11 kV (L-L) and a base MVA of 1.0, the feeder impedance is $0.025 + j0.1$ per unit and the load impedance is $0.5 + j2.0$ per unit. The prefault current in the steady state is 24.25 A (rms), i.e., 0.462 per unit.

Let us first consider the case in which the current limiter is not operational. If a fault occurs which short-circuits the load when the system is operating in the steady state, the fault current is only limited by the source and feeder impedance. The faulted load current is shown in Figure 6.3 in which the fault is applied at two different instants. It can be seen that depending on the instant of the fault, the fault current excursion may either be in the positive direction or in the negative direction. It can also be seen that the peak of the fault current is about 1200 A, i.e., more than 20 per unit. Now suppose an anti-parallel thyristor switch is used in the current limiting device rather than GTO switch. Then even if the thyristors are blocked once the fault is detected, the fault current will continue to flow till it reaches zero. Therefore, during this period the fault current will shoot up to an excessively high value. This obviously is unacceptable.

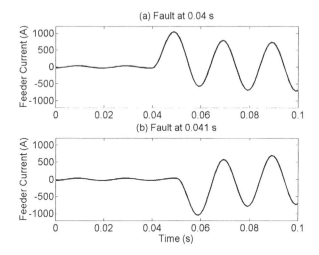

Figure 6.3. Feeder current before and after fault without current limiter

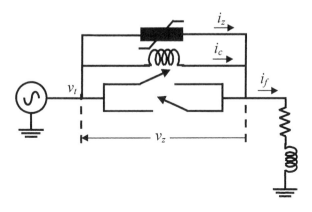

Figure 6.4. Radial distribution protected by current limiter

Let us now assume that the current limiter protects the feeder as shown in Figure 6.4. The current limiting inductance is chosen as 500 mH. We shall investigate the effect of the ZnO arrester on the fault current. Let us set the arrester voltage limit to 6.9 kV. The system response to the fault at the load bus is shown in Figure 6.5. It can be seen that the arrester voltage v_z is (nearly) zero before the fault. This voltage becomes equal to the terminal voltage v_t after the fault. However, since this voltage is clipped at 6.9 kV, the current through the limiting inductor cannot rise linearly. Therefore, the magnitude of this current (i_c) is considerably smaller than the fault current. To compensate this, the magnitude of the current i_z through the arrester becomes considerably large. The fault current i_f is the sum of the currents i_c and i_z. The arrester current is the dominant component of this current.

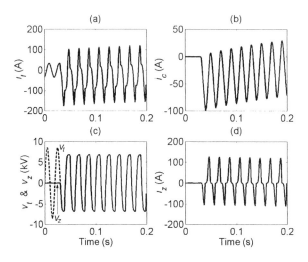

Figure 6.5. Current limited system response for an arrester voltage of 6.9 kV

Let us now increase the arrester voltage level to 11 kV such that it can accommodate full voltage across the limiter device. The system response is shown in Figure 6.6. It can be seen that the current through the arrester has reduced considerably and the fault current is almost same as the current through the limiting inductor. It is to be noted that the arrester current can be further reduced by increasing the arrester voltage level. This may however prove to be counterproductive as the arrester may then transiently allow this higher voltage to be applied across the anti-parallel GTO switch, thereby damaging it.

ΔΔΔ

6.2 Solid State Breaker (SSB)

The schematic diagram of a solid state breaker is shown in Figure 6.7 [4]. This circuit is almost similar to the circuit of the current limiter of Figure 6.1 except that an anti-parallel thyristor switch is added in series with the current limiting inductor. Given the present state of technology, an SSB cannot replace the standard circuit breaker. However, it can be applied to a large number of applications in which it can be used to provide a rapid interruption of a fault current.

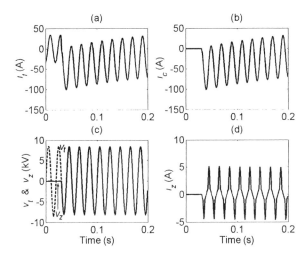

Figure 6.6. Current limited system response for an arrester voltage of 11 kV

In the circuit of Figure 6.7, the GTOs are the normal current carrying elements. At the onset of a fault, it goes through a number of sub-cycle auto reclose operations. If the fault does not clear, then the GTOs are turned off and the thyristors are turned on such that the fault current now starts flowing through the current limiting inductor [4]. The current through the limiting inductor is eventually cut off by blocking the thyristors. The ZnO arrester is also used here to protect the power electronics from lightning and switching surges and also to build up current through the inductor. In this figure a conventional switch is also placed in series with the SSB.

Let us consider the distribution system discussed in Example 6.1. The system response is shown in Figure 6.8. Note that the fault current clearance time is chosen arbitrarily in this figure to demonstrate the operating principle of the device. Also note that after the thyristor switch is blocked, the current i_c becomes equal to zero. However, there is a small amount of feeder current still flows through the ZnO arrester and the snubber circuit. This current can be interrupted by opening the conventional switch.

An alternate topology is proposed in [5] in which the anti-parallel GTO switch and the ZnO arrester are connected in parallel with a high-speed vacuum circuit breaker (VCB). This is shown in Figure 6.9. The current limiting inductor is missing in this topology. The current in the normal (unfaulted) state flows through the VCB. When a fault is detected, the GTOs are turned on and simultaneously an open signal is given to the VCB. The

VCB uses electromagnetic repulsion force to open the breaker at a high speed. The resulting arc produces a voltage that acts as a counter electromotive force. The fault current flowing through the VCB is reduced by this electromotive force and commutated to the GTO switch. When the fault current is completely commutated to the GTO switch, it can be interrupted by turning the GTO switch off. The ZnO arrester suppresses the over voltage that may occur. Note that the current limiting function is missing in this topology.

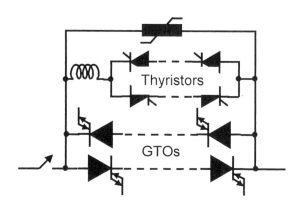

Figure 6.7. Schematic diagram of a solid state breaker

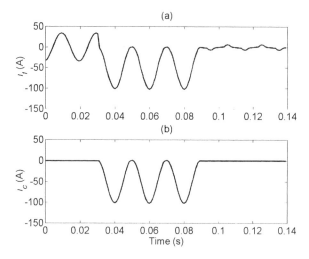

Figure 6.8. System response with a solid state breaker

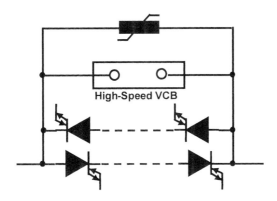

Figure 6.9. An alternate SSB topology

6.3 Issues in Limiting and Switching Operations

Electromechanical switchgears circuit breakers are still the most common forms of protective devices used in power systems. Usually two types of switchgears are used in power distribution systems – automatic reclosers and circuit breakers. Automatic reclosers operate for lower current ratings and are placed closer to loads, on feeders or at substations [2]. Circuit breakers are used for higher current ratings and are placed at substations. Reclosers can clear all faults of temporary nature. The circuit breakers are used as backup protective devices for cases in which the reclosers fail to clear faults. For ground faults in circuits without neutral earthing reactors, large currents flow through the networks. The protective devices must switch off the smallest portion of the circuit without affecting the majority of loads. This objective is accomplished through primary and backup protection using inverse-time overcurrent relays [6]. This basic principle of coordination must be kept in mind while placing solid state limiters and breakers.

To illustrate the idea let us consider the generic distribution system shown in Figure 6.10 in which the current limiting and breaking devices are labeled with numerals. This distribution system has two incoming transformers, each connected to one main bus. These two buses are connected by a bus-tie. As we have mentioned earlier, an SSB can interrupt a fault current much faster than an ordinary circuit breaker. Therefore even if the installation of an SSB in location 1 or 2 protects the downstream circuit from overcurrent, the coordination of the protective devices will be a major problem unless all the protective devices in the network are solid state. For example assume that 1 is an SSB while 4 is a conventional one. Then for a fault downstream from 4, breaker at 1 will operate before breaker at 4 thereby disconnecting both faulty and healthy feeder supplied by transformer (T_1). This clearly is unacceptable.

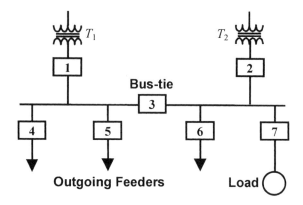

Figure 6.10. A distribution system protected by solid state devices

A potential installation point of an SSB is the bus-tie location 3. This will require no coordination with any other protection device. For example, for a fault in the transformer T_1 side of the system, the SSB will open the bus tie thereby preventing transformer T_2 from feeding the fault. This can be done irrespective of the action of the protective devices around the fault point. The other obvious potential application point of the SSB is location 7 that protects a load. A fault on the load side can be quickly isolated by the SSB without affecting any other part of the system.

The best position for the placement of a current limiter is at the output of the main incoming transformers, i.e., locations 1 and 2. This will limit the current for a fault at any part of the network. Furthermore, such an arrangement will not cause any coordination problem either. The current tap settings of the downstream overcurrent relays can then be set at lower values. Current limiters can also be put on the feeder locations 4, 5 and 6. These devices will have a lower current rating than those placed at 1 and 2. However, since these devices are more expensive than the conventional circuit breakers/reclosers, such an arrangement obviously will incur a much higher cost and may not be justifiable once the cost-benefit ratios are calculated.

A limiter at the bus tie location 3 can be most beneficial as it will have lower losses under normal operating conditions [2]. However since the current flowing through this position for a fault at any part of the circuit is maximum, the rating of the device at this location must be very high. From the above considerations a fault current limiter has the following requirements [2]

- It must limit the short circuit current such that it does not exceed the momentary or interrupting rating of any downstream protecting device.
- It must maintain the fault current within a specified limit till a downstream device clears the fault.
- The limiter must reset automatically after the fault is cleared.
- It must allow sufficient fault current to flow such that downstream overcurrent protection devices can isolate the fault.
- It must maintain normal performance throughout a normal sequence of operations of downstream protective devices.
- It must be relatively free of maintenance.

6.4 Solid State Transfer Switch (SSTS)

A solid state transfer switch can protect a sensitive load from voltage sag or interruption by quickly transferring the load to a healthy feeder in case of a voltage sag or interruption in the preferred supply feeder [7,8]. The schematic diagram of an SSTS protected sensitive load is shown in Figure 6.11. In this the SSTS includes a pair of anti-parallel thyristor switch and a pair of ZnO arresters. Usually the load is supplied by the preferred feeder and the load current flows through the switch Sw_1. When a deep voltage sag or interruption is detected in this feeder, the load is quickly transferred to the alternate feeder. This switching action is called make-before-break, i.e., before the switch Sw_1 is turned off, the switch Sw_2 is turned on. Once the load current starts flowing through the switch Sw_2 the switch Sw_1 is turned off.

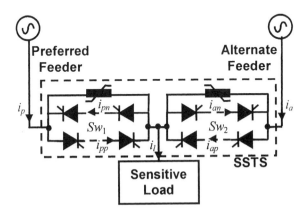

Figure 6.11. Schematic diagram of a sensitive load protected by SSTS

Figure 6.11 also shows the currents i_p, and i_a through the preferred and the alternate feeders respectively and the load current i_l. The currents i_{pp}

(in positive direction) and i_{pn} (in negative direction) through the switch Sw_1 and i_{ap} (in positive direction) and i_{an} (in negative direction) through the switch Sw_2 are also indicated in this figure. We shall now observe the behavior of these currents during the transfer operation with the help of the following example.

Example 6.2: Let us assume that the distribution system of Figure 6.11 is supplied by two 50 Hz, 6.35 kV (line-to-neutral, rms) sources. For simplicity we assume that both the sources are synchronized, i.e., the phase angle difference between the sources is zero. The resistance and inductance of both the preferred and alternate feeders are taken as 3.025 Ω and 38.5 mH respectively. The load resistance and inductance are given by 60.5 Ω and 770.3 mH respectively. We shall investigate what happens when the circuit breaker connecting the preferred feeder to the load opens accidentally.

Figure 6.12 depicts the transferring operation in which the load is transferred from the preferred feeder to the alternate feeder. It is assumed that the circuit breaker on the preferred feeder opens accidentally at 0.035 s. The transfer is initiated by closing the switch Sw_2 7 ms after the circuit breaker is opened. The switch Sw_1 is blocked 3 ms after the closing of the switch Sw_2. It can be seen that the current is smoothly switched from one feeder to another without any transient. Figure 6.13 depicts the currents through the switch during the transfer operation. Once the fault is detected and switch Sw_2 is enabled, the current i_{ap} starts conducting. However, the current i_{pn} still conducts despite the switch Sw_1 being blocked. The current through switch Sw_1 completely becomes zero at the next zero crossing of the current i_{pn} and thereafter the load is supplied only by the alternate feeder. It can be seen from Figure 6.12 that the load current for the make-before-break transfer is continuous.

ΔΔΔ

It is to be noted that the transferring time in the above example is chosen arbitrarily. The nature of the transfer however will depend on many factors. Some of these factors are: relative magnitude of the source voltages, phase angle difference between the preferred and alternate sources, feeder impedances, load impedance, fault detection time, whether the load is passive or active etc. The nature of the load current will depend on the combination of these parameters. The design of the transfer control must take into account all these factors for a satisfactory operation.

Example 6.2 illustrates the make-before-break switching action. It is however not always desirable that the SSTS operates in this fashion. For example consider the circuit of Figure 6.14 which depicts the circuit behavior when a fault occurs in the preferred feeder. The make-before-break

6. Solid State Limiting, Breaking and Transferring Devices

operation of the SSTS will cause the fault current to be temporarily supplied by the alternate feeder before the switch Sw_1 is cut off. The path for this current is indicated by the dotted line in Figure 6.14. The correct switching pattern is given in the following example.

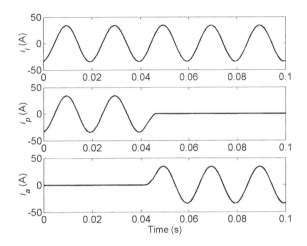

Figure 6.12. Make-before-break operation of SSTS

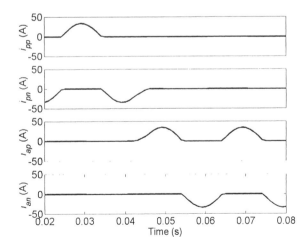

Figure 6.13. Switch currents during make-before-break operation

Example 6.3: Consider the same system given in Example 6.2. Let us assume that a fault has occurred in the preferred feeder at 0.035 s. The transfer is initiated by closing the switch Sw_2 7 ms after the occurrence of the fault and the switch Sw_1 is blocked 3 ms after the closing of the switch Sw_2. The fault is cleared by the opening of a circuit breaker in the primary feeder 0.65 s after the fault initiation. The currents in the preferred and alternate feeders and the load currents are shown in Figure 6.15. It can be seen that a large current flows in both the preferred and alternate feeders due to incorrect transfer operation. The path for this fault is as shown by the dotted line in Figure 6.14 and are also shown in Figure 6.16 which depicts the currents i_{pn} and i_{ap}.

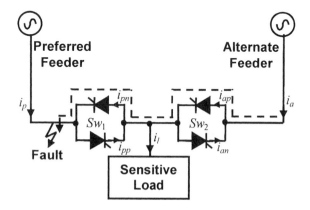

Figure 6.14. Incorrect transfer for a fault in the preferred feeder

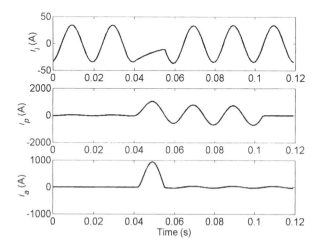

Figure 6.15. Currents during incorrect transfer for a fault in the preferred feeder

6. Solid State Limiting, Breaking and Transferring Devices

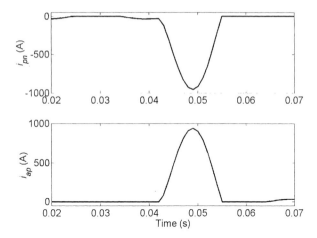

Figure 6.16. Currents through switches during incorrect transfer

To eliminate the possibility of such large current flowing in the switches, the transfer in this case has to be break-before-make [7]. This implies that the switching of Sw_2 has to be delayed till the current i_{pn} becomes zero. An alternate scheme in which the firing of Sw_1 is inhibited during the concurrence of the fault and enabling the firing of that thyristor in Sw_2 that allows the current to flow in the opposite direction of the fault current is given in [7].

The behavior of the circuit in which the firing of the switch Sw_2 is delayed is shown in Figures 6.17 and 6.18. The currents through the primary and alternate feeders and the load current are shown in Figure 6.17. It can be seen that even though the current through the preferred feeder becomes very large during the fault, the peak of the current through the alternate feeder does not exceed 50 A since firing of Sw_2 is now delayed. This switch is fired roughly after 35 ms after the occurrence of the fault when the current i_{pn} becomes zero. As a consequence the load current also becomes zero. The switch currents are shown in Figure 6.16. It can be seen that the delayed firing prevents the current i_{pn} to build up. It is however to be noted that if the switch Sw_2 is turned on before this current becomes zero, large current again will start flowing in the direction indicated in Figure 6.14.

ΔΔΔ

Example 6.3 shows that make-before-break switching scheme can lead to large currents to flow through a transfer switch in case of a fault in the preferred feeder. In such a case the transfer operation is delayed till the load current freewheels to zero to ensure that no current flows through the

230   Chapter 6

primary switch in case of a passive load. In case of a motor load, the motor will start feeding the fault and hence will go through periodic zero crossings. If the thyristors of the primary switch Sw_1 are blocked as soon as the fault is detected the fault current will be zero in the next available zero crossing. The secondary switch Sw_2 can then be turned on causing a minimum delay.

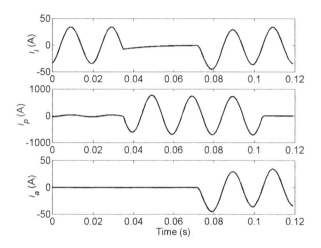

Figure 6.17. Currents during delayed transfer for a fault in the preferred feeder

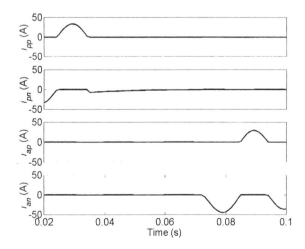

Figure 6.18. Currents through switches during correct transfer

An alternate topology of hybrid SSTS is given in [8]. The schematic diagram of this SSTS is given in Figure 6.19. In addition to the thyristors and the ZnO arresters this contains a pair of high-speed mechanical switches. During the normal operation the mechanical switch M_1 is closed and M_2 is open. Once a transfer is requested, M_1 is turned off and simultaneously the thyristor pair TH_1 is turned on. Consequently, the current is commutated to TH_1 and it is blocked at the next zero crossing of the current. As soon as the current is completely blocked, the thyristor pair TH_2 is turned on enabling the alternate feeder to supply the load. Once the current through TH_2 stabilizes, the mechanical switch M_2 is closed to bypass the current. The thyristor pair TH_2 is then blocked.

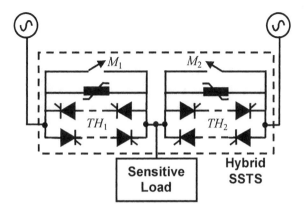

Figure 6.19. A hybrid SSTS topology

The main advantage of this hybrid switch system is that since the normal operation takes place through the mechanical switch, the power loss component is lower than the topology shown in Figure 6.11. Furthermore, since the thyristors are tuned on only during the transient and are not designed for continuous operation, the requirement on the cooling system will be less. However for this topology to be effective, two important factors must be considered. The main point is the operating time of the high-speed switch. The other point to ponder is the arc extinguishing mechanism once the contacts part. The fault current continues to flow through the mechanical switch till the arc is completely extinguished. Mitsubishi Electric Corporation has developed a magnetic repulsion based switch that can open and extinguish arc within 2 ms [8].

One of the primary objectives of a transfer switch is to protect a sensitive load from voltage sag/swell in the preferred feeder. The sags or swells must be detected quickly such that the transfer operation can be performed and in a sub-cycle time. This is discussed in the next section.

6.5 Sag/Swell Detection Algorithms

Ideally it is desirable to detect any voltage sag/swell instantaneously. This however cannot always be achieved. The next best option is to detect this with as little delay as possible. We present three detection algorithms below and also discuss their merits and demerits. It is to be noted here that for a fault in the preferred feeder, the main idea is to delay the transfer action. This can be accomplished by monitoring the current through the switch Sw_1 once the overcurrent relay picks up a fault in the preferred feeder.

6.5.1 Algorithm Based on Symmetrical Components

Let us assume that the transfer switch has to protect a load that can tolerate a minimum voltage sag of up to 70% of the nominal value in either of the three phases. We can then use the algorithm to extract the symmetrical components from samples presented in Chapter 3. This is illustrated by the following example.

Example 6.4: Let us assume that the voltage supplied by the preferred feeder at the load terminal is 1.0 per unit (peak) at a supply frequency of 50 Hz. We have to detect a sag in phase-c of the supply voltage in which the peak voltage of this phase drops to 0.7 per unit. The instantaneous voltages for the three phases are shown in Figure 6.20 (a).

The transferring algorithm here is based on the instantaneous symmetrical component extraction algorithm in which we have set a threshold level of 0.75 per unit (peak) for each phase. This implies that if the rms voltage of any phase drops below $0.75/\sqrt{2}$, the transferring operation is initiated.

The rms value of the phase-c voltage and the threshold line are shown in Figure 6.20 (b). It can be seen that it requires almost 9 ms for this algorithm to detect the fault. This is expected as the settling time of the symmetrical component extraction algorithm is half a cycle, i.e., 10 ms for a 50 Hz supply system. The detection time in this case increases with lowering the value of the threshold. Even though the maximum detection time will be restricted to 10 ms in this case, the detection time cannot be made significantly faster irrespective of the type of sag as it is completely dictated by the settling time of the symmetrical component extraction algorithm using a 10 ms window for averaging.

ΔΔΔ

6. Solid State Limiting, Breaking and Transferring Devices

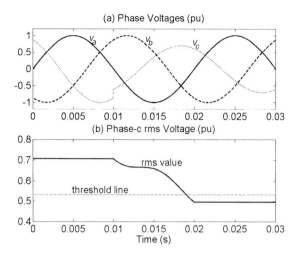

Figure 6.20. Sag detection based on symmetrical component extraction

6.5.2 Algorithm Based on Two-Axis Transformation

In this the instantaneous three phase voltages are transformed in a fixed two-axis coordinate system as [7]

$$\begin{bmatrix} v_\alpha \\ v_\beta \end{bmatrix} = \sqrt{\frac{2}{3}} \begin{bmatrix} 1 & -\frac{1}{2} & -\frac{1}{2} \\ 0 & \frac{\sqrt{3}}{2} & -\frac{\sqrt{3}}{2} \end{bmatrix} \begin{bmatrix} v_a \\ v_b \\ v_c \end{bmatrix} \quad (6.1)$$

The voltage vector obtained by the above transformation is further transformed into a rotating *dq*-coordinate using the equation

$$v_d + jv_q = e^{-j\theta}(v_\alpha + jv_\beta) \quad (6.2)$$

where θ is the transformation angle calculated for an initial value of θ_0 as

$$\theta = \theta_0 + \int_0^t \omega \, d\xi \quad (6.3)$$

The detection is achieved by comparing the voltage vector

$$V_{dq} = \sqrt{v_d^2 + v_q^2} \tag{6.4}$$

with a threshold value. Let us illustrate the idea with the help of the following example.

Example 6.5: Let us consider the same voltage and load tolerance level as given in Example 6.4. The voltage vector V_{dq} for sag in one phase, two phases and all three phases is shown in Figure 6.21. The instantaneous value of the vector is compared with that of a threshold value and the switching signal is generated when the value of V_{dq} falls below the threshold. The threshold value in this case is chosen as the minimum value of the sag in which the voltage of one phase falls to 75% of the nominal value. It can be seen that while the sag in all three phases can be detected almost instantaneously, a sag in either one or two phases can be made after a time delay. This time delay is dependent on the point of the voltage cycle at which the sag occurs. In fact it has been shown in [7] that depending on the threshold value and the occurrence instant of the sag, some single-phase sag will not even be detected.

ΔΔΔ

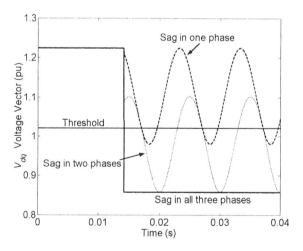

Figure 6.21. Sag detection based on two-axis transformation

6.5.3 Algorithm Based on Instantaneous Symmetrical Components

Let the voltages at the preferred terminal be given by v_a, v_b and v_c. Then the three instantaneous symmetrical component vectors can be obtained as

6. Solid State Limiting, Breaking and Transferring Devices

$$\begin{bmatrix} v_{a0} \\ v_{a1} \\ v_{a2} \end{bmatrix} = \frac{1}{\sqrt{3}} \begin{bmatrix} 1 & 1 & 1 \\ 1 & a & a^2 \\ 1 & a^2 & a \end{bmatrix} \begin{bmatrix} v_a \\ v_b \\ v_c \end{bmatrix} \qquad (6.5)$$

It has been shown in Chapter 3, that the magnitude of the vector v_{a1} is constant and equal to $\sqrt{3}/2$ when the three phase voltages are balanced with peak magnitude of 1.0 per unit. Thus the trajectory traced by the vector v_{a1} over a cycle is a circle. It has also been shown that the trajectory of this vector is an ellipse if the three phase voltages are not balanced.

Let us now assume that the supply voltage has a sag only in phase-a such that these voltages are given by

$$v_a = \kappa \sin \omega t, \; v_b = \sin(\omega t - 120°), \; v_c = \sin(\omega t + 120°)$$

where $\kappa < 1.0$. The vector v_{a1} is then given by

$$v_{a1} = \frac{1}{\sqrt{3}} \left[\kappa \sin \omega t + a \sin(\omega t - 120°) + a^2 \sin(\omega t + 120°) \right] \qquad (6.6)$$

The above equation can be simplified as

$$v_{a1} = \frac{1}{\sqrt{3}} \left[\left(\kappa + \frac{1}{2} \right) \sin \omega t - j \frac{3}{2} \cos \omega t \right] \qquad (6.7)$$

The path traced by the vector v_{a1} as ωt changes from 0 to 2π is an ellipse. We want to determine the maximum and minimum length of this vector. From (6.7) the magnitude of the vector v_{a1} is obtained as

$$|v_{a1}| = \frac{1}{\sqrt{3}} \left[\left(\kappa + \frac{1}{2} \right)^2 \frac{1 - \cos 2\omega t}{2} + \left(\frac{9}{4} \right) \frac{1 + \cos 2\omega t}{2} \right]^{1/2} \qquad (6.8)$$

We now take the derivative of (6.8) with respect to ωt and equate it to zero to obtain the extremal points. This yields

$$\sin 2\omega t = 0 \implies \omega t = 0, \pi/2, \pi, 3\pi/2 \qquad (6.9)$$

From (6.8) and (6.9) we see that the maximum length of the vector is obtained when $\omega t = 0$ or π and it is given by

$$|v_{a1}|_{max} = \frac{\sqrt{3}}{2} \tag{6.10}$$

Similarly, the minimum value of the vector occurs when $\omega t = \pi/2$ or $3\pi/2$ and the minimum value is

$$|v_{a1}|_{min} = \frac{1}{\sqrt{3}}\left(\kappa + \frac{1}{2}\right) \tag{6.11}$$

It can be seen that the maximum value obtained is the same as that of the length of the vector when all three phases are balanced with a peak of 1.0 per unit, while the minimum value changes linearly with the size of the voltage dip. In a similar way, it can be shown that for a magnitude sag only in the other two phases, the maximum and minimum values are the same as given in (6.10) and (6.11). However the occurrences of the maxima and minima differ for sags in different phases. Table 6.1 lists the points through which the vector goes through the extremals for a sag in any of the three phases. Note that for a sag in which the voltage of any one phase becomes 0.7 per unit, the maximum value as per (6.10) is 0.866, while the minimum value as per (6.11) is 0.6928. We shall use this concept for the transfer algorithm.

Table 6.1. Point of occurrence of maxima and minima for sag in difference phases

Sag in	Value of ωt	Value of the vector v_{a1}	Absolute Value of v_{a1}
Phase-a	0°	$0 - j0.866$	Maximum
	90°	$0.6928 - j0$	Minimum
	180°	$0 + j0.866$	Maximum
	270°	$-0.6928 - j0$	Minimum
Phase-b	30°	$0.3464 - j0.6$	Minimum
	120°	$0.75 + j0.433$	Maximum
	210°	$-0.3464 + j0.6$	Minimum
	300°	$-0.75 - j0.433$	Maximum
Phase-c	60°	$0.75 - j0.433$	Maximum
	150°	$0.3464 + j0.6$	Minimum
	240°	$-0.75 + j0.433$	Maximum
	330°	$-0.3464 - j0.6$	Minimum

Example 6.6: Let us again consider a load that can tolerate a maximum voltage sag of 0.7 per unit. The threshold value is arbitrarily chosen to be 0.7044. This is equal to the minimum value calculated as per (6.11) for a single-phase sag in which the phase voltage becomes 0.72 per unit. The complex plane plots of the periodic vector v_{a1} for different types of sag are shown in Figure 6.22. In these plots, the dotted line indicates the trace of a periodic vector with a magnitude of 0.7044. This implies that the threshold

6. Solid State Limiting, Breaking and Transferring Devices

value is represented by the dotted circle that indicates a balanced sag in which voltages of all the three phases have dropped to 0.8135 per unit.

It is assumed that the peak of all three phases of the system voltage is 1.0 per unit before the sag occurs. The pre-sag vector v_{a1}, as ωt changes from 0 to 2π is shown by the outer circle. This collapses into a smaller circle or an ellipse when the sag occurs. The transfer is initiated when the magnitude of the vector v_{a1} is less than 0.7044. A balanced sag in which the peak voltage of all the three phases becomes 0.7 per unit is shown in Figure 6.22 (a). It can be seen that this sag can be detected immediately. It is clear from Figure 6.22 (b) that the detection time can be at most half a cycle depending on the instant of occurrence of a single-phase sag. Similarly the maximum detection time for a two-phase sag is about one-quarter of a cycle.

The problem with this algorithm is however with unbalanced sags in which the voltage magnitudes of the phases are different. This algorithm tends to produce pessimistic results in such cases. Consider the complex plane plot of Figure 6.22 (d) in which the peak voltage of two phases has become 0.8 per unit while that of the third phase has dipped to 0.9 per unit. It is desirable that no transfer process should be initiated in this case. However as seen from Figure 6.22 (d) that the magnitude of the vector v_{a1} falls below the threshold value and this will cause a false transfer to be initiated for this sag. Similarly, this algorithm may also initiate false transfer for sags that are unbalanced in both magnitude and phase.

ΔΔΔ

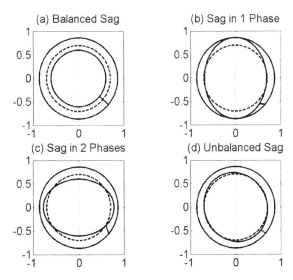

Figure 6.22. Periodic trajectory in complex for detection algorithm

In this section we have presented three different algorithms for the detection of voltage sag. Out of these three the first one is based on the symmetrical component extraction that has a settling time of half a cycle for alternating loads. This algorithm is very robust as it can tolerate and filter out voltage harmonics. Furthermore since averaging process is used in this algorithm, no false triggering is initiated for voltage spikes. However, the assumption with this method is that the frequency of the system is (almost) constant. This assumption might not be valid, especially when transients in other parts of the network cause the sag. The other two methods are similar with the maximum detection time being half a cycle. However being instantaneous in nature, none of these two methods can tolerate any spike in the voltage. Therefore further research is required for better sag detection strategies. Note that we have focused our attention on detection of voltage sags. However, all the algorithms mentioned above could be used for the detection of voltage swells as well. In that case, the transfer operation is initiated once the computed value exceeds an upper threshold level. Obviously this threshold level is higher than the normal operating voltage.

We can define the total transfer time as the duration from the inception of voltage sag/swell or fault in the preferred feeder to the load being completely transferred to the alternate feeder. In addition to the detection time, the circuit parameters also influence the total transfer time. Some of the factors that affect the transfer time are [9]

- The strategy of removing gating of the thyristors in the preferred feeder and triggering the thyristors in the alternate feeder.
- Load type, i.e., whether the load is active or passive.
- The magnitude and phase difference between supply voltages in the preferred and alternate feeders.
- Fault or sag/swell characteristics that affect the voltage difference between the sources.

The estimation of transfer time for RL loads is discussed in [9] for voltage sag and different types of faults. We must however remember that the accurate estimation of the transfer time is very difficult and we can at best get a rough estimate.

6.6 Conclusions

In this chapter we have discussed various power electronic based limiting, breaking and transferring devices. We have presented the topology and the operating principles of these devices. It is however to be kept in mind that the usage of the devices is still in its infancy. There is a

6. Solid State Limiting, Breaking and Transferring Devices

tremendous scope of building faster and more reliable devices. Also, there is scope for further research in the areas of placement, coordination and detection. We hope this chapter gives an overview based on which further research and development can be carried out.

6.7 References

[1] P. G. Slade, J. J. Wu, W. P. Stubler, E. J. Stacey, E. R. Voshall, J. J. Bonk, J. V. Porter and L. Hong, "The utility requirements for a distribution fault current limiter," *IEEE Trans. Power Delivery*, Vol. 7, No. 2, pp. 507-515, 1992.

[2] R. K. Smith, P. G. Slade, M. Sarkozi, E. J. Stacey, J. J. Bonk and H. Mehta, "Solid state distribution current limiter and circuit breaker: Application requirements and control strategies," *IEEE Trans. Power Delivery*, Vol. 8, No. 3, pp. 1155-1164, 1993.

[3] T. Ueda, M. Morita, H. Arita, J. Kida, Y. Kurosawa and Y. Yamagiwa, "Solid-state current limiter for power distribution system," *IEEE Trans. Power Delivery*, Vol. 8, No. 4, pp. 1796-1801, 1993.

[4] G. A. Taylor, "Power quality hardware solutions for distribution systems: Custom Power," *IEE North Eastern Center Power Section Symposium on Security and Power Quality of Distribution Systems*, 1995.

[5] T. Genji, O. Nakamura, M. Isozaki, M. Yamada, T. Morita and M. Kaneda, "400 V class high-speed current limiting circuit breaker for electric power system," *IEEE Trans. Power Delivery*, Vol. 9, No. 3, pp. 1428-1435, 1994.

[6] W. D. Stevenson, Jr., *Elements of Power Systems Analysis*, McGraw-Hill, New York, 1982.

[7] A. Sannino, "Static transfer switch: Analysis of switching conditions and actual transfer time," *IEEE Power Engineering Society Winter Meeting*, Columbus, Ohio, 2001.

[8] M. Takeda, H. Yamamoto, G. F. Reed, T. Aritsuka and I. Kamiyama, "Development of a novel hybrid switch device and application to a solid-state transfer switch," *IEEE Power Engineering Society Winter Meeting*, Vol. 2, pp. 1151-1156, New York, 1999.

[9] H. Mokhtari, S. B. Dewan and M. R. Iravani, "Analysis of a static transfer switch with respect to transfer time," *IEEE Trans. Power Delivery*, Vol. 17, No. 1, pp. 190-199, 2002.

Chapter 7

Load Compensation using DSTATCOM

In this chapter we shall discuss shunt compensation of distribution systems. The primary aims of a shunt compensator in a distribution system are to cancel or suppress

- the effect of poor load power factor such that the current drawn from the source has a near unity power factor.
- the effect of harmonic contents in loads such that current drawn from the source is nearly sinusoidal.
- the dc offset in loads such that the current drawn from the source has no offset.
- the effect of unbalanced loads such that the current drawn from the source is balanced.

In addition, as we shall discuss in Chapter 8, shunt compensators are also used to regulate voltages at a distribution bus thereby eliminating unbalance, harmonics and flicker in the bus voltage.

In Chapters 2 and 3 we have discussed the problems that are associated with a distribution system. It has been mentioned that some of the loads can cause significant problems for other loads in their vicinity. It is therefore desirable to compensate for such loads. However, there are economic considerations involved here. For example, whether a load should have a power factor correction depends on the amount of penalty that may be imposed on the customer for not compensating. On the other hand, the loads that cause fluctuations in the supply voltage may have to be compensated to achieve the required level of voltage regulation. The typical loads that cause voltage fluctuations are arc and induction furnaces, very large motors used in rolling mills etc. as they are continuously switched on or off. Again the loads that use power electronic devices such as adjustable speed drives, UPS etc. cause harmonic pollution. With these types of loads, a shunt compensator may be required to reduce the effects of the harmonics on the rest of the system, especially if the size of these loads is very large.

In this chapter we shall discuss how the four objectives mentioned before can be achieved using a shunt compensator. We shall assume that it is the responsibility of the customer to install and maintain these compensators. Some of the background materials required to follow the discussions are presented in Chapter 3. In this chapter we shall assume that the loads are supplied by stiff sources, i.e., there is no significant feeder impedance between the load and the source. This assumption will be relaxed in Chapter 8 in which we shall discuss more realistic DSTATCOM structures.

7.1 Compensating Single-Phase Loads

The schematic diagram of a single-phase load compensator is shown in Figure 7.1. In this diagram a voltage source is supplying a load that could be nonlinear as well. The point of connection of the load and the source is the point of common coupling (PCC). Since there is no feeder joining the source and the load, we shall designate the source to be stiff. Here the compensator consists of an H-bridge inverter and an interface inductor (L_f). The resistance R_f represents the resistance of the interface inductor due to its finite Q-factor as well as the losses in the inverter. One end of the compensator is connected at the PCC through the interface inductor while the other end is connected with the load ground. The dc side of the compensator is supplied by a dc capacitor C_{dc}. The inverter is expected to be controlled to maintain a voltage V_{dc} across this capacitor.

Let us assume that the load is nonlinear and draws a current that has a poor power factor. The instantaneous load current then can be decomposed as

$$i_l = i_{lp} + i_{lq} + i_{lh} \qquad (7.1)$$

where i_{lp} and i_{lq} are respectively the real and reactive parts of the current required by the load and i_{lh} is the harmonic current drawn by the load. The purpose of the compensator is to inject current i_f such that it cancels out the reactive and harmonic parts of the load current.

Now applying KCL at the PCC we get

$$i_l = i_s + i_f \Rightarrow i_s = i_l - i_f \qquad (7.2)$$

7. Load Compensation using DSTATCOM

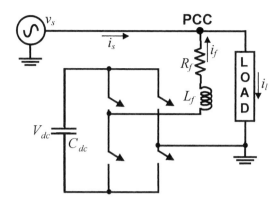

Figure 7.1. Schematic diagram of a single-phase compensator

We assume that the compensator operates in a hysteresis current control loop in which the compensator current tracks a reference current I_f^*. Let us now choose this reference current as

$$i_f^* = i_{lq} + i_{lh} \tag{7.3}$$

If the inverter accurately tracks this reference current, then the source current will be equal to the unity power factor current drawn by the load. Since the compensator does not draw or inject any real current, the average power consumed by the compensator is zero. Note that the above approach requires the on-line determination of the instantaneous reactive and harmonic components of the load current. There are however simpler approaches for the determination of the reference current. The following example illustrates one such approach.

Example 7.1: Let us assume that a 240 V (rms), 50 Hz source supplies a load that draws a current that has a fundamental and a harmonic part. The fundamental part of the load current has an rms value of 15 A at a power factor of 0.5 (lagging) and the harmonic part contains 5th and 7th harmonics. The instantaneous source voltage and the load current are given by

$$v_s = \sqrt{2} \times 240 \sin(\omega t)$$
$$i_s = \sqrt{2} \times 15 \left\{ \sin(\omega t - 60°) + \frac{1}{5}\sin 5(\omega t - 60°) + \frac{1}{7}\sin 7(\omega t - 60°) \right\}$$

where $\omega = 100\pi$. The load current is shown in Figure 7.2 (a).

Note that the source must supply only the real power required by the load. This can be accomplished by calculating the average power required by the load through a moving average filter that has an averaging window of half a cycle (10 ms). Let this average be denoted by p_{lav}. Since the desired source current must be unity power factor, its expression is then given by

$$i_s^* = \sqrt{2} \times \frac{p_{lav}}{240} \sin(\omega t) \qquad (7.4)$$

Since the source voltage is assumed to be a pure sinewave, the desired source current can be obtained through template matching, i.e., by taking samples of the instantaneous source voltage and scaling it by the factor $p_{lav}/240^2$. The reference compensator current is then given by the relation $I_f^* = I_l - I_s^*$.

The system response is shown in Figure 7.2. In this study, the system parameters chosen are

$$R_f = 0, \quad L_f = 20 \text{ mH}, \quad V_{dc} = 600 \text{ V}$$

It is assumed that the compensator is supplied by a constant voltage source and not a dc capacitor. Figure 7.2 (b) shows the scaled version of the source voltage and the source current. The source voltage is scaled by a factor of 10 such that its magnitude is comparable to that of the source current. The compensator is switched on once the p_{lav} is obtained after the first half cycle. It can be seen that the source current becomes sinusoidal and in phase with the source voltage after the first half cycle (10 ms). The current tracking error ($I_f^* - I_f$) is shown in Figure 7.2 (c). This error jumps as soon as the compensator is connected but then settles immediately. The compensator power is shown in Figure 7.2 (d). Even though the power is oscillating, its mean is zero. This implies that the compensator supplies the reactive and harmonic power required by the load, but no real power.

ΔΔΔ

The above example assumes that the compensator is supplied by a dc source. This however is an invalid assumption and in practical cases the source is replaced by a dc capacitor. Also we have assumed that the system is lossless (i.e., $R_f = 0$). This is also an invalid assumption. Therefore an additional loop must be incorporated for the control of the dc capacitor voltage control. Note that the losses in the system must be replenished by the supply itself. Thus the dc capacitor voltage can be held constant equal to a reference value if the current drawn from the source is higher than that given by (7.4). The additional amount of current is supplied to the capacitor to maintain its voltage constant. To accomplish this the dc capacitor voltage is averaged over one cycle. It is then compared with the reference voltage. The error is then put through an additional proportional-plus-integral (PI) or proportional-plus-differential (PD) loop. The output of this controller is added to the magnitude of the current calculated in (7.4).

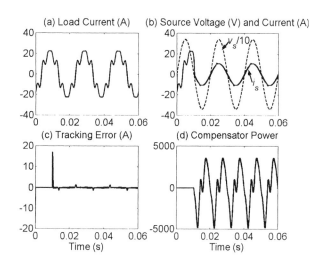

Figure 7.2. System response with single-phase compensator

The implementation aspects and dc capacitor loop control of a single-phase DSTATCOM are discussed in [1,2]. However since the power system is three-phase, the single-phase compensator has little value. Suppose we take the simplistic view that if we put three single-phase compensators then we can compensate a three-phase system. Even though this will enable us to cancel the reactive and harmonic currents in each phase, we shall not be able to balance an unbalanced load. The load balancing requires redistribution of real power equally between the phases. This will not be possible by three separate single-phase compensators. We shall therefore not consider this structure any further and will only discuss three-phase compensators.

7.2 Ideal Three-Phase Shunt Compensator Structure

To illustrate the functioning of shunt compensator, consider the three-phase, four-wire (3p4w) distribution system shown in Figure 7.3. All the currents and voltages that are indicated in this figure are instantaneous quantities. Here a three-phase balanced supply (v_{sa}, v_{sb}, v_{sc}) is connected across a star (Y) connected load. The loads are such that the load currents

(i_{la}, i_{lb}, i_{lc}) may not be balanced, may contain harmonics and dc offset. In addition, the power factor of the load may be poor. One implication of load not being balanced in this system is that there may be zero-sequence current i_{Nn} flowing in the 4th wire, i.e., in the path n-N as shown in Figure 7.3 [3,4].

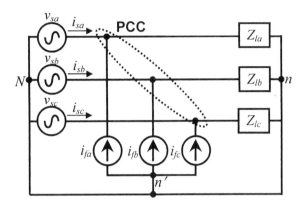

Figure 7.3. Schematic diagram of a shunt compensator for 3p4w distribution system that is supplying a Y-connected load

The shunt compensator is represented by three ideal current sources i_{fa}, i_{fb} and i_{fc}. The point of common coupling (PCC) is encircled in Figure 7.3. The current sources are connected in Y with their neutral n' being connected to the 4th wire. The purpose of the shunt compensator is to inject currents in such a way that the source currents (i_{sa}, i_{sb}, i_{sc}) are harmonic free balanced sinusoids and their phase angle with respect to the source voltages (v_{sa}, v_{sb}, v_{sc}) has a desired value. Let us illustrate the idea with the help of the following example. Note that in this chapter all the plots of load instantaneous real and reactive powers are shown in solid lines, the source powers are shown in dashed lines and the compensator powers are shown in dotted lines.

Example 7.2: Let the three phase instantaneous source voltages be given in per unit by

$$v_{sa} = \sqrt{2}\sin\omega t, \; v_{sb} = \sqrt{2}\sin(\omega t - 120°) \text{ and } v_{sc} = \sqrt{2}\sin(\omega t + 120°)$$

with $\omega = 100\pi$. Three unbalanced RL loads are connected across the supply. They are given in per unit as

$$Z_{la} = 6.0 + j3.0, \; Z_{lb} = 3.0 + j1.5 \text{ and } Z_{lc} = 7.5 + j1.5$$

7. Load Compensation using DSTATCOM

In addition, it is assumed that the load is drawing a 5th harmonic current of magnitude 0.05 per unit. The three load currents are then given in per unit by

$$i_{la} = 0.2108\sin(\omega t - 26.57°) + 0.05\sin 5\omega t$$
$$i_{lb} = 0.4216\sin(\omega t - 146.57°) + 0.05\sin 5(\omega t - 120°)$$
$$i_{lc} = 0.1849\sin(\omega t + 108.69°) + 0.05\sin 5(\omega t + 120°)$$

The load currents are shown in Figure 7.4 (a).

Let us now design a shunt compensator that does not supply any real power to the load. The entire amount of real power must then come from the supply. As per the discussion of the previous section, the real power supplied by the source is strictly utilized by the fundamental component of the load. The instantaneous power to the load is shown in Figure 7.4 (b). This power consists of a dc component and an oscillating component. The dc component with a mean value of 0.5282 per unit is the average real power supplied by the source. The average power in each phase will then be 0.1761 per unit.

It has been mentioned before that the shunt compensator can also correct the supply side power factor. Let us choose the desired supply power factor to be unity. The three source currents for this power level and power factor, are then given in per unit by

$$i_{sa} = \sqrt{2} \times 0.1761\sin \omega t = 0.249\sin \omega t$$
$$i_{sb} = 0.249\sin(\omega t - 120°)$$
$$i_{sc} = 0.249\sin(\omega t + 120°)$$

Applying KCL at the PCC we can write the following expression for the compensator currents

$$i_{f\beta} = i_{l\beta} - i_{s\beta}, \quad \beta = a, b, c \tag{7.5}$$

The various instantaneous quantities of the compensated system are shown in Figure 7.4. From Figure 7.4 (a) it can be seen that the load currents are distorted due to the presence of the 5th harmonic component. The three instantaneous powers are shown in Figure 7.4 (b). It can be seen that the instantaneous power drawn from the source (p_s) is constant even when the instantaneous power into the load (p_l) is oscillating and distorted. It is obvious then that the average component of the power comes from the source while the oscillating component comes from the compensator (p_f). It can also be seen that the compensator power (p_f) has mean of zero. This implies that the compensator does not supply any real power. The compensator currents are shown in Figure 7.4 (c). The source voltage and current of phase-a are shown in Figure 7.4 (d). In this figure the source voltage is scaled by a factor of 4 such that its magnitude is of the same order as the source current. Figure 7.4 (d) clearly demonstrates that the source voltage and current are in the same phase. This example clearly demonstrates that the shunt compensator balances the source currents, corrects their power factor to unity and eliminates harmonic from the source currents.

ΔΔΔ

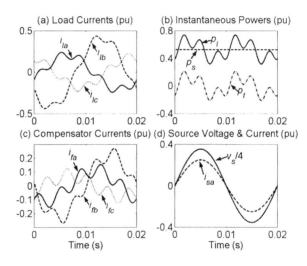

Figure 7.4. The system performance with the shunt compensator of Example 7.2

The compensator structure of Figure 7.3 can be used in a three-phase system when the load is Y-connected. When used in a three-phase, four-wire distribution system, the compensator balances the supply current thereby eliminating the neutral current. When the same structure is used in a three-phase, three-wire distribution system that supplies a Y-connected load, the link between the supply neutral (N) and the load neutral (n) is not present. However, the connection $n' - n$ is still important as this provides a path for the zero-sequence current to flow when the load is unbalanced.

The compensator structure of a three-phase, three-wire (3p3w) system supplying a Δ-connected load is shown in Figure 7.5. Like in the case of a Y-connected load, the compensator, represented by three current sources, are connected in parallel with the load. The only difference being that in the previous case, the compensator branches were connected between line and neutral and in this case they are connected between two lines [5].

7. Load Compensation using DSTATCOM

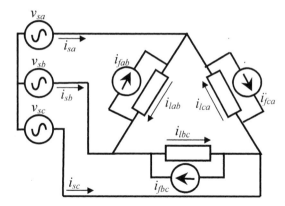

Figure 7.5. Schematic diagram of a shunt compensator for 3p3w distribution system that is supplying a Δ-connected load

The first step in shunt compensation, as illustrated in Example 7.2, is to generate a set of compensator currents i_{fa}, i_{fb} and i_{fc}. In actual practice the compensator is not made of three ideal current sources, but of a power electronic circuit that injects these currents in the distribution system. We shall therefore call these the reference currents of the compensator.

In Example 7.2 we generated the reference current by characterizing the load completely and extracting the component to be compensated. This however, is not desirable and may not even be possible as the load may change frequently. It is therefore imperative that the compensator reference currents be generated based on measurements of real-time quantities like voltages, currents and power. Again, as the compensator currents are instantaneous quantities, it will be desirable to generate them on an instant by instant basis. Below we present various techniques that can be used for the generation of the three-phase reference currents of the compensator.

7.3 Generating Reference Currents using Instantaneous PQ Theory

Hirofumi Akagi and his coworkers have described an instantaneous method of generating reference currents for shunt compensator in [6-8]. Since then various interpretations of this method have been presented [9-12]. This method is applicable to a three-phase, four-wire system. To begin with, we transform the three-phase voltages from *a-b-c* frame to *α-β-0* frame and vice versa using the following power invariant transformation

$$\begin{bmatrix} v_0 \\ v_\alpha \\ v_\beta \end{bmatrix} = \sqrt{\frac{2}{3}} \begin{bmatrix} 1/\sqrt{2} & 1/\sqrt{2} & 1/\sqrt{2} \\ 1 & -1/2 & -1/2 \\ 0 & \sqrt{3}/2 & -\sqrt{3}/2 \end{bmatrix} \begin{bmatrix} v_a \\ v_b \\ v_c \end{bmatrix} \qquad (7.6)$$

$$\begin{bmatrix} v_a \\ v_b \\ v_c \end{bmatrix} = \sqrt{\frac{2}{3}} \begin{bmatrix} 1/\sqrt{2} & 1 & 0 \\ 1/\sqrt{2} & -1/2 & \sqrt{3}/2 \\ 1/\sqrt{2} & -1/2 & -\sqrt{3}/2 \end{bmatrix} \begin{bmatrix} v_0 \\ v_\alpha \\ v_\beta \end{bmatrix} \qquad (7.7)$$

We can also use the same transform matrix for transforming currents.

The instantaneous three-phase power is then given by

$$p_{3\phi} = v_a i_a + v_b i_b + v_c i_c = v_\alpha i_\alpha + v_\beta i_\beta + v_0 i_0 = p + p_0 \qquad (7.8)$$

where p is the total instantaneous real power in the three phase wires and $p_0 = v_0 i_0$ is the instantaneous power in the zero-sequence network. Let us define the following variable

$$q = v_\alpha i_\beta - v_\beta i_\alpha = \frac{1}{\sqrt{3}}\{i_a(v_c - v_b) + i_b(v_a - v_c) + i_c(v_b - v_a)\} \qquad (7.9)$$

We now investigate the fundamental frequency equivalent of the variable q through the following example.

Example 7.3: Let us consider the following balanced three-phase voltages and currents

$$v_a = V_m \sin\omega t \qquad i_a = I_m \sin(\omega t - \phi)$$
$$v_b = V_m \sin(\omega t - 120^\circ), \quad i_b = I_m \sin(\omega t - 120^\circ - \phi)$$
$$v_c = V_m \sin(\omega t + 120^\circ) \quad i_c = I_m \sin(\omega t + 120^\circ - \phi)$$

We can then write

$$v_a - v_b = \sqrt{3} V_m \sin(\omega t + 30^\circ)$$
$$v_b - v_c = \sqrt{3} V_m \sin(\omega t - 90^\circ)$$
$$v_c - v_a = \sqrt{3} V_m \sin(\omega t + 150^\circ)$$

7. Load Compensation using DSTATCOM

Using the above relations we get

$$i_a(v_b - v_c) = -\sqrt{3}V_m I_m \{\cos\omega t \sin(\omega t - \phi)\}$$
$$= -\frac{\sqrt{3}}{2}V_m I_m \{\sin(2\omega t - \phi) - \sin\phi\}$$
$$i_b(v_c - v_a) = -\sqrt{3}V_m I_m \{\cos(\omega t - 120°)\sin(\omega t - 120° - \phi)\}$$
$$= -\frac{\sqrt{3}}{2}V_m I_m \{\sin(2\omega t - 240° - \phi) - \sin\phi\}$$
$$i_c(v_a - v_b) = -\sqrt{3}V_m I_m \{\cos(\omega t + 120°)\sin(\omega t + 120° - \phi)\}$$
$$= -\frac{\sqrt{3}}{2}V_m I_m \{\sin(2\omega t + 240° - \phi) - \sin\phi\}$$

Adding the above three terms together we get

$$i_a(v_b - v_c) + i_b(v_c - v_a) + i_c(v_a - v_b) = \frac{3\sqrt{3}}{2}V_m I_m \sin\phi = \sqrt{3}\,Q = -\sqrt{3}\,q$$

where Q is the reactive power required by the circuit.

△△△

We thus see that the quantity q given in (7.9) is the reactive power absorbed by a circuit when both voltages and currents contain only the fundamental frequency. However, this quantity can be used in a much broader context when either voltages or currents or both have many frequency components. Akagi et al called this term the instantaneous imaginary power [7]. We can write from (7.8) and (7.9)

$$\begin{bmatrix} p \\ q \end{bmatrix} = \begin{bmatrix} v_\alpha & v_\beta \\ -v_\beta & v_\alpha \end{bmatrix} \begin{bmatrix} i_\alpha \\ i_\beta \end{bmatrix} \qquad (7.10)$$

This is equivalent to writing

$$\begin{bmatrix} i_\alpha \\ i_\beta \end{bmatrix} = \frac{1}{v_\alpha^2 + v_\beta^2} \begin{bmatrix} v_\alpha & -v_\beta \\ v_\beta & v_\alpha \end{bmatrix} \begin{bmatrix} p \\ q \end{bmatrix}$$
$$= \frac{1}{v_\alpha^2 + v_\beta^2} \begin{bmatrix} v_\alpha & -v_\beta \\ v_\beta & v_\alpha \end{bmatrix} \left\{ \begin{bmatrix} p \\ 0 \end{bmatrix} + \begin{bmatrix} 0 \\ q \end{bmatrix} \right\} = \begin{bmatrix} i_{\alpha p} \\ i_{\beta p} \end{bmatrix} + \begin{bmatrix} i_{\alpha q} \\ i_{\beta q} \end{bmatrix} \qquad (7.11)$$

The following components of current can then be defined from the above equation

α - axis instantaneous active current : $i_{\alpha p} = \dfrac{v_\alpha}{v_\alpha^2 + v_\beta^2} p$

α - axis instantaneous reactive current : $i_{\alpha q} = -\dfrac{v_\beta}{v_\alpha^2 + v_\beta^2} q$

β - axis instantaneous active current : $i_{\beta p} = \dfrac{v_\beta}{v_\alpha^2 + v_\beta^2} p$

β - axis instantaneous reactive current : $i_{\beta q} = \dfrac{v_\alpha}{v_\alpha^2 + v_\beta^2} q$

Let the instantaneous powers in α-axis and β-axis be denoted respectively by p_α and p_β. We can then write from (7.10) and (7.11)

$$\begin{bmatrix} p_\alpha \\ p_\beta \end{bmatrix} = \begin{bmatrix} v_\alpha i_\alpha \\ v_\beta i_\beta \end{bmatrix} = \begin{bmatrix} v_\alpha i_{\alpha p} \\ v_\beta i_{\beta p} \end{bmatrix} + \begin{bmatrix} v_\alpha i_{\alpha q} \\ v_\beta i_{\beta q} \end{bmatrix} = \begin{bmatrix} p_{\alpha p} \\ p_{\beta p} \end{bmatrix} + \begin{bmatrix} p_{\alpha q} \\ p_{\beta q} \end{bmatrix} \quad (7.12)$$

We now define the following quantities

α - axis instantaneous active power : $p_{\alpha p} = v_\alpha i_{\alpha p}$
α - axis instantaneous reactive power : $p_{\alpha q} = v_\alpha i_{\alpha q}$
β - axis instantaneous active power : $p_{\beta p} = v_\beta i_{\beta p}$
β - axis instantaneous reactive power : $p_{\beta q} = v_\beta i_{\beta q}$

Let us now expand these expressions

$$p_{\alpha p} = \dfrac{v_\alpha^2}{v_\alpha^2 + v_\beta^2} p, \quad p_{\beta p} = \dfrac{v_\beta^2}{v_\alpha^2 + v_\beta^2} p$$

Adding the above two expressions we get

$$p = p_{\alpha p} + p_{\beta p} \quad (7.13)$$

Similarly adding the reactive power components we get

7. Load Compensation using DSTATCOM

$$p_{\alpha q} + p_{\beta q} = 0 \qquad (7.14)$$

We can then conclude the following

- The sum of $p_{\alpha p}$ and $p_{\beta p}$ is equal to the instantaneous real power. Therefore they are referred to as instantaneous active powers.
- The instantaneous powers $p_{\alpha q}$ and $p_{\beta q}$ cancel each other and do not contribute to the real power. They are thus called instantaneous reactive powers.

The instantaneous three-phase power is then given by

$$p_{3\phi} = p_{\alpha p} + p_{\beta p} + p_0 \qquad (7.15)$$

Let us consider the following example.

Example 7.4: Let the following balanced three-phase voltages

$$v_a = V_m \sin\omega t, \quad v_b = V_m \sin(\omega t - 120°) \text{ and } v_c = V_m \sin(\omega t + 120°)$$

be supplying a non-linear load. The load currents contain 3rd and 5th harmonics in addition to the fundamental. These currents are given by

$$i_a = \sum_{n=1,3,5} \frac{I_m}{n} \sin(n\omega t - \phi_n)$$

$$i_b = \sum_{n=1,3,5} \frac{I_m}{n} \sin\{n(\omega t - 120°) - \phi_n\}$$

$$i_c = \sum_{n=1,3,5} \frac{I_m}{n} \sin\{n(\omega t + 120°) - \phi_n\}$$

Transforming the voltages and currents into $\alpha\text{-}\beta\text{-}0$ frame we get

$$v_0 = \frac{1}{\sqrt{3}}\{v_a + v_b + v_c\} = 0$$

$$v_\alpha = \sqrt{\frac{2}{3}}\left\{v_a - \frac{1}{2}v_b - \frac{1}{2}v_c\right\} = \sqrt{\frac{3}{2}}v_a = \sqrt{\frac{3}{2}}V_m \sin\omega t$$

$$v_\beta = \frac{1}{\sqrt{2}}\{v_b - v_c\} = -\frac{2}{\sqrt{2}}V_m \cos\omega t \sin 120° = -\sqrt{\frac{3}{2}}V_m \cos\omega t$$

$$i_0 = \frac{1}{\sqrt{3}}\{i_a + i_b + i_c\} = \frac{I_m}{\sqrt{3}}\sin(3\omega t - \phi_3)$$

$$i_\alpha = \sqrt{\frac{2}{3}}\left\{i_a - \frac{1}{2}i_b - \frac{1}{2}i_c\right\} = \sqrt{\frac{3}{2}}I_m \sin(\omega t - \phi_1) + \sqrt{\frac{3}{2}}\frac{I_m}{5}\sin(5\omega t - \phi_5)$$

$$i_\beta = \frac{1}{\sqrt{2}}\{i_b - i_c\} = -\sqrt{\frac{3}{2}}I_m \cos(\omega t - \phi_1) + \sqrt{\frac{3}{2}}\frac{I_m}{5}\cos(5\omega t - \phi_5)$$

It can be seen from the above expression that the 3rd harmonic current is present only in the zero-sequence. Furthermore, the zero-sequence power is given by $p_0 = v_0 i_0 = 0$. We can then get from (7.10)

$$p = v_\alpha i_\alpha + v_\beta i_\beta = \frac{3}{2}V_m I_m \left\{\cos\phi_1 - \frac{1}{5}\cos(6\omega t - \phi_5)\right\}$$

$$q = v_\alpha i_\beta - v_\beta i_\alpha = -\frac{3}{2}V_m I_m \left\{\sin\phi_1 - \frac{1}{5}\sin(6\omega t - \phi_5)\right\}$$

ΔΔΔ

The above example clearly demonstrates that there are two components of real and reactive power present in a system when the load contains harmonics. We can then write

$$p = p_{av} + p_{osc}$$
$$q = q_{av} + q_{osc}$$
(7.16)

where the subscript *av* indicates the mean or dc value and the subscript *osc* indicates the oscillating component. We have already observed the average (dc) power and the oscillating power in Figure 7.4 of Example 7.2. The reactive power will also have two similar components. We shall now discuss the reference current generation scheme for the compensator using the above-mentioned theory.

Refer to the compensator structure shown in Figure 7.3. Suppose we want to compensate only for the reactive power of the load such that the current drawn from the source is unity power factor. The compensator then must supply the entire reactive power requirement of the load. Let $i_{f\alpha}$ and $i_{f\beta}$ be the α-β components of the compensator and $v_{s\alpha}$ and $v_{s\beta}$ be the α-β

7. Load Compensation using DSTATCOM

components of the source voltage v_s. Note from Figure 7.3 that the source voltage is applied both across the load and the compensator. Now since the compensator must supply the entire load reactive power and no real power, we can write from (7.11)

$$\begin{bmatrix} i_{f\alpha} \\ i_{f\beta} \end{bmatrix} = \frac{1}{v_{s\alpha}^2 + v_{s\beta}^2} \begin{bmatrix} v_{s\alpha} & -v_{s\beta} \\ v_{s\beta} & v_{s\alpha} \end{bmatrix} \begin{bmatrix} 0 \\ q_l \end{bmatrix} \qquad (7.17)$$

where q_l is the instantaneous imaginary power of the load. A reverse transformation using (7.7) will yield the compensator currents in a-b-c plane. In a similar way we can also compensate for both the imaginary power and the oscillating component of the real power. In that case we can modify (7.17) to get

$$\begin{bmatrix} i_{f\alpha} \\ i_{f\beta} \end{bmatrix} = \frac{1}{v_{s\alpha}^2 + v_{s\beta}^2} \begin{bmatrix} v_{s\alpha} & -v_{s\beta} \\ v_{s\beta} & v_{s\alpha} \end{bmatrix} \begin{bmatrix} p_{losc} \\ q_l \end{bmatrix} \qquad (7.18)$$

where p_{losc} is the oscillating component of the load power.

Let us now consider the following example on the application of this instantaneous power analysis. Note that in this example and all the subsequent examples of this chapter it is assumed that the compensator is represented by three ideal current sources as shown in Figure 7.3 (or Figure 7.5 for Δ-connected loads). This implies that these current sources inject the exact reference currents obtained through the different algorithms. Practical compensator implementation will be discussed in Chapter 8 where, in particular, it will be shown how the reference currents can be tracked using voltage source inverters.

Example 7.5: Let a Y-connected balanced RL load be connected to a balanced three-phase supply with an rms value of 1.0 per unit. The instantaneous source voltages and load are given in per unit by

$$v_{sa} = \sqrt{2} \sin \omega t$$
$$v_{sb} = \sqrt{2} \sin(\omega t - 120°)$$
$$v_{sc} = \sqrt{2} \sin(\omega t + 120°)$$
$$Z_{la} = Z_{lb} = Z_{lc} = 5.0 + j5.0$$

with $\omega = 100\pi$. In addition to the RL load, the source also supplies a nonlinear load that is drawing square wave current of peak 0.1 per unit. The load currents are shown in Figure 7.6 (a). Figure 7.6 (b-d) shows the system plots when only q_l is compensated as per (7.17). It is obvious from Figure 7.6 (b) that the source current is in phase with the source voltage indicating that the entire amount of reactive power is supplied by the compensator. In this figure the source voltage is scaled by a factor of three. The source currents are shown in Figure 7.6 (c). It can be seen that they are distorted and this is totally undesirable. The instantaneous powers are shown in Figure 7.6 (d). It can be seen that the compensator power is zero while the load and source powers are equal.

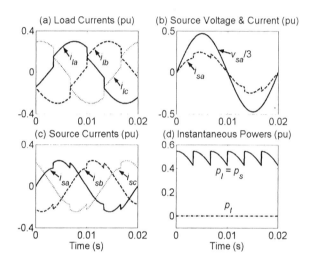

Figure 7.6. System response with balanced nonlinear load when only q is compensated

We now use the compensation algorithm given in (7.18) instead. The system plots are shown in Figure 7.7. It can be seen that the source currents are all balanced sinusoids and are in phase with the source voltage. The compensator in this case however supplies a zero-mean oscillating power such that the source supplies the average (dc) power required by the load. The compensator currents, shown in Figure 7.7 (a), are balanced but distorted, as they have to cancel the distortion of the load currents.

ΔΔΔ

In the above example the average power consumed by the load is obtained by a moving average (MA) filter rather than to a low-pass Butterworth filter suggested by Akagi [8]. An MA filter continuously calculates the average over a certain number of past consecutive samples. To illustrate the idea, let us assume that the signal is uniformly sampled with a

7. Load Compensation using DSTATCOM

sampling time of 100 μs. This means that there are 100 samples in a half cycle if the fundamental frequency is 50 Hz. Then, at any given instant, the average of last 100 consecutive samples is taken to produce the power average. Since the fundamental power fluctuates at 100 Hz, the average of any consecutive 100 values corresponds to a full cycle of the fundamental power waveform. Thus the starting point need not be synchronized with the zero crossing of the waveform. This has the advantage that any change in the instantaneous power is reflected in the power average just after half a cycle.

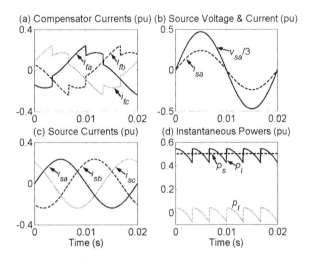

Figure 7.7. System response with balanced nonlinear load when both q and p_{losc} are compensated

Example 7.6: Let us now investigate how this algorithm performs when the load is unbalanced. Let a Y-connected unbalanced RL load be connected to a balanced three-phase supply with an rms voltage of 1.0 per unit. The source voltages are the same as given in Example 7.5 and the loads are given in per unit by

$$Z_{la} = 5.0 + j5.0, \quad Z_{lb} = 5.0 + j1.0 \text{ and } Z_{lc} = 3.0 + j2.0$$

In addition three single-phase rectifier loads that are drawing a current of magnitude of 0.1 per unit are connected in parallel with the RL load. The load currents are shown in Figure 7.8 (a).

We first use the compensation algorithm given in (7.18) in which both q and p_{losc} are compensated. The system plots are shown in Figure 7.8. It is evident from Figure 7.8 (b) that the source current is not unity power factor. Furthermore, as can be seen from Figure 7.8 (c), the source currents are neither harmonic-free, nor are they balanced. The reason for this unwanted behavior is the presence of zero-sequence current in the source. This is shown in Figure 7.8 (d) from which it can be seen that this current contains harmonics as well.

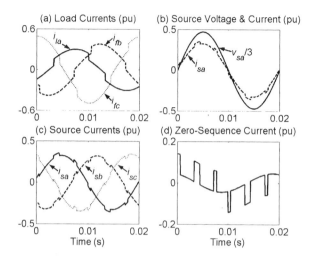

Figure 7.8. System response with unbalanced nonlinear load when both q and p_{losc} are compensated, but the zero-sequence is not compensated

To correct this unwanted behavior, the compensation algorithm of (7.18) is used again with the added stipulation that the zero-sequence of the load current be taken into consideration while obtaining the compensator currents in the a-b-c frame. Refer to Figure 7.3 in which it is shown that the compensator neutral (n') is connected to the load neutral (n). Thus in order to prevent the source from drawing neutral current, the load neutral current must flow through the compensator neutral. Thus, once the compensator currents are obtained in α-β plane using (7.18), we generate these currents in the a-b-c frame using (7.6) and (7.7) as

$$\begin{bmatrix} i_{fa}^* \\ i_{fb}^* \\ i_{fc}^* \end{bmatrix} = \sqrt{\frac{2}{3}} \begin{bmatrix} 1/\sqrt{2} & 1 & 0 \\ 1/\sqrt{2} & -1/2 & \sqrt{3}/2 \\ 1/\sqrt{2} & -1/2 & -\sqrt{3}/2 \end{bmatrix} \begin{bmatrix} (i_{la}+i_{lb}+i_{lc})/\sqrt{3} \\ i_{f\alpha} \\ i_{f\beta} \end{bmatrix} \quad (7.19)$$

where the superscript '*' denotes the instantaneous reference values.

The system response with the zero-sequence compensation is shown in Figure 7.9. It can be seen from Figure 7.9 (a) & (b) that the source currents are balanced with unity power factor. The three different powers are shown in Figure 7.9 (c), while the instantaneous imaginary powers are shown in Figure 7.9 (d). It can be seen that imaginary power out of the source is zero, which implies that the compensator supplies the total load imaginary power.

ΔΔΔ

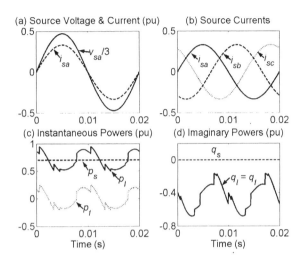

Figure 7.9. System response with unbalanced nonlinear load when both q and p_{losc} are compensated along with the zero-sequence compensation

It is important to note here that if the compensator neutral is not available, or in other words, we have a three-wire compensator, the source currents cannot be balanced if the distribution system is three-phase, four-wire and the load is Y-connected. On the other hand, if the distribution system contains only three wires, the zero-sequence current will not have a path to flow. This will ensure that the sum of the three source currents is zero and hence the compensator will be able to balance these currents.

7.4 Generating Reference Currents using Instantaneous Symmetrical Components

We have discussed the theory of the instantaneous symmetrical components in Chapter 3. In this section we shall utilize this theory for generating instantaneous reference currents. The compensation scheme presented here can be applied to either a three-phase three-wire system or a three-phase four-wire system. The loads can be either connected in Y or in Δ. We shall discuss the compensation of Y-connected loads first.

7.4.1 Compensating Star Connected Loads

The objective in either three or four wire system compensation is to provide balanced supply current such that its zero sequence component is zero. We therefore have

$$i_{sa} + i_{sb} + i_{sc} = 0 \tag{7.20}$$

In the method discussed in Section 7.3, there is no direct control over the power factor angle from the source and the algorithm forces the source current to be unity power factor. In the method under consideration, this angle can be set to have any desired value [4]. Let us assume that the source voltages are balanced and are given by

$$v_{sa} = \sin\omega t,\ v_{sb} = \sin(\omega t - 120°),\ v_{sc} = \sin(\omega t + 120°)$$

Then from (3.16) we get

$$v_{sa1} = \frac{1}{\sqrt{3}}\{v_{sa} + av_{sb} + a^2 v_{sc}\}$$

The angle of the vector is then given by

$$\phi = \angle(v_{sa1}) = \tan^{-1}\left\{\frac{\frac{\sqrt{3}}{2}v_{sb} - \frac{\sqrt{3}}{2}v_{sc}}{v_{sa} - \frac{1}{2}v_{sb} - \frac{1}{2}v_{sc}}\right\} = \tan^{-1}\left\{\frac{\frac{\sqrt{3}}{2}(v_{sb} - v_{sc})}{\frac{3}{2}v_{sa}}\right\} \tag{7.21}$$

Substituting the values of the instantaneous voltages in (7.21) we get

$$\phi = \tan^{-1}\left\{\frac{-\cos\omega t}{\sin\omega t}\right\} = \omega t - \frac{\pi}{2} \tag{7.22}$$

It can thus be seen from (7.22) that the angle of the vector v_{sa1} will change linearly as t changes. We can then easily force another vector to follow (or lead) this vector by an arbitrary angle. If we now assume that the phase of the vector i_{sa1} lags that of v_{sa1} by an angle ϕ, we get

$$\angle\{v_{sa} + av_{sb} + a^2 v_{sc}\} = \angle\{i_{sa} + ai_{sb} + a^2 i_{sc}\} + \phi \tag{7.23}$$

7. Load Compensation using DSTATCOM

Substituting the values of a and a^2 (7.23) can be expanded as

$$\angle\left\{\left(v_{sa} - \frac{1}{2}v_{sb} - \frac{1}{2}v_{sc}\right) + j\frac{\sqrt{3}}{2}(v_{sb} - v_{sc})\right\}$$

$$= \angle\left\{\left(i_{sa} - \frac{1}{2}i_{sb} - \frac{1}{2}i_{sc}\right) + j\frac{\sqrt{3}}{2}(i_{sb} - i_{sc})\right\} + \phi$$

Equating the angles, we can write from the above equation

$$\tan^{-1}(K_1/K_2) = \tan^{-1}(K_3/K_4) + \phi \qquad (7.24)$$

where

$$K_1 = \frac{\sqrt{3}}{2}(v_{sb} - v_{sc}), \quad K_2 = v_{sa} - \frac{1}{2}v_{sb} - \frac{1}{2}v_{sc},$$

$$K_3 = \frac{\sqrt{3}}{2}(i_{sb} - i_{sc}) \text{ and } K_4 = i_{sa} - \frac{i_{sb}}{2} - \frac{i_{sc}}{2}$$

Using the formula

$$\tan(\alpha + \beta) = \frac{\tan\alpha + \tan\beta}{1 - \tan\alpha\tan\beta}$$

(7.24) can be expanded as

$$\frac{K_1}{K_2} = \tan\left\{\tan^{-1}(K_3/K_4) + \phi\right\} = \frac{K_3/K_4 + \tan\phi}{1 - (K_3/K_4)\tan\phi}$$

Solving the above equation we get

$$(v_{sb} - v_{sc} - 3\beta v_{sa})i_{sa}$$
$$+ (v_{sc} - v_{sa} - 3\beta v_{sb})i_{sb} + (v_{sa} - v_{sb} - 3\beta v_{sc})i_{sc} = 0 \qquad (7.25)$$

where

$$\beta \equiv \tan\phi/\sqrt{3}. \qquad (7.26)$$

It is interesting to note the implication of (7.25). When the power factor angle is assumed to be zero, (7.25) implies that the instantaneous reactive power supplied by the source is zero. On the other hand, when this angle is non-zero, the source supplies a reactive power that is equal to β times instantaneous power.

As we have seen before that the instantaneous power in a balanced three-phase circuit is constant while for an unbalanced circuit it has a double frequency component in addition the dc value. In addition, the presence of harmonics adds to the oscillating component of the instantaneous power. The objective of the compensator is to supply the oscillating component such that the source supplies the average value of the load power. Therefore we obtain

$$v_{sa}i_{sa} + v_{sb}i_{sb} + v_{sc}i_{sc} = p_{lav} \tag{7.27}$$

where p_{lav} is the average power drawn by the load. Since the harmonic component in the load does not require any real power, the source only supplies the real power required by the load.

Combining (7.20), (7.25) and (7.27) we get

$$\begin{bmatrix} 1 & 1 & 1 \\ v_{sb} - v_{sc} - 3\beta v_{sa} & v_{sc} - v_{sa} - 3\beta v_{sb} & v_{sa} - v_{sb} - 3\beta v_{sc} \\ v_{sa} & v_{sb} & v_{sc} \end{bmatrix} \begin{bmatrix} i_{sa} \\ i_{sb} \\ i_{sc} \end{bmatrix} = \begin{bmatrix} 0 \\ 0 \\ p_{lav} \end{bmatrix} \tag{7.28}$$

Assuming that the current are tracked without error, the KCL at PCC can be written in terms of the reference currents as

$$i_{fk}^* = i_{lk} - i_{sk}, \quad k = a,b,c$$

Substituting the above equation in (7.28) and solving we get

$$\left.\begin{aligned} i_{fa}^* &= i_{la} - \frac{v_{sa} + (v_{sb} - v_{sc})\beta}{v_{sa}^2 + v_{sb}^2 + v_{sc}^2} p_{lav} \\ i_{fb}^* &= i_{lb} - \frac{v_{sb} + (v_{sc} - v_{sa})\beta}{v_{sa}^2 + v_{sb}^2 + v_{sc}^2} p_{lav} \\ i_{fc}^* &= i_{lc} - \frac{v_{sc} + (v_{sa} - v_{sb})\beta}{v_{sa}^2 + v_{sb}^2 + v_{sc}^2} p_{lav} \end{aligned}\right\} \tag{7.29}$$

7. Load Compensation using DSTATCOM

Example 7.7: To illusrate the working principle of the compensator, let us consider the load and the supply voltage given in Example 7.6. When source currents are required to be in phase with the source voltage, the results are identical to those given in Figure 7.9 as the requirements that

– the source supplies only the average power to the load,
– the compensator supplies the instantaneous reactive power and oscillating power to the load and
– no zero-sequence current flows out of the source neutral

are identical in both the algorithms. The system response when leading or lagging current required from the source is shown in Figure 7.10. We have considered two cases – Case-a when the desired source currents lead the source voltages by 30° and Case-b when the currents lag the source voltages by 30°. The results for Case-a are shown in Figure 7.10 (a & b) and that of Case-b are shown in Figure 7.10 (c & d). Note that since the load remains unchanged, the average power consumed by the load and the instantaneous load power remain unchanged. Therefore, the compensator power also is identical in both cases. However, to accommodate the same amount of power flow, the amplitude of the source current will vary. In fact, the peak of the source current for unity power factor operation is 0.3311 per unit, while for the cases shown in Figure 7.10 it is 0.3823 per unit. This peak value is the same for both the cases as the power factor is the same.

ΔΔΔ

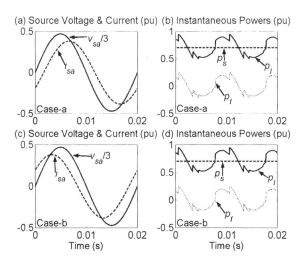

Figure 7.10. System response with unbalanced load when the power factor required is (a & b) 30° leading and (c & d) 30° lagging

It can be shown by simplifying (7.29) that if the load is balanced and free of distortion and if ϕ is the same as that of the phase angle of the load current, the compensator currents become zero. This implies that the load currents are equal to the source currents and the compensator is not required to perform any task. Note that the sum of the source currents in a properly compensated system is zero as per (7.20). Therefore we get the following equation for exact tracking

$$\sum_{k=a,b,c} i_{lk} = \sum_{k=a,b,c} i_{fk} \qquad (7.30)$$

The above equation has three possible implications. These are discussed below.

1. In a three-phase, four-wire distribution system, the summation in either side of (7.30) will be non-zero when any unbalance is present in the load. In this case the zero sequence load current will flow between the compensator and the load in the path $n - n'$ of Figure 7.3.
2. In a three-phase, three-wire distribution system, the connection $N - n$ is absent. In this case if the path $n - n'$ is present, the zero sequence current will again circulate in this path for an unbalanced load.
3. In a three-phase, three-wire (3p3w) distribution system, if the neutral path $n - n'$ is also absent, then the terms on the both sides of the above equation will be zero, i.e.,

$$\sum_{k=a,b,c} i_{lk} = \sum_{k=a,b,c} i_{fk} = 0$$

Hence there will be no zero sequence current at any point of the network. However the voltages between $N - n'$ and $n - n'$ will now oscillate. The following example illustrates this.

Example 7.8: Let us consider the same source and the same RL load as given in Example 7.6 except that a 3p3w distribution system is considered here. The nonlinear load is not considered here. It is assumed that the system is operating in the steady state. The compensator is connected after the first half cycle once the value of p_{lav} is obtained through the moving average filter. Subsequently the load is changed to a balanced RL load at the end of the 3rd cycle. This balanced load is given by $5 + j5$ per unit per phase. The system response is shown in Figure 7.11. It can be seen that the source supplies a unity power factor current to the compensated load. Also the source currents are balanced all through. Once the load is changed to the

7. Load Compensation using DSTATCOM

balanced load the power supplied by the source (p_s) becomes equal to the power consumed by the load (p_l). As a consequence the compensator power (p_s) becomes zero, as the compensator now supplies purely reactive power required for the unity power factor operation. It is to be noted that the compensator currents are non zero as they supply the reactive component of the load current. The neutral voltage (v_{Nn}), shown in Figure 7.11 (d), oscillates when the load is unbalanced but becomes zero as soon as the load becomes balanced.

ΔΔΔ

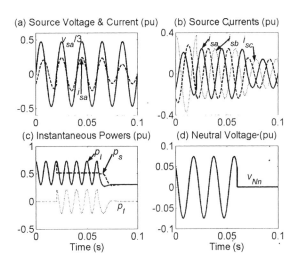

Figure 7.11. Compensator performance in a 3p3w distribution system

It is to be noted that the algorithms presented in Sections 7.3 and 7.4 can eliminate any dc offset in the load since both of them force the sum of the sources current to be zero. This is possible only when the compensator neutral is available for connection and no transformer is present in the compensator realization. If the compensator is realized using transformers, then the neutral current cannot be compensated. Otherwise the presence of dc will saturate the transformers.

7.4.2 Compensating Delta Connected Loads

The basic scheme is shown in Figure 7.5. The aim of the scheme is to generate the three reference current waveforms for i_{fab}, i_{fbc} and i_{fca} from the measurements of source voltages and load currents such that the supply sees a balanced load. The requirements for the compensating currents in this case are same as that of the previous case with Y-connected load. Therefore the requirements (7.20), (7.23) and (7.27) are valid in this case also. As a result, (7.24) and (7.25) are also valid.

Now we can write the following from Figure 7.5

$$
\left.\begin{array}{l}
i_{sa} = i_{lab} - i_{fab} - i_{lca} + i_{fca} \\
i_{sb} = i_{lbc} - i_{fbc} - i_{lab} + i_{fab} \\
i_{sc} = i_{lca} - i_{fca} - i_{lbc} + i_{fbc}
\end{array}\right\} \tag{7.31}
$$

For zero circulating current inside the delta, we can write from Figure 7.5

$$i_{lab} + i_{lbc} + i_{lca} - (i_{fab} + i_{fbc} + i_{fca}) = 0 \tag{7.32}$$

Substituting (7.31) in (7.25) and solving we get

$$
\begin{aligned}
(i_{lab} - i_{fab})\{v_{sc} + \beta(v_{sa} - v_{sb})\} + (i_{lbc} - i_{fbc})\{v_{sa} + \beta(v_{sb} - v_{sc})\} \\
+ (i_{lca} - i_{fca})\{v_{sb} + \beta(v_{sc} - v_{sa})\} = 0
\end{aligned} \tag{7.33}
$$

Finally substituting (7.31) in (7.27) and solving, we get

$$
\begin{aligned}
(i_{lab} - i_{fab})(v_{sa} - v_{sb}) + (i_{lbc} - i_{fbc})(v_{sb} - v_{sc}) \\
+ (i_{lca} - i_{fca})(v_{sc} - v_{sa}) = p_{lav}
\end{aligned} \tag{7.34}
$$

Combining (7.32)-(7.34) we get

$$
\begin{bmatrix}
1 & 1 & 1 \\
v_{sc} + \beta(v_{sa} - v_{sb}) & v_{sa} + \beta(v_{sb} - v_{sc}) & v_{sb} + \beta(v_{sc} - v_{sa}) \\
(v_{sa} - v_{sb}) & (v_{sb} - v_{sc}) & (v_{sc} - v_{sa})
\end{bmatrix}
\begin{bmatrix}
i_{lab} - i_{fab} \\
i_{lbc} - i_{fbc} \\
i_{lca} - i_{fca}
\end{bmatrix}
= \begin{bmatrix} 0 \\ 0 \\ p_{lav} \end{bmatrix} \tag{7.35}
$$

From the above equation we get the reference currents as

7. Load Compensation using DSTATCOM

$$\left. \begin{array}{l} i_{fab} = i_{lab} - \dfrac{v_{sab} - 3\beta v_{sc}}{\Delta} P_{lav} \\[6pt] i_{fbc} = i_{lbc} - \dfrac{v_{sbc} - 3\beta v_{sa}}{\Delta} P_{lav} \\[6pt] i_{fca} = i_{lca} - \dfrac{v_{sca} - 3\beta v_{sb}}{\Delta} P_{lav} \end{array} \right\} \quad (7.36)$$

where

$$\Delta = (v_{sa} - v_{sb})^2 + (v_{sb} - v_{sc})^2 + (v_{sc} - v_{sa})^2$$

Example 7.9: In this example an unbalanced RL Δ-connected load is connected to the supply that is same as given in Example 7.8. The per unit values of the load impedances are

$$Z_{ab} = 5 + j4, \ Z_{bc} = 3 + j4 \ \Omega \text{ and } Z_{ca} = 7 + j1.$$

In addition, three single-phase full-wave uncontrolled rectifiers are connected, one across each phase and neutral. The rectifiers are drawing uneven square wave currents of amplitude 0.15 per unit, 0.1 per unit and 0.2 per unit in phases ab, bc and ca respectively. The load currents are shown in Figure 7.12 (a). The compensator is connected to the system in such a way that the source currents lag the source voltage by an angle of 30°. The desired angle is changed over to unity power factor angle at the end of the 1st cycle. The three source voltages and currents are shown in Figure 7.12 (b-d). It is obvious from these figures that the source currents become unity power factor at the beginning of the 2nd cycle. The average real power drawn from the source however remains unchanged during this transition. The magnitude of the source current varies to accommodate the change in the power factor.

$\Delta\Delta\Delta$

While correcting the power factor, it may not always be economical compensate for unity power factor. The two methods presented in this section leave the choice of power factor angle to the customer. In this regard this is more versatile than the method presented in the previous section. Another important aspect to be noted here is that even though methods presented in both the previous section and this require the instantaneous measurements of variables, to achieve perfect compensation they rely on the average value of power. The average power is obtained through a moving average filter based on half cycle averaging. Hence, the compensator can compensate for any change in the load only half a cycle after its occurrence. Note that a compensator that uses half cycle averaging will not be able to compensate any sub-harmonic component in the load current.

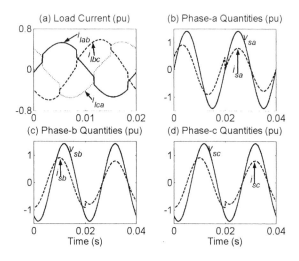

Figure 7.12. Compensating Δ-connected load with a change in the desired power factor angle at the end of 1st cycle

7.5 General Algorithm for Generating Reference Currents

We have discussed the instantaneous real and reactive power in Section 3.2.3. In this section we shall use them to generate the compensator reference currents. Using the definitions given in Section 3.2.3 it is possible to write the instantaneous vector of the filter reference currents in terms of its active and reactive components as

$$i_f^* = i_{fp}^* + i_{fq}^* = \frac{p_f v_s}{v_s \cdot v_s} + \frac{q_f \times v_s}{v_s \cdot v_s} \qquad (7.37)$$

where p_f is the instantaneous scalar power drawn by the filter, v_s, q_f and i_f^* are vectors defined by

$$v_s = \begin{bmatrix} v_{sa} \\ v_{sb} \\ v_{sc} \end{bmatrix}, \quad q_f = \begin{bmatrix} q_{fa} \\ q_{fb} \\ q_{fc} \end{bmatrix}, \quad i_f^* = \begin{bmatrix} i_{fa}^* \\ i_{fb}^* \\ i_{fc}^* \end{bmatrix}$$

7. Load Compensation using DSTATCOM

Similarly the instantaneous power drawn from the source can also be written in terms source real and reactive powers as

$$i_s^* = i_{sp}^* + i_{sq}^* = \frac{p_s v_s}{v_s \cdot v_s} + \frac{q_s \times v_s}{v_s \cdot v_s} \qquad (7.38)$$

where p_s is the scalar power supplied by the source and the vector source reactive power is given by $q_s = [q_{sa}\ q_{sb}\ q_{sc}]^T$.

Expanding (7.37) and (7.38) using the definitions of p and q and noting that $i_f^* = i_l - i_s$, we get the following general algorithm [13,14]

$$\left.\begin{aligned}
i_{fa}^* &= i_{la} - i_{sa} = i_{la} - \frac{1}{\sum_{k=a,b,c} v_{sk}^2}\left(p_s v_{sa} + q_{sb} v_{sc} - q_{sc} v_{sb}\right) \\
i_{fb}^* &= i_{lb} - i_{sb} = i_{lb} - \frac{1}{\sum_{k=a,b,c} v_{sk}^2}\left(p_s v_{sb} + q_{sc} v_{sa} - q_{sa} v_{sc}\right) \\
i_{fc}^* &= i_{lc} - i_{sc} = i_{lc} - \frac{1}{\sum_{k=a,b,c} v_{sk}^2}\left(p_s v_{sc} + q_{sa} v_{sb} - q_{sb} v_{sa}\right)
\end{aligned}\right\} \qquad (7.39)$$

The generation of the reference filter currents through (7.39) requires the measurements of the desired source powers. The algorithm given in [15] is based on (7.37) and it requires the desired filter powers. Considering filter current sources of Figure 7.3 to be ideal, the appropriate selection of source power terms, p_s, q_{sa}, q_{sb} and q_{sc} based on load powers yields different kinds of compensation schemes. The implementation of these schemes involves continuous measurement of system voltages and load currents and real time calculation of various active and reactive load power components.

7.5.1 Various Compensation Schemes and Their Characteristics Based on the General Algorithm

It has been discussed before that both real and reactive powers have an average component and an oscillating component. In the discussion given below we shall denote the average quantities by the subscript *av* and the oscillating quantities by the subscript *osc*. We can then write from (3.34)

$$q_l = q_{lav} + q_{losc} = \begin{bmatrix} q_{la} \\ q_{lb} \\ q_{lc} \end{bmatrix} = \begin{bmatrix} q_{la\_av} \\ q_{lb\_av} \\ q_{lc\_av} \end{bmatrix} + \begin{bmatrix} q_{la\_osc} \\ q_{lb\_osc} \\ q_{lc\_osc} \end{bmatrix} \qquad (7.40)$$

$$p_l = p_{lav} + p_{losc} \qquad (7.41)$$

As defined in Section 3.2.3, the scalar instantaneous reactive power can be given by the algebraic sum as per (3.23) $q_{lsum} = (q_{la} + q_{lb} + q_{lc})/\sqrt{3}$. Therefore the scalar average and zero mean oscillating values of the instantaneous reactive powers over three phases are,

$$\left. \begin{aligned} q_{lsum}{}^{av} &= \frac{1}{\sqrt{3}} \left(q_{la\_av} + q_{lb\_av} + q_{lc\_av} \right) \\ q_{lsum}{}^{osc} &= \frac{1}{\sqrt{3}} \left(q_{la\_osc} + q_{lb\_osc} + q_{lc\_osc} \right) \end{aligned} \right\} \qquad (7.42)$$

This implies that while q_{lav} and q_{losc} are vectors, $q_{lsum}{}^{av}$ and $q_{lsum}{}^{osc}$ are scalars. Let the desired power factor angle between the source voltage and the source current be ϕ. Then we define a factor γ as

$$\gamma = \frac{p_{lav}}{q_{lsum}{}^{av}} \tan \phi \qquad (7.43)$$

The range of γ usually varies between 0 and 1.

The different cases of load compensation possible by stipulating that the compensator supplies various portions of active and reactive powers are given in Table 7.1. This table gives the values of p_s, q_{sa}, q_{sb} and q_{sc} chosen in (7.39) to obtain seven different types of compensation.

7.5.2 Discussion of Results

In the following examples, we discuss each case in details. For all the cases except Case-2 we consider the same source and load as given in Example 7.6. Note that in Figures 7.13 to 7.19 the notation q denotes the instantaneous scalar reactive power, which is actually q_{sum}.

Example 7.10 (Case-1): For this case, the compensator supplies the total reactive power (q_{lsum}) and also the zero mean oscillating active power (p_{losc})

7. Load Compensation using DSTATCOM

of the load. The source supplies only the average load power (p_{lav}). The result is shown in Figure 7.13. It can be seen that the compensated source current is in phase with the source voltage. In fact the results are identical to those shown in Figure 7.9. It can also be shown that the relations given in (7.29) with the power factor angle ϕ taken as 0 are identical to those given in (7.39) for this case, i.e., when the source supplies the average real power and no reactive power.

△△△

Table 7.1. The instantaneous real and reactive power to be supplied by the source for different types of compensation

Case	p_s	q_{sa}	q_{sb}	q_{sc}
1	p_{lav}	0	0	0
2	p_{lav}	$\gamma q_{lsum}^{av}/\sqrt{3}$	$\gamma q_{lsum}^{av}/\sqrt{3}$	$\gamma q_{lsum}^{av}/\sqrt{3}$
3	p_l	0	0	0
4	p_l	$q_{lsum}^{av}/\sqrt{3}$	$q_{lsum}^{av}/\sqrt{3}$	$q_{lsum}^{av}/\sqrt{3}$
5	p_{lav}	$q_{lsum}^{osc}/\sqrt{3}$	$q_{lsum}^{osc}/\sqrt{3}$	$q_{lsum}^{osc}/\sqrt{3}$
6	p_{lav}	$q_{lsum}/\sqrt{3}$	$q_{lsum}/\sqrt{3}$	$q_{lsum}/\sqrt{3}$
7	p_l	$q_{lsum}^{osc}/\sqrt{3}$	$q_{lsum}^{osc}/\sqrt{3}$	$q_{lsum}^{osc}/\sqrt{3}$

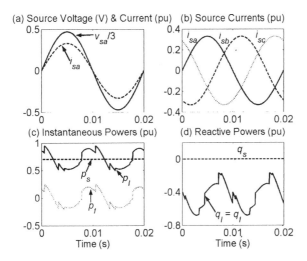

Figure 7.13. System response for Case-1

Example 7.11 (Case-2): This is the same as Case 1 except that the source now supplies a part of the total reactive power q_{lsum} required by the load. The reactive power to be supplied by the source can be fixed by choosing the desired power factor angle ϕ. For this we have assumed a balanced distortion-free load that has an impedance of $5 + j5$ per unit. This implies that the power factor angle of the load is 45° (lagging).

The system response is shown in Figure 7.14 for two different desired source power factor angles. First the desired angle is chosen as 30° (lagging). This is evident from Figure 7.14 (a) which depicts the phase-a source voltage and current. The instantaneous reactive powers are shown in Figure 7.14 (b). It can be seen that part of the reactive power requirement of the load is supplied by the source and the compensator supplies the rest. Figure 7.14 (c & d) shows the response when the desired power factor angle is 60° (lagging). For this case the source supplies more reactive power than required by the load. Therefore the compensator must absorb the rest of the reactive power. This is evident from Figure 7.14 (d) where the sign of q_f is positive. It is to be noted that this mode is operation is exactly the same as discussed in Section 7.4.

ΔΔΔ

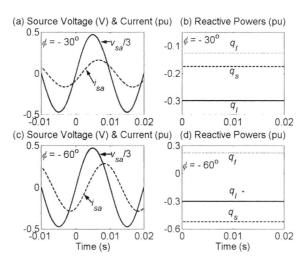

Figure 7.14. System response for Case-2

Example 7.12 (Case-3): In this mode the source supplies the total instantaneous power requirement of the load (p_l) and the compensator supplies the total reactive power required by the load (q_{lsum}). Since the compensator fully compensates the load reactive power, the source current becomes unity power factor. However, since the source supplies both the average and oscillating real power, the distortion and unbalance in the load current is passed on to the source current. The result for this case is shown in Figure 7.15. The operation of the compensator in this mode is exactly the same as the algorithm proposed in [9,11].

ΔΔΔ

7. Load Compensation using DSTATCOM

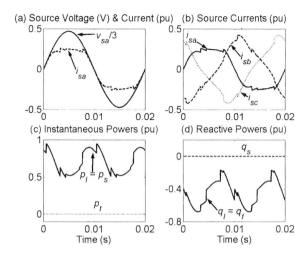

Figure 7.15. System response for Case-3

Example 7.13 (Case-4): This is a combination of cases 2 and 4. In this mode the source supplies the entire real power and a part of the average reactive power. The compensator supplies the rest of the reactive power and no real power. These facts are evident from Figure 7.16, which depicts the result when the desired power factor angle is chosen to be 30° (lagging). It can be seen that even though the source currents are unbalanced and distorted, the phase angle difference between the source voltage and current is as stipulated.

ΔΔΔ

Example 7.14 (Case-5): The source, in this case, supplies the average real power and the oscillating reactive power required by the load. The compensator then supplies the oscillating real power and average reactive power required by the load. The result is shown in Figure 7.17. It can be seen that the source current is in phase with the source voltage. However they are neither balanced nor sinusoidal as the source is required to supply oscillating component of the reactive power.

ΔΔΔ

Example 7.15 (Case-6): In this case the source supplies the average real power and the total reactive power, while the compensator supplies the zero-mean oscillating real power and no reactive power. The system response is shown in Figure 7.18. It can be seen that the source current is not unity power factor, as the source is required to supply the entire reactive power. In fact the phase angle of the source current will be same as that of the load current in this case.

ΔΔΔ

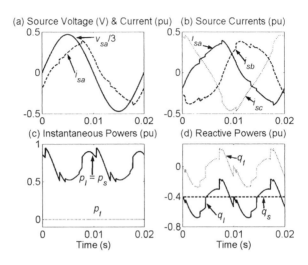

Figure 7.16. System response for Case-4

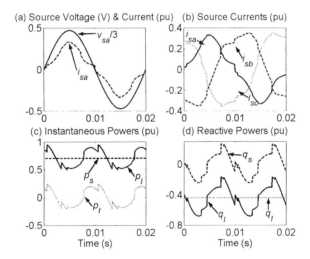

Figure 7.17. System response for Case-5

Example 7.16 (Case-7): In this case the source supplies the total real power and oscillating component of the reactive power required by the load,

7. Load Compensation using DSTATCOM

while the compensator supplies only the average reactive power. The system response for this case is shown in Figure 7.19.

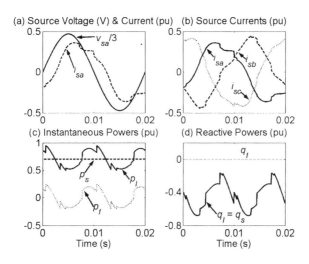

Figure 7.18. System response for Case-6

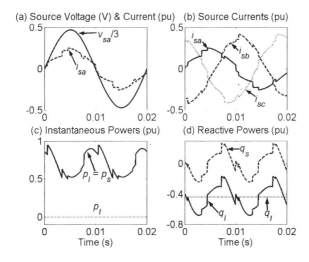

Figure 7.19. System response for Case-7

Thus we can make the compensator to take action on any of the various combination of the power terms. However the average load power p_{lav} must always come from the source. Similarly in an actual compensator, the losses must also be supplied by the source. While we can make several other combinations, the most commonly used combinations (Cases 1, 2, 3) have been included above. The simulation for the Cases 2, 3, 4 and 6 in Table 7.1, using the *pq* theory, has been reported in [11]. For these cases, the simulated results given above using generalized instantaneous reactive power theory are found to be similar.

The different compensation schemes given in Table 7.1 give rise to different characteristics. Based on above simulation results, the effects of compensation for each case are summarized in Table 7.2. In this table, UPF and UDF stand for unity power factor and unity displacement factor respectively. As given in (3.63), the displacement factor is related to the power factor angle between the fundamental waveforms of the distorted voltage and current signals. It is observed from simulation results that in several cases the source currents are unbalanced and have notches. The unbalance results when the source supplies zero-mean oscillating powers. The notches result from discontinuities in the oscillating components of real and reactive powers due to the ideal rectifier diode bridge. In addition to ones given in Table 7.1, alternate choices of real and reactive powers are reported in [16]. These are however not included here as they result in more distorted waveforms than shown in Figures 7.15 to 7.19.

Table 7.2. Various compensation strategies and their characteristics

Case	The effects of compensation on the source currents
1	UPF, sinusoidal balanced
2	Non-UPF, sinusoidal balanced
3	UDF, notches, unbalanced
4	Non-UDF, notches, unbalanced
5	UDF, notches, unbalanced
6	Non-UDF, notches, unbalanced
7	UDF, notches, unbalanced

7.6 Generating Reference Currents when the Source is Unbalanced

We have so far assumed that the sources are stiff and balanced. In practice, these assumptions may not true. In this section we investigate the implication of having unbalanced voltage sources. We shall however assume that the voltage sources are undistorted, i.e., at fundamental frequency. Let us begin our discussion with the following example.

Example 7.17: Let us consider the same load as used in Example 7.6. The source voltages are unbalanced and are given in per unit by

7. Load Compensation using DSTATCOM

$$v_{sa} = \sqrt{2} \sin \omega t$$
$$v_{sb} = 0.8 \times \sqrt{2} \sin(\omega t - 120°)$$
$$v_{sc} = 1.2 \times \sqrt{2} \sin(\omega t + 120°)$$

The source voltage and load currents are shown in Figure 7.20 (a) and Figure 7.20 (b) respectively. We use the same algorithm as given by (7.19). The system plots are shown in Figure 7.20. From Figure 7.20 (c) it can be seen that the source currents are both unbalanced and distorted. Furthermore, the real power drawn from the source is not a steady dc value. This can be seen from Figure 7.20 (d). The imaginary power drawn from the source however remains zero. Thus the compensator supplies the entire imaginary power requirement of the load.

ΔΔΔ

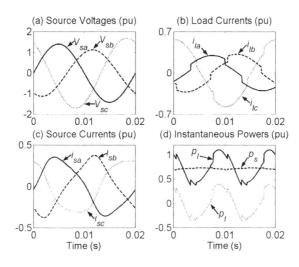

Figure 7.20. System response with unbalanced nonlinear load and unbalanced source when both q and p_{osc} are compensated along with the zero-sequence compensation

In a similar way, the currents will be distorted when the reference currents are generated based on the algorithm given in Section 7.4.1. It is to be noted that the algorithm given in (7.29), which assumes a balanced source voltage, cannot be directly used here. These equations can however be easily modified to consider the general case. It can be shown that for unity power factor operation these equations are

$$i_{fa}^* = i_{la} - i_{sa} = i_{la} - \frac{v_{sa} - v_0}{\Delta} p_{lav}$$
$$i_{fb}^* = i_{lb} - i_{sb} = i_{lb} - \frac{v_{sb} - v_0}{\Delta} p_{lav} \qquad (7.44)$$
$$i_{fc}^* = i_{lc} - i_{sc} = i_{lc} - \frac{v_{sc} - v_0}{\Delta} p_{lav}$$

where

$$v_0 = \frac{1}{3} \sum_{k=a,b,c} v_{sk} \text{ and } \Delta = \left[\sum_{k=a,b,c} v_{sk}^2\right] - 3v_0^2$$

Note that when the system is balanced, v_0 will be zero and (7.44) will be identical to (7.29) for unity power factor operation.

Using the modified algorithm of (7.44) it can be shown that the source currents and voltages are in phase. The source currents are however distorted and their magnitudes are not equal. Even though the algorithm can find a solution, the unequal phase currents added up to zero only with severe and unacceptable distortions. Therefore alternate formulations are imperative.

In general, the compensating currents for this problem can be generated to achieve one of the following three objectives [14]:

- Equal equivalent resistance in all three phases.
- Equal source current magnitude in all three phases
- Equal sharing of average power by all three phases

The algorithms for generating compensator currents for each of the above cases are discussed below.

7.6.1 Compensating to Equal Resistance

The aim of this scheme is to inject the compensator currents in such a way that the supply sees a balanced resistive load. This means that the ratios of instantaneous source voltage and current in each phase are equal, i.e.,

$$\frac{v_{sa}}{i_{sa}} = \frac{v_{sb}}{i_{sb}} = \frac{v_{sc}}{i_{sc}} = R_{eq} \qquad (7.45)$$

7. Load Compensation using DSTATCOM

This gives the following two equations

$$v_{sa}i_{sb} - v_{sb}i_{sa} = 0 \qquad (7.46)$$

$$v_{sb}i_{sc} - v_{sc}i_{sb} = 0 \qquad (7.47)$$

Furthermore, the three-phase average power supplied by the source must be equal to the average power drawn by the load. Since the compensated load is resistive, from this consideration we get

$$|V_{sa}||I_{sa}| + |V_{sb}||I_{sb}| + |V_{sc}||I_{sc}| = P_{lav} \qquad (7.48)$$

where V_{sa} and I_{sa} are respectively the phasor source voltage and current of phase-a. Now if we denote the peak of the phase-a voltage and current by V_{sam} and I_{sam} respectively, we can then write

$$|V_{sa}||I_{sa}| = \frac{V_{sam}I_{sam}}{2} = \frac{V_{sam}^2 I_{sam}}{2V_{sam}} = \frac{V_{sam}^2 i_{sa}}{2v_{sa}}$$

Similar expressions can also be written for the other two phases. We then rewrite (7.48) as

$$\frac{V_{sam}^2}{2v_{sa}}i_{sa} + \frac{V_{sbm}^2}{2v_{sb}}i_{sb} + \frac{V_{scm}^2}{2v_{sc}}i_{sc} = P_{lav} \qquad (7.49)$$

Combining (7.46), (7.47) and (7.49) and rearranging, we get the following expressions for the reference compensator currents

$$\left.\begin{aligned} i_{fa}^* &= i_{la} - i_{sa} = i_{la} - \frac{2v_{sa}}{\Delta_1}P_{lav} \\ i_{fb}^* &= i_{lb} - i_{sb} = i_{lb} - \frac{2v_{sb}}{\Delta_1}P_{lav} \\ i_{fc}^* &= i_{lc} - i_{sc} = i_{lc} - \frac{2v_{sc}}{\Delta_1}P_{lav} \end{aligned}\right\} \qquad (7.50)$$

where

$$\Delta_1 = V_{sam}^2 + V_{sbm}^2 + V_{scm}^2$$

Assuming that the peaks of the source voltages remain constant, the instantaneous compensator reference currents can be computed from (7.50). From (7.45) and (7.48) we can write

$$\frac{|V_{sa}|^2 + |V_{sb}|^2 + |V_{sc}|^2}{R_{eq}} = p_{lav}$$

Solving the above equation we get

$$R_{eq} = \frac{\Delta_1}{2 p_{lav}} \qquad (7.51)$$

Example 7.18: Let us consider the same system as given in Example 7.17. The source voltage and load currents are the same as shown in Figure 7.20. We shall now use the compensator of (7.50).

The system response is shown in Figure 7.21. It can be seen that the source currents are in phase with the source voltages. Furthermore, the magnitude of the source currents is a fixed proportion (R_{eq}) of the source voltages. The proportionality constant (R_{eq}) in this case has a value of 4.202. This compensation will however not result in balanced source current. As a consequence, the power drawn from the source contains both its ac and dc components and oscillates at 100 Hz. The compensator however supplies a zero mean oscillating power as evident from Figure 7.21 (d).

△△△

7.6.2 Compensating to Equal Source Currents

Let us consider the general case in which both the magnitudes and phase angles of the supply voltage are unbalanced. The supply voltage is given by

$$\begin{aligned} v_{sa} &= V_{sam} \sin \omega t \\ v_{sb} &= V_{sbm} \sin(\omega t - 2\pi/3 + \theta_b) \\ v_{sc} &= V_{scm} \sin(\omega t + 2\pi/3 + \theta_c) \end{aligned} \qquad (7.52)$$

In (7.52), the magnitudes V_{sam}, V_{sbm} and V_{scm} are unequal. The phase angles θ_b and θ_c contribute to the phase unbalance. Let us define a set of fictitious set of voltages v'_{sa}, v'_{sb} and v'_{sc} as

7. Load Compensation using DSTATCOM

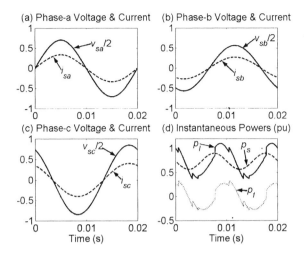

Figure 7.21. System response when compensated to equal resistance

$$v'_{sa} = V'_{sm} \sin \omega t$$
$$v'_{sb} = V'_{sm} \sin(\omega t - 2\pi/3) \qquad (7.53)$$
$$v'_{sc} = V'_{sm} \sin(\omega t + 2\pi/3)$$

We shall use this balanced set of voltages in the compensator algorithm of (7.39). The use of balanced voltages in the algorithm produces balanced compensated source currents. For balanced compensated source currents, both sets of voltages given by (7.52) and (7.53) above, should yield equal average real power, p_{sav}. From this requirement, we obtain the magnitude V'_{sm} as,

$$V'_{sm} = \frac{1}{3}(v_{sa} + v_{sb}\alpha_b + v_{sc}\alpha_c) \qquad (7.54)$$

where

$$\alpha_b = \frac{\cos(\phi + \theta_b)}{\cos \phi} \quad \text{and} \quad \alpha_c = \frac{\cos(\phi + \theta_c)}{\cos \phi} \qquad (7.55)$$

ϕ being the desired phase angle between the source voltage v_{sa} and the desired compensated source current i_{sa}. In the case of unbalance in magnitudes only $\theta_b = \theta_c = 0$. The factors α_b and α_c in (7.55) will then reduce to unity and consequently V'_{sm} becomes the average of the unequal magnitudes v_{sa}, v_{sb} and v_{sc}.

Based on above considerations, The modified algorithm for filter reference currents is given from (7.39) as [17]

$$\left.\begin{aligned} i^*_{fa} &= i_{la} - \frac{1}{\Delta_2}\left(p_s v'_{sa} + q_{sb} v'_{sc} - q_{sc} v'_{sb}\right) \\ i^*_{fb} &= i_{lb} - \frac{1}{\Delta_2}\left(p_s v'_{sb} + q_{sc} v'_{sa} - q_{sa} v'_{sc}\right) \\ i^*_{fc} &= i_{lc} - \frac{1}{\Delta_2}\left(p_s v'_{sc} + q_{sa} v'_{sb} - q_{sb} v'_{sa}\right) \end{aligned}\right\} \quad (7.56)$$

where

$$\Delta_2 = V'^2_{sa} + V'^2_{sb} + V'^2_{sc}$$

Example 7.19: Let us consider the same system as discussed in Examples 7.17 and 7.18. The power terms in (7.56) are obtained from Case 1 in Table 7.1 and measured load powers by the same procedure as in the case of the balanced voltages as discussed before. The simulated results for unity power factor operation, i.e., $\gamma = 0$ are shown in the Figure 7.22. It is seen from Figure 7.22 (b) that the compensated source currents obtained are balanced sinusoids. Moreover, since the source voltage does not have any phase unbalance, the source currents are unity power factor. Despite the source currents being balanced, the real and reactive power supplied by the source oscillates at 100 Hz, since the source voltages are unbalanced.

△△△

7.6.3 Compensating to Equal Average Power

In this case, the source voltages are assumed to be same as those given in (7.52) and the source currents are given by

$$\begin{aligned} i_{sa} &= I_{sam} \sin \omega t \\ i_{sb} &= I_{sbm} \sin(\omega t - 2\pi/3 + \theta_b) \\ i_{sc} &= I_{scm} \sin(\omega t + 2\pi/3 + \theta_c) \end{aligned} \quad (7.57)$$

7. Load Compensation using DSTATCOM

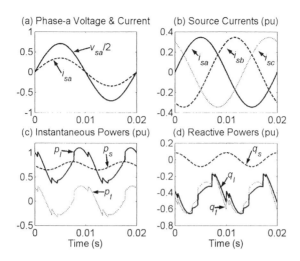

Figure 7.22. System response when compensated to equal current

Combining (7.52) with (7.57), the phase-a power is given by

$$p_{sa} = \frac{V_{sam} I_{sam}}{2}(1 - \cos 2\omega t)$$

Similar expressions can also be derived for the other two phases. The first term in the above expression is the average component, and this must be equal for all three phases, i.e.,

$$\frac{V_{sam} I_{sam}}{2} = \frac{V_{sbm} I_{sbm}}{2} = \frac{V_{scm} I_{scm}}{2} = \frac{P_{lav}}{3} \qquad (7.58)$$

The phase-a source current is then given by

$$i_{sa} = I_{sam} \sin \omega t = \frac{2 P_{lav}}{3} \times \frac{v_{sa}}{V_{sam}^2}$$

Similar expressions can also be derived for the other two phases. Therefore the reference compensator currents are

$$\left.\begin{array}{l}i_{fa} = i_{la} - i_{sa} = i_{la} - \dfrac{2v_{sa}}{3V_{sam}^2}P_{lav} \\[6pt] i_{fb} = i_{lb} - i_{sb} = i_{lb} - \dfrac{2v_{sb}}{3V_{sbm}^2}P_{lav} \\[6pt] i_{fc} = i_{lc} - i_{sc} = i_{lc} - \dfrac{2v_{sc}}{3V_{scm}^2}P_{lav}\end{array}\right\} \quad (7.59)$$

It is to be noted that only when $\theta_b = \theta_c = 0$, i.e., the source voltages have unbalance in magnitude only, the ac power from the source is zero. Let us consider the following example.

Example 7.20: For the same system as discussed the results of the compensation (7.59) are shown in Figure 7.23. Even though the source currents are in phase with the source voltages, their magnitudes are not proportional with that of the source voltages. Furthermore, since the source voltages have magnitude unbalance only, the power drawn from the source however contains only its dc component. The compensator however supplies a zero mean oscillating power.

△△△

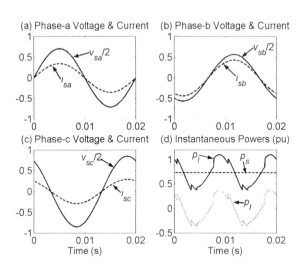

Figure 7.23. System response when compensated to equal power

Other strategies that are similar to the equal current strategy have been reported in the literature. These are termed as sinusoidal current source strategy in [18] and equal current criteria in [19]. However in [18], to find the fictitious balanced set of voltages for control algorithm, the complex α–β–0 transformation and elaborate computations are used. In [19] the synchronous detection method is used but it is limited to the case of unity power factor and magnitude unbalance only. The algorithm given in Section 7.6.2 using generalized theory is simple to use and considers unbalances in magnitudes and/or phase angles. The algorithm also provides the facility of setting the desired power factor angle, ϕ.

7.7 Conclusions

In this chapter we have presented various methods of load compensation in a power distribution system. Numerical examples presented assume that the compensators are realized by ideal current sources. We shall discuss structures of practical compensators (DSTATCOMs) in the next chapter where the reference currents generated using use some of the above algorithms will then tracked using voltage source inverters.

In our discussions in this chapter we have assumed that the sources are stiff, whether they are balanced or unbalanced. This, however, may not be a valid assumption. Since the source voltage may not be available for measurements, we may have to depend on the local measurements that are distorted due to the presence of the nonlinear load to begin with. We must therefore eliminate the effect of the load harmonics from the bus voltage before we can use this voltage for compensation. This issue and the other practical issues will be discussed in the next chapter.

7.8 References

[1] D. A. Torrey and A. M. A. M. Al-Zamel, "Single phase active power filters for multiple nonlinear loads," *IEEE Trans. Power Electronics*, Vol. 10, No. 3, pp. 263-272, 1995.

[2] J. C. Wu and H. L. Jou, "Simplified control method for single phase active power filter," *Proc. IEE*, Pt. B, Vol. 143, No. 2, pp. 219-224, 1996.

[3] A. Ghosh and A. Joshi, "A new method for load balancing and power factor correction using instantaneous symmetrical components," *PE Letters in IEEE Power Engg. Rev*, vol. 18, no. 9, pp. 60-62, 1998.

[4] A. Ghosh and A. Joshi, "A new approach to load balancing and power factor correction in power distribution system," *IEEE Trans. on Power Delivery*, Vol. 15, No. 1, pp. 417-422, 2000.

[5] A. Ghosh and A. Joshi, "The use of instantaneous symmetrical components for balancing a delta connected load and power factor correction," *Electric Power Systems Research*, Vol. 54, pp. 67-74, 2000.

[6] H. Akagi, Y. Kanazawa, K. Fujita and A. Nabae, "Generalized theory of the instantaneous reactive power and its application," *Electrical Engineering in Japan*, Vol. 103, No. 4, pp. 58-65, 1983.

[7] H. Akagi, Y. Kanazawa and A. Nabae, "Instantaneous reactive power compensators comprising switching devices without energy storage components," *IEEE Trans. Industry Applications*, Vol. IA-20, No. 3, pp. 625-630, 1984.

[8] H. Akagi, A. Nabae and S. Atoh, "Control strategy of active power filters using multiple voltage-source PWM converters," *IEEE Trans. Industry Applications*, Vol. IA-22, No. 3, pp. 460-465, 1986.

[9] T. Furuhashi, S. Okuma, Y. Uchikawa, "A study on the Theory of Instantaneous Reactive Power," *IEEE Trans. Industrial Electronics*, Vol. 37, No. 1, pp. 86-90, Feb. 1990.

[10] J. L. Willems, "A new interpretation of the Akagi-Nabae power components for nonsinusoidal three phase situations," *IEEE Trans. on Instrumentation & Measurements*, Vol. 41, No. 4, pp. 523-527, 1992.

[11] E. H. Watanabe, R. M. Stephan and M. Aredes, "New concepts of instantaneous active and reactive powers in electrical systems with generic loads," *IEEE Trans. Power Delivery*, Vol. 8, No. 2, pp. 697-703, 1993.

[12] A. Ferrero and G. Supeti-Furga, "A new approach to the definition of power components in three-phase systems under nonsinusoidal conditions," *IEEE Trans. Instrumentation & Measurements*, Vol. 40, No. 3, pp. 568-577, 1991.

[13] M. K. Mishra, A. Joshi and A. Ghosh, "A new algorithm for active shunt filters using instantaneous reactive power theory," *IEEE PE Letters in IEEE Power Engg. Review*, Vol. 20, No. 12, pp. 56-58, 2000.

[14] M. K. Mishra, A. Joshi and A. Ghosh, "Unified shunt compensator algorithm based on generalised instantaneous reactive power theory," *Proc. IEE – Generation, Transmission & Distribution*, Vol. 148, No. 6, pp. 583-589, 2001.

[15] F. Z. Peng and J. S. Lai, "Generalized instantaneous reactive power theory for three-phase power systems," *IEEE Trans. Instrumentation & Measurements*, Vol. 45, No. 1, pp. 293-297, 1996.

[16] F. Z. Peng, G. W. Ott and D. J. Adams, "Harmonic and reactive power compensation based on the generalized instantaneous reactive power theory for three-phase four-wire systems," *IEEE Trans. Power Electronics*, Vol. 13, No. 6, pp. 1174-1181, 1998.

[17] M. K. Mishra, A. Joshi and A. Ghosh, "A new compensation algorithm for balanced and unbalanced distribution systems using generalized instantaneous reactive power theory," *Electric Power systems Research*, Vol. 60, pp. 29-37, 2001.

[18] M. Ardes, J. Hafner and K. Heumann, "Three-phase four-wire shunt active filter control strategies," *IEEE Trans. Power Electronics*, Vol. 12, No. 2, pp. 311-318, 1997.

[19] C. L. Chen C. E. Lin and C. L. Huang, "Reactive and harmonic current compensation for unbalanced three-phase systems using the synchronous detection method," *Electric Power Systems Research*, Vol. 26, pp. 163-170, 1993.

Chapter 8

Realization and Control of DSTATCOM

In Chapter 7 we have restricted our discussion mainly on the generation of reference currents. We have assumed that the shunt compensator, which tracks the reference currents, is represented by three ideal current sources. In practice however these current sources are implemented using voltage source inverters. The inverter circuit along with interface transformers/inductors is called a distribution static compensator (DSTATCOM). In our discussions in Chapter 7 we have seen that a DSTATCOM may have to inject a set of three unbalanced currents that may also contain harmonics. Therefore, the VSI associated with a DSTATCOM must be able to inject currents in one phase independent of the other two phases. From this point of view the structure of a DSTATCOM attains significance. A DSTATCOM operating as a current source has been termed as DSTATCOM in current control mode in Chapter 4.

Furthermore in Chapter 7 we have assumed that the load is connected to a stiff voltage source. Therefore the shunt compensator can measure the PCC voltages and use them in the reference current generation algorithms without any problem as these voltages are pure sinusoids. This however may not be possible in actual systems where the loads are connected at the end of the feeder. The PCC voltage in this case will be unbalanced if the load is unbalanced. In addition, the PCC voltage will be distorted by both the harmonics generated by a nonlinearity in the load and by the switching frequency harmonics generated by the DSTATCOM. Furthermore there will be switching and resistive losses in the DSTATCOM circuit. These losses must be supplied by the source. We must therefore suitably modify the reference current generation algorithm to accommodate all these factors.

Finally, to provide a path for the harmonic current generated by the VSI realizing the DSTATCOM to flow, we must place additional filters in the circuit. The presence of the filer and feeder impedances will then increase the system order. As we have seen in Chapter 5 that a simple output feedback switching controller may result in an unstable closed-loop system when applied to higher order systems. We must therefore choose the DSTATCOM controller structure carefully such that the closed-loop control convergence is guaranteed.

The DSTATCOM is a shunt device. It should therefore be able to regulate the voltage of a bus to which it is connected. The operating principle of a DSTATCOM in this mode has been termed as the DSTATCOM in voltage control mode. We shall show in this chapter that even though the structure of DSTATCOM used in both current control and voltage control modes is the same, its operating principle is different. In the current control mode it is required to follow a set of reference currents while in the voltage control mode it is required to follow a set of reference voltages. We shall also discuss the reference voltage generation scheme and the control of DSTATCOM in the voltage control mode in this chapter.

8.1 DSTATCOM Structure

One form of DSTATCOM structure is shown in Figure 8.1. It contains three H-bridge VSIs that are connected to a common dc storage capacitor. In this figure each switch represents a power semiconductor device and an anti-parallel diode combination. Each VSI is connected to the network through a transformer. Six output terminals of the transformer are connected in star. These six terminals can also be connected in delta to compensate a Δ-connected load. In that case, each transformer is connected in parallel with the corresponding load [1]. The purpose of including the transformers is to provide isolation between the inverter legs. This prevents the dc storage capacitor from being shorted through switches in different inverters. The inductance L_f in this figure represents the leakage inductance of each transformer and additional external inductance, if any. The switching losses of an inverter and the copper loss of the connecting transformer are represented by a resistance R_f. The iron losses of the transformer are neglected. For star connected load, the neutral point of the three transformers is connected to the load neutral. The dotted line indicates the 4th wire and is connected to the system neutral N, if available.

A three-phase full-bridge inverter is not suitable for a DSTATCOM application. A well known constraint of such inverters is that the sum of current through its three legs must be zero. It will not thus be possible to compensate for the zero-sequence current that might be flowing in the load, nor can it eliminate any dc current from flowing into the source from the load. This will result in distortion in the source currents. The topology

8. Realization and Control of DSTATCOM

shown in Figure 8.1 allows three independent current injections as this contains three separate H-bridge inverters. This however has the drawback that it will fail if the load draws any dc current. The dc current will saturate the transformers causing heating and increased losses thereby reducing the life of the transformers. We however retain the option of not compensating the dc component.

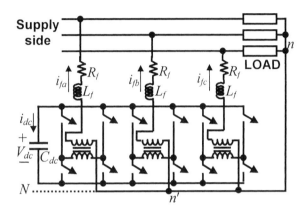

Figure 8.1. A typical compensator structure in which three separate VSIs are supplied from a common dc storage capacitor

An alternate topology that allows the injection of three independent currents including any dc current that the load may draw is proposed in [2]. This topology contains two dc storage capacitors as shown in Figure 8.2. In this circuit the junction (n') of the two capacitors is connected to the neutral of the load. This neutral clamped topology allows a path for the zero-sequence current and therefore the three injected currents can be independently controlled. Note that in this configuration there is no transformer and each leg of the VSI is connected to the point of common coupling through an interface reactor. The inductance L_f and the resistance R_f in Figure 8.2 represent this interface reactor. Another important component of the topology of Figure 8.2 is a chopper circuit that is represented by the switches S_{ch1} and S_{ch2}, the inductance L_p and the resistance R_p. The purpose of this circuit is to balance the voltages in the two capacitors.

Consider the circuit of Figure 8.3, which is a magnification of the chopper circuit of Figure 8.2. In this the anti-parallel diodes are also indicated for each chopper switch. Let us define the voltages across C_{dc1} and C_{dc2} by V_{dc1} and V_{dc2} respectively. Normally the switches S_{ch1} and S_{ch2} are kept open and the voltages V_{dc1} and V_{dc2} are equal. Now suppose the voltage V_{dc1} drops and V_{dc2} rises. The switch S_{ch2} is then closed such that a current is built up in the inductor L_p. Once the current reaches a certain level, the switch S_{ch2} is opened. The inductor current then discharges through the diode D_{ch1} to bring up the voltage V_{dc1} to the desired level. Similarly, the charge can be transferred from the capacitor C_{dc1} to the capacitor C_{dc2} by closing the switch S_{ch1} to build current in L_p and then charging C_{dc2} through the diode D_{ch2} by opening the switch S_{ch1}. The feedback control of this chopper circuit is essential for the success of the scheme. Various chopper control schemes are discussed in [2-5] where many simulation and experimental results are given.

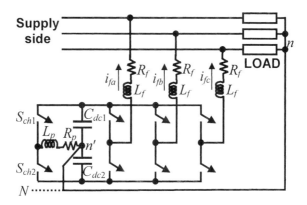

Figure 8.2. A compensator structure in which a neutral-clamped three-phase VSI is used

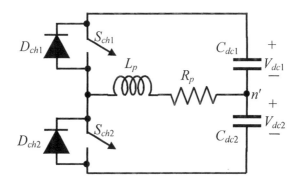

Figure 8.3. The chopper circuit of the neutral clamped DSTATCOM of Figure 8.2

An alternate topology is shown in Figure 8.4 in which a VSI with four legs are used. This requires only one dc storage unit. Three of its legs are used for phase connection while the fourth leg is connected to the load neutral and the supply neutral, if available, through a reactance [6-8]. The reference current for the fourth leg is the negative sum of three phase load currents. This nullifies the effect of dc component of load current. To maintain the adequate charge on dc-side capacitor a PI regulator is used to control the flow of real power from ac side towards dc side of the converter.

When the compensator is working, zero sequence current is routed to path $n\rightarrow n'$ containing switching frequency harmonics. Using fourth leg of inverter, the negative of zero sequence current $-i_0$ is tracked. Certainly it needs a higher bandwidth VSI to track negative of neutral current ($-i_0$) as i_0 contains harmonics due to non-linear loads. This increases the switching losses. If this current is not tracked properly, it will leave high switching frequency current components in the $N\rightarrow n$ path, which is not desirable. The advantage of the topology shown in Figure 8.4 over the one shown in Figure 8.2 is that it requires one less capacitor. However, the topology of Figure 8.2 is superior as it intrinsically incorporates an LC filter for the neutral path.

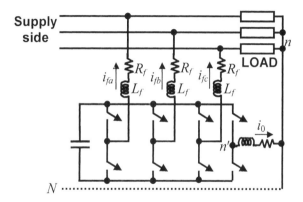

Figure 8.4. A compensator structure which uses a four-leg VSI

In the simplest form of current tracking, each leg of the VSIs of Figures 8.1, 8.2 and 8.4 are operated in the hysteresis-band current control mode to track the reference currents generated. This method is simple and will produce desired results when the source is stiff, i.e., it is assumed that there is no impedance between the source and the point of common coupling. There can be various other tracking methods as explained in Chapter 5. We shall also use one such method later in the chapter when we discuss current tracking with non-stiff sources.

8.2 Control of DSTATCOM Connected to a Stiff Source

In the discussions to follow, we shall assume the topology of Figure 8.1 unless stated otherwise. An important aspect that must be considered during any practical implementation of DSTATCOM is the losses in the inverter and connecting transformer or interface reactor, modeled by R_f in Figure 8.1. The dc voltage stored (V_{dc}) in the storage capacitor C_{dc} must be maintained in order for perfect current tracking. This however will not be possible unless the loss due to R_f is somehow replenished by drawing additional power from the source. In order to accommodate this, the real power supplied by the source is not equal to the average load power as given by (7.27) but the total of average load power and power lost in the DSTATCOM circuit, i.e.,

$$v_{sa}i_{sa} + v_{sb}i_{sb} + v_{sc}i_{sc} = p_{lav} + p_{loss} \tag{8.1}$$

where p_{loss} is the loss due to R_f. We can then modify (7.29) as

$$\begin{aligned} i_{fa}^* &= i_{la} - \frac{v_{sa} + (v_{sb} - v_{sc})\beta}{v_{sa}^2 + v_{sb}^2 + v_{sc}^2}(p_{lav} + p_{loss}) \\ i_{fb}^* &= i_{lb} - \frac{v_{sb} + (v_{sc} - v_{sa})\beta}{v_{sa}^2 + v_{sb}^2 + v_{sc}^2}(p_{lav} + p_{loss}) \\ i_{fc}^* &= i_{lc} - \frac{v_{sc} + (v_{sa} - v_{sb})\beta}{v_{sa}^2 + v_{sb}^2 + v_{sc}^2}(p_{lav} + p_{loss}) \end{aligned} \tag{8.2}$$

We now have to generate p_{loss} through a suitable feedback control such that the dc voltage (V_{dc}) across the storage capacitor C_{dc} is maintained.

The feedback should be able to correct the deviation of the average value of V_{dc} from a reference value V_{ref}. The average value can be obtained at the end of each cycle or even running average can be used as feedback signals. Alternatively, we can also use end of a cycle value of the dc capacitor voltage. Refer to Figure 8.1 in which i_{dc} is the current through the capacitor C_{dc}. This current charges or discharges the capacitor C_{dc}. Thus the average capacitor voltage is held constant when the average value of the dc capacitor current i_{dc} over a cycle is zero. Now we know that

$$V_{dc} = \frac{1}{C_{dc}}\int i_{dc}dt \tag{8.3}$$

Thus the deviation of V_{dc} from the reference value V_{ref} at the end of each fundamental frequency cycle gives a good indication of the deviation of the average value of capacitor current i_{dc} from zero. From this consideration we can also correct for the deviation of the end of cycle value of V_{dc} from V_{ref}.

8. Realization and Control of DSTATCOM

Let us define the following error

$$e = V_{ref} - V_{dc}^{cyl} \qquad (8.4)$$

where V_{dc}^{cyl} is either the average value of the dc capacitor voltage or the value of the capacitor voltage at the end of a cycle. In the simplest form, we can then use a proportional-plus-integral (PI) controller to correct for any discharge in the capacitor voltage. The controller is then given by

$$P_{loss} = K_p e + K_I \int e\, dt \qquad (8.5)$$

The value of p_{loss} is then substituted in (8.2) to draw an additional amount of real power from the source to maintain the dc capacitor voltage.

Example 8.1: Let us consider a three-phase, four-wire system in which a set of three balanced voltages is supplying a load. The peak of these voltages is √2 per unit. These voltage are supplying three unbalanced RL loads given in per unit as

$$Z_{la} = 2.0 + j1.5,\ Z_{lb} = 2.55 + j1.25\ \text{and}\ Z_{lc} = 1.0 + j2.3$$

In addition to this, the load contains a three-phase rectifier that is drawing a peak current of 0.15 per unit. The load currents are shown in Figure 8.5 (a). We assume a DSTATCOM topology of Figure 8.1. The interface circuit parameters in per unit are

$$R_f = 0.625\ \text{and}\ X_f = \omega L_f = 1.25$$

The dc capacitor value and its reference voltage are given by

$$X_{dc} = \frac{1}{\omega C_{dc}} = 0.1 \text{ per unit and } V_{ref} = 2.0 \text{ per unit}$$

It is assumed that the turns ratio of the interface transformers is 1:1. The DSTATCOM is controlled by a PI controller of the form (8.5) with the following parameters

$$K_P = 2.0\ \text{and}\ K_I = 10.0$$

In this loop V_{dc}^{cyl} is taken as the average value of the dc capacitor voltage over one cycle. Once the reference currents are generated, they are tracked in a hysteresis band current control scheme. The hysteresis band is chosen as 0.01 per unit. The switching frequency for this choice of hysteresis band is around 4.5 kHz.

The simulation results are shown in Figure 8.5. It is assumed that the DSTATCOM dc capacitor is precharged to a dc voltage of 2.0 per unit and the DSTATCOM is connected to the system after the first half cycle once the first average value of the load power (p_{lav}) is obtained. The desired power factor angle is chosen to be 0°. The PI controller output (p_{loss}) and the voltage across the dc capacitor are shown in Figure 8.5 (b) and (c) respectively. It can be seen that p_{loss} settles within about 0.12 s (6 cycles). This implies that the source currents reach their steady state values within about 6 cycles. It is to be noted that the tracking performance depends on forcing a current through the inductor L_f. Therefore the tracking performance will be good, as long as the capacitor voltage is higher than the peak of the ac voltage at PCC. Since the excursion in the capacitor voltage is less than 0.1 per unit from the reference value, the tracking performance will not be affected by the slow convergence of the capacitor voltage. The compensated source currents are shown in Figure 8.5 (d) along with the source voltage of phase-a for one cycle after the controller convergence. It can be seen that they are balanced and have unity power factor.

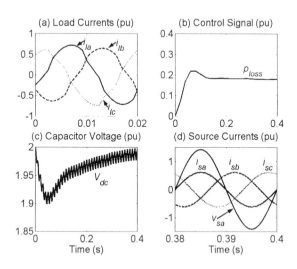

Figure 8.5. System response with DSTATCOM when the load is connected to a stiff source

To observe the transient response of the circuit, the system loads are suddenly changed. It is assumed that the system has been started from a fully

8. Realization and Control of DSTATCOM

uncharged state and is operating for 0.4 s when the load change occurs. The system load and other parameters considered are the same as above and the system response for the first 0.4 s is the same as that shown in Figure 8.5. At the end of 0.4 s, the RL load is made balanced with the load impedance given by

$$Z_{la} = Z_{lb} = Z_{lc} = 2.5 + j2.25 \text{ per unit}$$

The nonlinear component of the load remains unchanged.

The system response is shown in Figure 8.6. In this we have only shown the PI controller output and the dc capacitor voltage. It can be seen from Figure 8.6 (a) that the p_{loss} term again settles within about 6 cycles. Also note from Figure 8.6 (b) that the capacitor voltage takes about 0.2 s to come back to its steady state following the disturbance. Again the excursion of this voltage from its reference value during this transient is very small and hence it will have not much bearing on the tracking performance.

ΔΔΔ

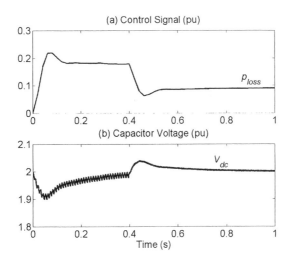

Figure 8.6. Transient response with DSTATCOM when the load is connected to a stiff source

In the above example the average power demand of the load reduces after it has been made balanced. Then following observations can be made from Figure 8.6.

− As the load changes at the end of 20 cycles, the average power demand of the load reduces. Therefore the excess power supplied by the source during the transient enters the DSTATCOM making the dc capacitor

voltage to rise. This rise in the capacitor voltage is arrested by the control action which brings the voltage to its nominal value.
- The p_{loss} term settles to a lower value after the load change. Since the average power demand of the load reduces, the source current magnitude also reduces. This will lead to reductions in the magnitudes of the fundamental component of the currents injected by the DSTATCOM. Obviously, this will lead to a lower value of power loss in the DSTATCOM circuit.

The above example clearly demonstrates one practical aspect of the shunt compensation. However, in our discussion we have assumed that the supply voltage is stiff, i.e., connected at the point of common coupling without any source impedance. In the discussion given below we develop it further to handle the case when the compensator is connected to a source that is not stiff.

8.3 DSTATCOM Connected to Weak Supply point

Consider the system shown in Figure 8.7. In this an unbalanced and nonlinear load is supplied by a balanced voltage source (v_s) through a feeder. The feeder has a resistance R and inductance L. Let us denote the PCC voltage as the terminal voltage (v_t). Then this voltage is applied across the load as well as across the DSTATCOM. The load is compensated by a DSTATCOM, the reference signal of which is generated using any of the methods given in Chapter 7. However, all these methods will suffer from the problem of harmonic contamination due the VSI of DSTATCOM. Let us illustrate this with the following example.

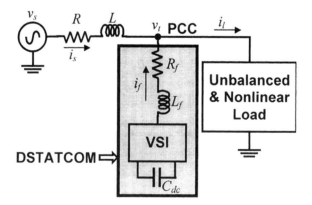

Figure 8.7. Single-line diagram of shunt compensation of a load supplied through a feeder

8. Realization and Control of DSTATCOM

Example 8.2: Consider the system of Figure 8.7. Assume that it is supplied by a balanced voltage source with a peak of $\sqrt{2}$ per unit. It supplies an unbalanced RL load given in per unit by

$$Z_{la} = 2.0 + j1.5, \ Z_{lb} = 2.55 + j1.25 \text{ and } Z_{lc} = 1.0 + j2.3$$

The feeder impedance is given in per unit by

$$R = 0.05 \text{ and } X = \omega L = 0.3$$

It is assumed that the VSIs are lossless and supplied by a fixed dc voltage such as a battery. The DSTATCOM parameters are given in per unit by

$$V_{dc} = 2.0, \ R_f = 0 \text{ and } X_f = \omega L_f = 1.0$$

Since the DSTATCOM is supplied by a fixed dc source, we do not require the capacitor voltage control loop discussed in Section 8.2. We then use the algorithm of (7.29), i.e., (8.2) with p_{loss} taken as zero. It is assumed that β is zero, i.e., the desired source power factor is unity.

It is assumed that the shunt compensator takes the measurements of local quantities only. We connect the compensator at the end of first half cycle after the average power is available. The results are shown in Figure 8.8. The load currents are shown in Figure 8.8 (a). Figure 8.8 (b) depicts the source currents. It can be seen that the source currents become distorted as soon as the compensator is pressed into action. Figure 8.8 (c) depicts phase-a voltage at the point of common coupling or the terminal voltage (v_t). This voltage increases in magnitude as soon as the compensator is connected. As a result, the magnitude of the load currents increases. This is evident from Figure 8.8 (a). The phase-a of the injected compensator current (i_f) is shown in Figure 8.8 (d). The reason for the distortion in the source current is simple. The injected current contains high frequency harmonics. These harmonic currents, when injected into the feeder, corrupt the terminal voltage. Now since the terminal voltage is not sinusoidal, the reference current generation scheme based on this voltage will not function properly. As a result of this, the injected compensator currents will be erroneous thereby injecting more harmonics in the feeder. Furthermore, we have chosen a hysteresis band of 0.01 per unit and this has resulted in a switching frequency of around 15 kHz. Obviously this is very high for power switches and will result in excessive switching losses.

△△△

The above example clearly demonstrates that we cannot use the algorithm given in (7.29) without suitable modification due to the distortion of the terminal voltage. This distortion is caused by the VSI realizing the DSTATCOM. Since the terminal voltage is the locally measured variable, it should be used in the algorithm of (7.29). One way of alleviating this problem of a distorted reference is to use the fundamental positive sequence of the terminal voltage. This can be accomplished using the fundamental positive sequence extraction algorithm given in Section 3.3.1.

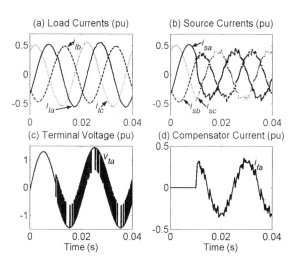

Figure 8.8. Imperfect compensation of a system with a non-stiff source

To achieve this we measure the three-phase terminal voltage v_{ta}, v_{tb} and v_{tc} through voltage transducers. These are then used in (3.56) to obtain the phasor positive sequence of the fundamental component of the terminal voltage V_{taf}, V_{tbf} and V_{tcf}. Note that for this, the reference time is taken as the zero-crossing of the phase-a of the terminal voltage. This reference can be obtained through a phase-locked loop (PLL). Once these quantities are obtained we can use the instantaneous positive sequence fundamental component of the terminal voltage v_{taf}, v_{tbf} and v_{tcf} in the same way as explained in Example 3.7. We shall now use these voltage terms instead of the actual measured voltages in the algorithm given in (7.29).

Example 8.3: Let us consider the same system as given in Example 8.2 and use the fundamental extraction algorithm as discussed above. The compensator reference currents are then extracted based on this fundamental voltage. The results are shown in Figure 8.9. The source currents are shown in Figure 8.9 (a). It can be seen that they are balanced and relatively free of low frequency harmonics. The phase-a of the actual terminal voltage is shown in Figure 8.9 (b), while the extracted positive sequence voltages of all the phases are shown in Figure 8.9 (c). It is evident that even though the source currents are almost free of harmonics, the terminal voltage still contains a very large amount of harmonics. This of course will cause distortion in any other load that may be connected at the terminal bus. The phase-a of the compensator current is shown in Figure 8.9 (d). The hysteresis band in this case is also chosen as 0.01 per unit. However the switching frequency in this case is below 5 kHz as opposed to around 15 kHz in the previous case.

ΔΔΔ

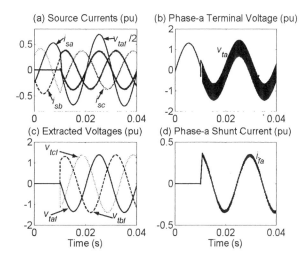

Figure 8.9. Acceptable compensation but unacceptable terminal voltage with a non-stiff source

8.3.1 DSTATCOM Structure for Weak Supply Point Connection

Example 8.3 clearly demonstrates that through the extraction of the fundamental component of the terminal voltage we can eliminate any distortion in the source current. However, the switch-frequency component in the terminal voltage still remains. We thus must provide a path for the switching harmonic current to flow thorough a filter. To facilitate this, we modify the circuit of Figure 8.7 by placing a filter capacitor (C_f) in parallel with the DSTATCOM at the point of common coupling. This is shown in Figure 8.10. With respect to this figure we define the injected shunt current as

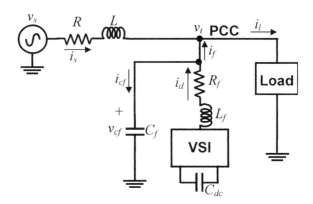

Figure 8.10. Single-line diagram of shunt compensation in the presence of a filter capacitor

$$i_f = i_d - i_{cf} \tag{8.6}$$

This choice enables us to use the reference current generation relations used before since the KCL at PCC, i.e., $i_l = i_f + i_s$ remains undisturbed. Further note that the terminal voltage in this case is the voltage across the capacitor C_f, i.e., $v_t = v_{cf}$.

The single phase equivalent circuit of the compensated system of Figure 8.10 is shown in Figure 8.11 (a) while its Thevenin equivalent looking left of the points $\alpha\beta$ is shown in Figure 8.11 (b). In this $Z = R + j\omega L$ and $X_{cf} = 1/\omega C_f$. The Thevenin equivalent circuit parameters are as follows

$$v_{th} = \frac{-jX_{cf}}{Z - jX_{cf}} v_s \text{ and } Z_{th} = \frac{-jZX_{cf}}{Z - jX_{cf}} \tag{8.7}$$

Assuming the feeder to be lossless, i.e. $R = 0$, the Thevenin voltage and feeder impedance are given as

$$v_{th} = -\frac{X_{cf}}{\omega L - X_{cf}} v_s \text{ and } Z_{th} = -j\frac{\omega L X_{cf}}{\omega L - X_{cf}} \tag{8.8}$$

It must be ensured that the feeder inductance L and the capacitor C_f do not resonate at the fundamental frequency of 50 or 60 Hz. Let a capacitor C_{fn} be of a value such that it would resonate with the feeder reactance at a frequency ω_n. We can then write

8. Realization and Control of DSTATCOM

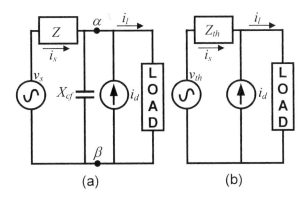

Figure 8.11. (a) Single-phase equivalent of the circuit of Figure 8.10 and (b) its Thevenin equivalent

$$C_{fn} = \frac{1}{\omega_n^2 L} \tag{8.9}$$

Let us assume that this value is equal to C_{f0} when ω_n is equal to the fundamental frequency (ω). The filter capacitor C_f should never be chosen near C_{f0}. We then have to choose either $C_f \ll C_{f0}$ or $C_f \gg C_{f0}$. However if C_f is very large, the impedance between PCC and ground becomes very small resulting in excessive currents through filter capacitors. Therefore the choice of $C_f \gg C_{f0}$ is invalid. We must therefore restrict C_f to be much smaller than C_{f0}. For example the feeder inductance (L) in Examples 8.2 and 8.3 is given by 9.55×10^{-4}. Therefore from (8.9) C_{f0} is 0.0106. Again note that the load may be supplied through a feeder that has load buses upstream. In such a case, the feeder impedance may be the Thevenin equivalent looking at the distribution system from the PCC. Therefore we can make a guess about the feeder impedance value which may change anytime. From this consideration we must choose the value of C_f that is significantly less than 0.0106. Let us consider the following example.

Example 8.4: Let us consider the same feeder and load parameters as given in Example 8.2. We however use different leakage impedance for the interface transformers of the DSTATCOM. We maintain the assumption that the DSTATCOM is lossless, i.e., $R_f = 0$, and the VSIs are supplied by a fixed dc source. The filter capacitor and transformer leakage inductance are given in per unit as

$$X_{cf} = \frac{1}{\omega C_{filt}} = 7.02 \text{ and } X_f = \omega L_f = 0.2$$

Once the fundamental positive sequence terminal voltages are obtained using the extraction algorithm, the reference compensator currents are generated using the algorithm given in (7.29). They are then tracked using a hysteresis band current control scheme with a band of 0.01 in which the compensator is turned on at 0.01 s after the average power and the fundamental positive component of the terminal voltage are obtained. Figure 8.12 shows phase-a source current for this case. It can be seen that the source current diverges as soon as the compensator is pressed into action. As has been already explained in Chapter 5 (see Section 5.7) that a hysteresis controller can be viewed as a high gain proportional controller. Also the hysteresis band adds a small phase lag in the tracking operation. The combined effect of the proportional control with a phase lag is detrimental on the second order system of Figure 8.10 especially when the variable to be tracked (i.e., i_f) is not directly controlled by the inverter. It is thus imperative that this system needs a stabilizing controller.

ΔΔΔ

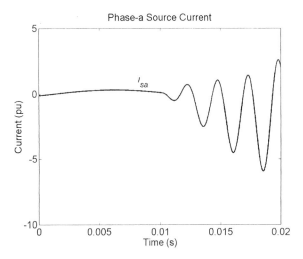

Figure 8.12. Effect of hysteresis tracking of the reference compensator current

8.3.2   Switching Control of DSTATCOM

The equivalent circuit of the compensated system is shown in Figure 8.13. In this figure, the load is represented by an RL load. Note that the inverter control u can have values ± 1. However to derive a control law, we

8. Realization and Control of DSTATCOM

assume for the time being that u is equal to a continuous signal u_c. Comparing with Figure 8.10 we can summarize the current relations as

$$i_s = i_1, \quad i_l = i_3$$
$$i_{cf} = i_1 - i_2 - i_3 \qquad (8.10)$$
$$i_f = i_3 - i_1$$

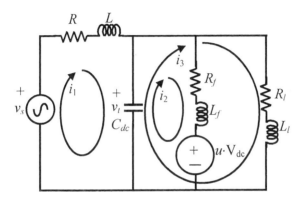

Figure 8.13. Equivalent circuit of a DSTATCOM compensated distribution system

With respect to this figure, let us define the following state vector

$$x^T = \begin{bmatrix} i_1 & i_2 & i_3 & v_t \end{bmatrix}$$

The state-space equation of the circuit then can be written as

$$\dot{x} = \begin{bmatrix} -R/L & 0 & 0 & -1/L \\ 0 & -R_f/L_f & 0 & -1/L_f \\ 0 & 0 & -R_l/L_l & -1/L_l \\ 1/C_f & -1/C_f & -1/C_f & 0 \end{bmatrix} x + \begin{bmatrix} 1/L \\ 0 \\ 0 \\ 0 \end{bmatrix} v_s + \begin{bmatrix} 0 \\ -V_{dc}/L_f \\ 0 \\ 0 \end{bmatrix} u_c$$

$$= Ax + B_1 v_s + B_2 u_c$$

(8.11)

The above state equation contains the feeder and load impedances. While the former is measurable, the latter may change at any time. Moreover, any feedback controller must rely only on the locally measured variables. From Figure 8.10 it can be seen that the local variables for the DSTATCOM are terminal voltage, source current, filter current and the filter capacitor current. Based on the above observation and from (8.10), we make the following state transformation

$$z = \begin{bmatrix} i_f \\ i_{cf} \\ v_t \\ i_l \end{bmatrix} = \begin{bmatrix} -1 & 0 & 1 & 0 \\ 1 & -1 & -1 & 0 \\ 0 & 0 & 0 & 1 \\ 0 & 0 & 1 & 0 \end{bmatrix} x = Px \qquad (8.12)$$

The state equation (8.11) then can be transformed into

$$\dot{z} = PAP^{-1}z + PB_1 v_s + PB_2 u_c = \Lambda z + \Gamma_1 v_s + \Gamma_2 u_c \qquad (8.13)$$

Assuming we have full control over u_c, we can design a state feedback controller of the form

$$u_c = -K(z - z_{ref}) \qquad (8.14)$$

where K is the feedback gain matrix and z_{ref} is the desired state vector. Once a suitable gain matrix is selected based on a control law, the switching control is obtained from

$$u = -hys(u_c) \qquad (8.15)$$

where the *hys* function is defined by

$$\begin{aligned} &\text{if } w > \lim \text{ then } hys(w) = 1 \\ &\text{else if } w < -\lim \text{ then } hys(w) = -1 \end{aligned} \qquad (8.16)$$

The selection of lim determines the switching frequency while tracking the reference. In this control law the switching decision is based on a linear combination of multiple states. Hence we call this control a *switching band tracking control*. Let us demonstrate its working principle based on the following examples.

Example 8.5: Let us consider the same system given in Example 8.4. In addition to the load given in Example 8.1, we also add a rectifier load that is drawing a peak current of 0.35 per unit. We now use the pole shift controller discussed in Section 5.7.6 of Chapter 5 to obtain the control signal u_c. To implement this controller we first choose a suitable sampling time and then determine the discrete-time equivalent of the system of (8.13) using this sampling time. We then obtain the gain matrix K by radially shifting the open-loop poles by the pole shift factor λ.

Note that in a three-phase unbalanced system the system equation (8.11) is different for different phases. However to avoid finding three different gain matrices, we choose phase-a for the computation of the gain matrix. We now choose a pole shift factor (λ) of 0.98. This gives the following feedback gain matrix

$$K = [15.359 \quad 3.609 \quad 3.312 \quad -55.355]$$

Using this gain matrix all the eigenvalues of the closed-loop system are found to be within the circle $|z| = 0.9797$. Now the load can change any time. Therefore the values of the load parameters used in (8.11) is just a guess. Also note that we have added a nonlinear load with the RL load. Therefore to avoid the complexity of forming a reference for the load current, the gain matrix is restricted to

$$K = [15.359 \quad 3.609 \quad 3.312 \quad 0]$$

This reduction in state feedback results in a minimal shift in the closed-loop eigenvalues as all the closed-loop eigenvalues are inside the circle $|z| = 0.9951$. Thus stability is not at all endangered by the reduced feedback, but the computation is reduced considerably. Once u_c is obtained using the pole shift design, the switching logic u is obtained from (8.15) and (8.16). The value of lim chosen is 0.45.

The response is shown in Figure 8.14. From this it can be seen that both load currents and the terminal voltages become balanced and free of harmonics once the compensator action settles. Figure 8.14 (d) shows the three instantaneous powers. It can be seen that the power drawn by the load changes dramatically with the introduction of the compensator as it balances the terminal voltage as well. As expected, the source power becomes constant and the compensator power becomes zero-mean. The switching frequency with this value of pole shift factor is found to be 3.75 kHz.

Note that the system is very sensitive to value of the pole shift factor chosen. In fact it was observed that the closed-loop system becomes unstable when this value is chosen to be 0.97. On the other hand the switching frequency of the inverter reduces to 2.1 kHz when a pole shift factor 0.99 is chosen. However as a result of this, the tracking becomes inferior. The terminal voltages and instantaneous powers are shown in Figure 8.15. A high frequency ripple visible in the voltage and power waveforms.

ΔΔΔ

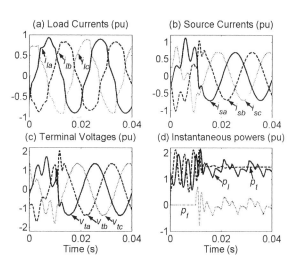

Figure 8.14. System response with a pole shift controller with $\lambda = 0.98$

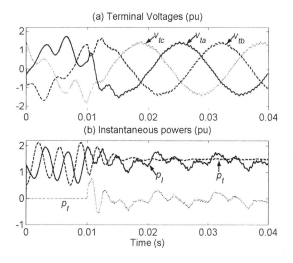

Figure 8.15. System response with a pole shift controller with $\lambda = 0.99$

The above example shows that we are quite restricted in our application of the pole shift controller. It is thus imperative that we must design a more robust controller. Below we present a LQR design example. It is to be noted

Example 8.6: Let us consider the same system as given in Example 8.5. We now design an LQR to obtain u_c with the following parameters

$$Q = \begin{bmatrix} 20 & & & \\ & 1 & & \\ & & 5 & \\ & & & 0 \end{bmatrix}, \quad r = 0.1$$

The choice enables us to put maximum weighting (20) on the shunt injected current i_f. The capacitor voltage has been given a weighting of 5 and the current through the filter capacitor has been given the weighting of 1. The load current has not been weighted at all. With these parameters the feedback gain matrix is found to be

$$K = \begin{bmatrix} 13.659 & 6.252 & 10.356 & -2.732 \end{bmatrix}$$

The eigenvalues of the system for this gain matrix are to the left of the line $\text{Re}(s) = -356$ on the s-plane. If we now discard the load current for a reduced feedback, the eigenvalues are found to be on the left of the line $\text{Re}(s) = -344$. Therefore the stability of the system is guaranteed with this reduced feedback. The system results are shown in Figure 8.16 in which only the source currents and the terminal voltages are shown. It can be seen that they become balanced sinusoids after the compensator is pressed into action at 0.01 s. The switching frequency is found to be around 5.8 kHz.

△△△

It is to be noted that in last two examples, the state feedback needs the reference waveforms of the compensator current, the terminal voltage and the current through the filter capacitor (i_{cf}). Of these, the first one is generated using the algorithm of (7.29) and the reference terminal voltage is taken to be the fundamental positive sequence of the terminal voltage. The rational behind this choice lies in the fact that if the source currents are balanced then voltage at the PCC terminal will also be balanced provided the source is balanced [9]. Once the terminal voltage reference is obtained, the reference for the current through the filter capacitor is easily obtained by taking the derivative of the terminal voltage reference. The issues involved in the tracking reference convergence are discussed in Section 5.7.4.

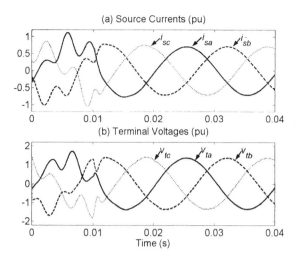

Figure 8.16. System response with linear quadratic controller (LQR)

8.3.3 DC Capacitor Control

Let us illustrates the performance of the DSTATCOM and filter capacitor combination when the DSTATCOM is supplied through a dc capacitor rather than a dc source using the following example.

Example 8.7: Consider the same system as given in Example 8.6. The DSTATCOM is not assumed to be lossless in this case. We thus have added the following per unit quantities with the configuration of Example 8.6 where the capacitor value is quoted at the system frequency of 50 Hz to maintain the per unit system

$$R_f = 0.1 \text{ and } X_{dc} = \frac{1}{\omega C_{dc}} = 0.106$$

The dc capacitor is controlled using (8.4) and (8.5) and the reference currents are generated using (8.2). The PI controller parameters chosen are

$$K_p = 3 \text{ and } K_I = 2$$

The system is started from rest, i.e., the compensator is connected at the end of first half cycle after the average load power is obtained. The reference quantities are tracked using the LQR state feedback controller with the same design parameters as given in Example 8.6. The capacitor is assumed to be precharged to 2.0 per unit and this value is also chosen as the set point for this voltage. The dc voltage controller is started from a zero initial condition. The results are shown in Figure 8.17 in which the control signal and the capacitor voltage are plotted. It can be seen that after the initial transient, the capacitor voltage returns to its desired value. It settles in less than 0.2 s.

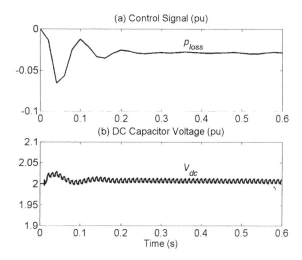

Figure 8.17. Response of the control signal and dc capacitor voltage when the control system is started from zero initial condition

To investigate the effect of load change on the compensator system, the RL component of the load is suddenly changed to (all in per unit)

$$Z_{la} = 1.2 + j2.5, \ Z_{lb} = 2.0 + j3.25 \text{ and } Z_{lc} = 10.0 + j2.0$$

when the system is operating in the steady state. The harmonic currents drawn by the load however remains unchanged. The system response is shown in Figure 8.18. In Figure 8.18 (a) the instantaneous powers are shown. Even if the load power settles within a couple of cycles, the source power, and, as a consequence, the compensator power, take longer to settle. The capacitor voltage settles within 0.15 s. This is shown in Figure 8.18 (b). It is to be noted that during this transient the capacitor voltage does not deviate much from its nominal value. Therefore the tracking remains perfect apart from the half cycle delay in formulation of correct references.

ΔΔΔ

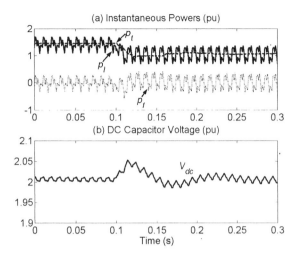

Figure 8.18. Changes in instantaneous powers and capacitor voltage with load change

8.4 DSTATCOM Current Control through Phasors

So far in our discussions we have assumed that the source is balanced. This however may not be always true. Consider, for example, the distribution system shown in Figure 8.19. In this a stiff source supplies four buses. Of the four buses, only the stiff source supplies Bus-1 and all the other buses supplied by the feeder. Assume that we have to install a shunt compensator in Bus-4. Then the total distribution system including the loads can be represented by a Thevenin voltage and impedance looking left at Bus-4. Therefore even if the voltage source v_s is balanced, the Thevenin equivalent voltage source may not be balanced if any of the loads is unbalanced. Similarly this voltage source can be distorted if any of the loads is distorted. We must therefore resort to operating the DSTATCOM in such a way that it takes into account these factors. Note that since we are referring to unbalance and distortion, we shall use the fundamental sequence component extraction algorithm of Section 3.3.1. We shall call this operation as the DSTATCOM in phasor current control mode since we shall generate the reference currents through the phasor analysis.

Let us, for the time being, consider that the Thevenin equivalent results in three balanced impedances even if the equivalent source is unbalanced or distorted. We can then have the following three conditions [10]

1. When both load and source are unbalanced.
2. When both load and source are unbalanced and the load is distorted.
3. When both load and source are unbalanced as well as distorted.

8. Realization and Control of DSTATCOM

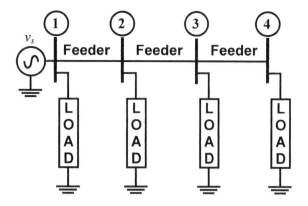

Figure 8.19. A typical distribution system

Below we shall present the design approaches for the three cases mentioned above based on which the customer can draw sinusoidal current from the feeder.

8.4.1 Case-1: When Both Load and Source are Unbalanced

The reference for the terminal voltage is obtained through the Fourier extraction of the fundamental voltage, i.e., $v_t^* = v_t^{fund}$ where the superscript *fund* denotes the fundamental quantity. Once this reference is obtained it is easy to obtain a reference for the current through the filter capacitor. To extract a reference for the injected shunt current we note that the desired source current must be fundamental positive sequence phasors. Since $i_l = i_s + i_f$, the compensator must supply the negative and zero sequence required by the load current. We can then write in phasor domain

$$\left. \begin{array}{l} I_{f0}^* = I_{l0} \\ I_{f2}^* = I_{l2} \\ I_{f1}^* = 0 \end{array} \right\} \qquad (8.17)$$

where the subscripts 0, 1 and 2 indicate zero, positive and negative sequences respectively. We can then use the algorithm presented in Section 3.3.1 to extract the sequence components from their samples. Once they are obtained we can use inverse sequence transform to set the reference value.

Assuming that the compensator tracks the reference currents accurately we get the following power equations

$$P_l + jQ_l = V_{t1}I_{l1}^* + V_{t2}I_{l2}^* + V_{t0}I_{l0}^*$$
$$P_t + jQ_t = V_{t1}I_{s1}^* = V_{t1}I_{l1}^* \tag{8.18}$$
$$P_f + jQ_f = V_{t2}I_{l2}^* + V_{t0}I_{l0}^*$$

This implies that the source supplies the positive sequence real and reactive powers required by the load and the compensator supplies the negative and zero sequence powers of the load. However the power supplied by the compensator is small and the dc capacitor voltage can be maintained through a feedback control.

Example 8.8: Let us consider a system whose fundamental frequency is 50 Hz. It is assumed that the Thevenin equivalent voltages of the source are unbalanced and are given in per unit by

$$v_{sa} = \sqrt{2}\sin(100\pi t)$$
$$v_{sb} = \sqrt{2} \times 1.25 \sin(100\pi t - 120°)$$
$$v_{sc} = \sqrt{2} \times 0.85 \sin(100\pi t + 120°)$$

The Thevenin equivalent source impedance is given in per unit by

$R = 0.05$ and $X = \omega L = 0.3$

Let us assume that the load buses to the right of the PCC contain passive RL loads. We shall refer the combined feeder and load impedances to the right of PCC as load impedances for simplicity. They are unbalanced are given in per unit by

$$Z_{la} = R_{la} + jX_{la} = 2.0 + j1.5$$
$$Z_{lb} = R_{lb} + jX_{lb} = 2.55 + j1.25$$
$$Z_{lc} = R_{lc} + jX_{lc} = 1.0 + j2.3$$

For simplicity in discussion at this stage, it is assumed that the VSIs are lossless and supplied by a fixed dc source instead of the dc storage capacitor. The DSTATCOM parameters are given in per unit are

$$V_{dc} = 2.0,\ R_f = 0,\ X_f = \omega L_f = 0.2 \text{ and } X_{filt} = \frac{1}{\omega C_{filt}} = 7.02$$

8. Realization and Control of DSTATCOM

The reference waveforms are tracked by an LQR that is designed with the following parameters

$$Q = \begin{bmatrix} 30 & & & \\ & 0 & & \\ & & 1 & \\ & & & 0 \end{bmatrix}, \quad r = 0.05$$

Note that since the system configuration and control law remain the same we use the equations (8.11) to (8.16) for the tracking controller. Also note that the above choice of weighting matrix puts maximum weight on the shunt current and much less weight on the terminal voltage. The full state feedback gain matrix obtained through the above choice is

$$K = \begin{bmatrix} 23.97 & 5.67 & 11.47 & -3.39 \end{bmatrix}$$

We however use a reduced order feedback in which the last element is equated to zero. For this reduced order feedback, the dominant closed-loop eigenvalues are found to be on the left of the line $Re(s) = -2476$ as opposed to the line $Re(s) = -2529$ for the full state feedback. For the hysteresis band control of (8.16) the value of lim chosen is 0.45.

The result of this compensation is shown in Figure 8.20 in which the DSTATCOM is connected at the end of the first half cycle. In this we only show the source currents and the terminal voltages. It can be seen that the source currents become balanced within about 1.5 cycles. The terminal voltage however will remain unbalanced, as the source voltage is still unbalanced. As a consequence the power entering the PCC will be oscillating at 100 Hz.

ΔΔΔ

8.4.2 Case-2: When Both Load and Source are Unbalanced and Load Contains Harmonics

The reference for the terminal voltage and current through the filter capacitor is obtained in the same way as Case-1. However, to extract the reference for the injected current we must remember that the compensator should not only supply the sequence currents given in (8.17) but should also supply the harmonic content of the load. To facilitate this, let us assume that we have obtained the instantaneous three-phase reference current i_f^* from the inverse transformation of i_{f0}^*, i_{f1}^* and i_{f2}^* given in (8.17). We then use the following relation to modify the existing i_f^*

$$i_f^*\big|_{new} = i_f^*\big|_{old} + i_l - i_l^{fund} \qquad (8.19)$$

This implies that the compensator cancels the harmonic current drawn by the load. The following example illustrates the idea.

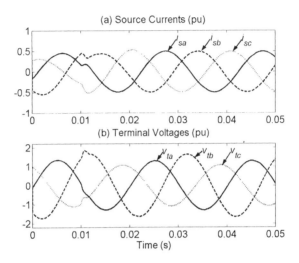

Figure 8.20. Compensated source currents and terminal voltages for Case-1

Example 8.9: Let us consider the same system as given in Example 8.8 except that in this case we have included a rectifier type load that is drawing a peak current of 0.35 per unit in addition to the RL load. Using the same feedback gain matrix as given in Example 8.8, we obtain the results that are shown in Figure 8.21. It can be seen that the terminal voltages become free of distortion as the compensator is connected at the end of the first half cycle. They however remain unbalanced, as the source is unbalanced.

ΔΔΔ

8.4.3 Case-3: Both Load and Source are Unbalanced and Distorted

This is the case in which we assume that both the source voltages and load currents are distorted by harmonics. This has to be treated differently than the previous two cases as the generation of references is of a completely different nature. Consider the single-line diagram of Figure 8.10. If we want

8. Realization and Control of DSTATCOM

the source current to be balanced and harmonic free, then the terminal voltage must contain exactly the same amount of harmonics as the source. The reference for the injected current must be extracted in the same way as given by (8.17) and (8.19). However, the reference for terminal voltage must then be

$$v_t^* = v_t^{fund} + v_t^{harm} \qquad (8.20)$$

where v_t^{harm} is the harmonic content of the source. Typically, the source voltage does not contain any even harmonics and for phase-a is

$$v_{sa}^{harm} = \sum_{n=3,5,7,\cdots} V_{san} \sin(n\omega t) \qquad (8.21)$$

Again note that

$$\dot{v}_t = \frac{1}{C_f} i_{cf}$$

Then the reference for the current through the filter-capacitor is then easily obtained from (8.20) and (8.21).

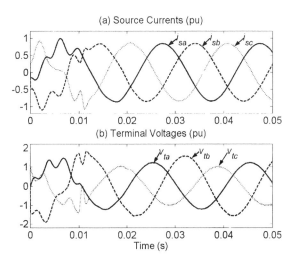

Figure 8.21. Compensated source currents and terminal voltages for Case-2

Example 8.10: Let us consider the same system as in Example 8.9. We assume that the source is distorted with 5th and 7th harmonics in addition to the unbalance given in Example 8.8. The reference for the terminal voltage is generated assuming that the source voltage is completely measurable. The state feedback controller gains used in Example 8.8 are used in this case also. The results are shown in Figure 8.22 in which the compensator is connected at the end of first half cycle. The source currents become balanced before the end of the second cycle as evident from Figure 8.22 (a). The terminal voltages are shown in Figure 8.22 (b). The phase-a source and terminal voltage are plotted in Figure 8.22 (c). It can be seen that they both contain the same harmonics of similar magnitude since for zero harmonic current to flow through the feeder, the DSTATCOM must create a harmonic voltage equal to the voltage source harmonics. To illustrate the tracking performance, the reference and actual injected current waveforms for the phase-a are plotted in Figure 8.22 (d). It can be seen that the injected current tracks the reference within half a cycle.

ΔΔΔ

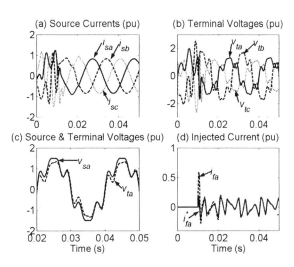

Figure 8.22. Compensated source currents and terminal voltages for Case-3

In the above example we have assumed that the measurement of the source voltage is completely available to us. This is not a valid assumption as the source may be remotely located. We can at most get a Thevenin equivalent voltage in a radial distribution system. We must therefore estimate the harmonic content of the source voltage from the measured data. This can be easily done when the feeder impedance is completely known. Referring to Figure 8.10, we can determine the 5th harmonic component of the source voltage through the measurement of 5th harmonic component of terminal voltage and source current in phasor notation as

8. Realization and Control of DSTATCOM

$$V_{s5} = V_{t5} + (R + j5\omega L)I_{s5}$$

Thus assuming that the feeder impedance is completely known, we can estimate the 5th harmonic component of source voltage as follows:

- Using the Fourier coefficient extraction using half cycle averaging, estimate the 5th harmonic component of terminal voltage and source current and use them in the above relation to estimate the 5th harmonic of source voltage.
- Use this value to set a reference for terminal voltage. Reference current through the filter capacitor is also set using this.
- Once the controller tracking system converges, the 5th harmonic component of the source current will be zero and we shall have $V_{s5} = V_{t5}$.

Similar operation must also be performed for all odd harmonic components up to a defined order.

It is further unrealistic to expect that the feeder impedance of Figure 8.10 be completely known as it may also be a Thevenin impedance. A straightforward use of the algorithm given above will then not be possible. Fortunately however, estimation of the entire feeder impedance is not necessary for the extraction of terminal voltage reference. The circuit to the point of common coupling can be viewed as a uniformly distributed feeder that is supplied by a stiff source. We thus need the voltage at any point on this feeder and the corresponding impedance. We shall call the procedure of estimating the voltage using a fractional value of the Thevenin impedance as partial back projection. For example a 50% back projection will mean using $0.5(R + jn\omega)$ in computing the source voltage of nth harmonic. The proof of this reference extraction convergence is given in Section 5.7.4 of Chapter 5 under the assumption that the harmonic values are used after the circuit reaches steady state.

Example 8.11: As developed in Section 5.7.4, the stability of the reference formulation depends on the eigenvalues of

$$P\left[jn\omega I - \left(I - \frac{\Gamma_2 K}{K\Gamma_2}\right)\Lambda\right]^{-1}\left\{jn\omega \frac{\Gamma_2 K}{K\Gamma_2}\right\}$$

being inside the unit circle. Ignoring the load, the three states of Figure 8.10 are i_f, v_t and i_{cf}. In current tracking mode, the reference for i_f is known. The reference for v_t should be found by back projection of measured harmonic voltage v_t back to the source. The matrix P in the above equation expresses this estimate of the back projected voltage and the corresponding filter current at a particular harmonic in terms of the measured harmonic voltage.

Let us consider the same system as given in Example 8.10 with the same set of controller gains. The worst damped eigenvalues for two different amounts of back projection are shown in Figure 8.23. In this figure, the dominant eigenvalues for different odd harmonics are indicated by '*'. The unit circle is also indicated in the figure. It can be seen that all harmonic eigenvalues of the system are stable for a back projection of 75%. However, for a back projection of 250%, all eigenvalues are unstable except for the fundamental and the 3rd harmonic.

We now use partial back projection in the simulation studies. Figure 8.24 shows the source currents obtained with four different back projections. The feedback system is not stable and the source currents never become sinusoid for 25% back projection. For 50% back projection, the source currents settle in about 3 cycles, while they settle down within 2 cycles for 75% back projection. The source currents take considerably longer to settle for 150% back projection. In fact, through the simulation studies it was observed that the system attains steady state with perfect tracking when the amount of back projection is limited between 40% to 160%. It is to be noted that there is a slight discrepancy between the results obtained using the simulation studies and that obtained using the convergence proof of Section 5.7.4. In the proof we have assumed that the system first reaches steady state before we extract the harmonic components. In the simulation however the harmonic components are extracted as we move along in time using the moving averaging. The result obtained through the simulation will show a lower range of stability than the convergence formula.

ΔΔΔ

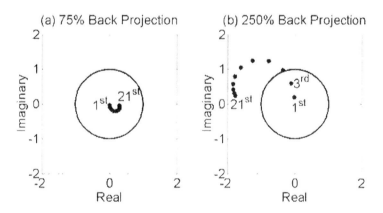

Figure 8.23. System eigenvalues for two different back projections

8. Realization and Control of DSTATCOM

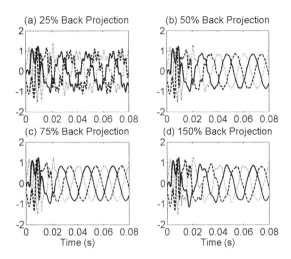

Figure 8.24. Source currents with four different back projections

Another important point to be noted if the back projection is below 40%, then the instability in the reference quantities are limited by the operation of pulse width modulated VSI. This will result in distortion in source currents but they will not diverge because of the finite gain of the inverter. However, if the back projection is over 250%, the unstable 5th and 7th harmonic will swamp the fundamental and the whole system will collapse. The above example however clearly demonstrates that if we use a reasonable estimate of the feeder impedance (including Thevenin impedance, if any), the compensator will be able to correct for current harmonics at the load side as well as voltage harmonics on the source side.

8.4.4 DC Capacitor Control

So far in the discussion we have assumed that a dc source, rather than a dc capacitor is supplying the VSIs of the DSTATCOM. We have assumed further that the compensator is lossless. We shall now relax these assumptions and put a feedback loop to regulate the dc capacitor voltage. Refer to (8.17), which gives the relation for reference shunt currents in current control mode. The equation is derived assuming that the fundamental positive sequence of the source current is equal to that of the load current. This assumption will no longer be strictly true since the positive sequence of the source current must be higher than its load current counterpart in order to supply the real power requirement of the dc capacitor. The difference in two currents will be forced through the shunt path. We can then modify (8.17) as

$$\left.\begin{array}{l}I_{f0}^* = I_{l0}\\I_{f2}^* = I_{l2}\\I_{f1}^* = I_{loss}\end{array}\right\} \tag{8.22}$$

where I_{loss} is the current required by the dc capacitor to maintain its charge and it must be obtained from the dc capacitor feedback loop.

The feedback should be able to correct the deviation of the average value of V_{dc} from a reference value V_{ref}. In the simplest form of feedback, we have used a proportional-plus-integral (PI) controller to correct for any discharge in the capacitor voltage. The controller is then given by

$$z_c = K_P \left(V_{ref} - V_{dc}^{cyl}\right) + K_I \int \left(V_{ref} - V_{dc}^{cyl}\right) dt \tag{8.23}$$

where the average value of the dc capacitor voltage V_{dc}^{cyl} can either be measured at the end of a cycle or can be the moving average. The amount of z_c required to sustain the dc capacitor voltage must then be drawn equally from the three phases. As a result of which we can substitute in (8.22)

$$I_{loss} = -z_c/3 \tag{8.24}$$

Example 8.12: Let us consider the same system as discussed before. We have assumed that the load side is harmonically distorted, while the source side is free of harmonics. We have removed the assumption that the compensator is lossless and have chosen the following per unit quantities

$$R_f = 0.1, \; X_{dc} = \frac{1}{\omega C_{dc}} = 0.106 \text{ and } V_{dc} = 2.0$$

The PI controller parameters chosen are

$$K_P = 8 \text{ and } K_I = 25$$

The system is started from rest, i.e., the compensator is connected at the end of first half cycle after the average load power is obtained. The capacitor is assumed to be precharged to 2.0 per unit and this value is also chosen as the set point for this voltage. The PI controller is started from zero initial condition. The results are shown in Figure 8.25. It can be seen that the dc capacitor voltage settles within about 0.3 s. In this figure the steady state waveforms of the source current are also plotted. It can be seen that these are balanced without any distortion. This implies that the additional amount of real current required is distributed equally in the three phases.

△△△

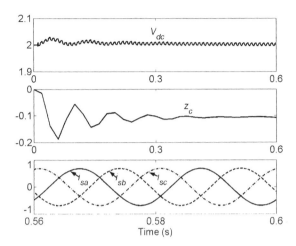

Figure 8.25. Performance of a DSTATCOM with dc capacitor control system

8.5 DSTATCOM in Voltage Control Mode

In a distribution system there may be several compensating devices of different kinds. However, in a radial distribution system, the voltage of a particular bus can be distorted or unbalanced if the loads in any part of the system are nonlinear or unbalanced. The customers connected to that bus would be supplied by a set of unbalanced and distorted voltages, even when their loads are not contributing to the bus voltage pollution. Therefore a DSTATCOM can be used at this bus to reduce harmonics and balance the bus voltages.

Consider the three-phase, four-wire radial system shown in Figure 8.19. Let us assume that we would like to correct the voltage of Bus-3. The single-phase Thevenin equivalent of the system is the same as that shown in Figure 8.11 (b). In this v_s, R and L constitute the Thevenin equivalent looking towards left into the network, while the equivalent load is the impedance looking towards right into the network, at Bus-3. We now have to use the DSTATCOM in the voltage control mode at Bus-3. Below we present two different control schemes of DSTATCOM in voltage control mode.

8.5.1 State Feedback Control of DSTATCOM in Voltage Control Mode

For the state feedback control of the DSTATCOM operating as a voltage regulator we need to specify the following references with respect to Figure 8.10: terminal voltage v_t, injected current i_f, current through the filter capacitor i_{cf} and load current i_l. Of these, the load current is load dependent that may change at any time. Just as we have done in Examples 8.5 and 8.6, we have eliminated the load current from the feedback signal.

We can choose an arbitrary magnitude for the reference signal of v_t. For simplicity, let us denote this magnitude by $|V_t^*|$. The instantaneous phase-a terminal reference voltage is then given by $v_{ta}^* = |V_t^*|\sin(\omega t + \phi_t)$. Since we require a balanced voltage at the terminal, the reference voltage for the other two phases can be obtained by phase shifting this waveform by 120°. However, the angle of the reference signal must be chosen such that the real power drawn by the DSTATCOM is zero in the steady state. To facilitate this, we use the following feedback signal [10]

$$\phi_t(k+1) = \phi_t(k) + C_P p_{fav} \qquad (8.25)$$

where C_P is a gain, p_{fav} is the average of the shunt power and k is the discrete-time index. Once ϕ_t is obtained from (8.25), the reference voltages for the three phases are calculated. The instantaneous three-phase reference voltages for the terminal can then be generated with respect to any phase locked point that provides the reference time setting. Once this voltage reference is generated, the corresponding current through the filter-capacitor can be generated as follows. Since the reference voltage for phase-a is $|V_t^*|\sin(\omega t + \phi_t)$, the reference current for this phase is then given by $|V_t^*|\omega C_f \cos(\omega t + \phi_t)$.

To generate i_f, it is to be noted that $i_l = i_s + i_f$. Hence the reference for i_f could be the instantaneous difference between these two currents. However, in order to prevent distortion creeping in, it is desirable to generate this reference using the fundamental value of the source current, i.e.,

$$i_f^* = i_l - i_s^{fund} \qquad (8.26)$$

Example 8.13: In this example we demonstrate the performance of a voltage regulating DSTATCOM. The system parameters are the same as that given in Example 8.8. The equivalent load, in addition to the RL load, also contains a rectifier load that draws a current with the peak of 0.35 per unit. For simplicity in discussion at this stage, it is assumed that the VSIs are

8. Realization and Control of DSTATCOM

lossless and supplied by a fixed dc source instead of the dc storage capacitor. The control gain (C_P) of (8.25) is chosen as -0.15 and $|V_t^*|$ is chosen as 1.2 per unit.

The control is designed with

$$Q = \begin{bmatrix} 1 & & & \\ & 0 & & \\ & & 50 & \\ & & & 0 \end{bmatrix}, \quad r = 0.05$$

The two variables that are most important for control are the terminal voltage v_t followed by the current i_f. The weighting matrix Q reflects the importance of these variables. For this set of parameters, the gain matrix for phase-a is then given by

$$K = \begin{bmatrix} 3.25 & 9.23 & 30.35 & -1.79 \end{bmatrix}$$

Using this feedback, the oscillatory closed-loop eigenvalues are to the left of the line $\text{Re}(s) = -7193$. To avoid the complexity of forming a reference for the load current, we use a reduced order feedback in which the last element of the gain matrix is chosen as zero. This reduction in state feedback results in a minimal shift in the closed-loop eigenvalues to the left of $\text{Re}(s) = -7187$. Thus stability is not at all endangered by the reduced feedback, but the computation is reduced considerably. Furthermore, the same feedback matrix that is designed for phase-a is used for the other two phases as well.

The DSTATCOM performance is shown in Figure 8.26 in which Figure 8.26 (a) depicts the load currents. The compensator is connected at the end of first half cycle (10 ms) after the first fundamental extraction is completed or first power average is obtained. The terminal voltages are shown in Figure 8.26 (b). It can be seen that the terminal voltages become balanced sinusoids within about one cycle of the connection of the compensator. The controller output is shown in Figure 8.26 (c). This quantity is plotted a little longer than the other quantities. It can be seen as soon as the DSTATCOM is pressed into service this quantity goes through a transient. However, it settles down quickly to an angle of around $-6°$. Note that this angle is measured with respect to the source voltage as it determines the power flow over the line. We may therefore need the timing information from the source for the successful implementation of this algorithm. The fast convergence of the algorithm is proved from the tracking error plot of Figure 8.26 (d).

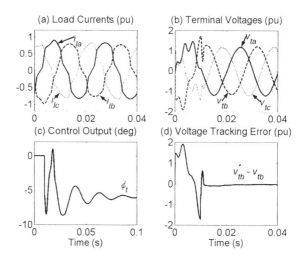

Figure 8.26. Performance of voltage controlled DSTATCOM when source is unbalanced and load is both unbalanced and distorted

We now distort the source voltage by adding 5th and 7th harmonics to it with the magnitudes of the harmonic components being inversely proportional to their harmonic number. The source voltage is shown in Figure 8.27 (a). The terminal voltage becomes balanced sinusoid within about one cycle of connecting the compensator. This is shown in Figure 8.27 (b). The controller output and voltage tracking error for phase-b are also shown in Figure 8.27 (c) and (d) respectively.

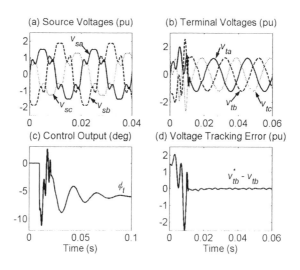

Figure 8.27. Performance of voltage controlled DSTATCOM when both source and load are unbalanced and distorted

8. Realization and Control of DSTATCOM

Let us now consider the case when there are random disturbances in the source voltage. We assume that the source is unbalanced and load is both unbalanced and distorted. The source voltage, as shown in Figure 8.28 (a), contains spikes of irregular height that appear randomly. The phase-b of the uncompensated terminal voltage and source current are shown in Figure 8.28 (b) and (c) respectively. Due to the presence of the feeder inductance, these waveforms are not sinusoidal. The compensator is connected at the end of the first half cycle and the compensated terminal voltages are shown in Figure 8.28 (d). It can be seen that the terminal voltages become balanced sinusoids within 2 cycles of the connection of the compensator. It is however to be noted that the glitches in the supply voltage are smoothed by the feeder inductance, thereby providing opportunity for the DSTATCOM to achieve correction. If the flicker voltage source were very close to the terminal bus itself, it would have been impossible for the DSTATCOM to correct it fully due to the limited bandwidth of the VSIs.

ΔΔΔ

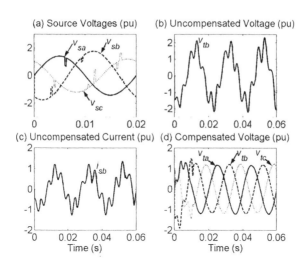

Figure 8.28. Elimination of random voltage variation using a DSTATCOM

In the above example we have assumed that the DSTATCOM is supplied by a dc voltage source. We shall now relax this assumption and include a feedback controlled dc capacitor. Refer to (8.25) through which the angle of the reference terminal voltage of the DSTATCOM is computed in order to force the real power drawn by the DSTATCOM to zero. Since the aim is to draw real power from the source, we now modify this equation. In order to maintain the capacitor voltage as constant, a negative feedback of this voltage is added in the angle computation as

$$\phi_t(k+1) = \phi_t(k) + C_p p_{fav} + C_V \left(V_{ref} - V_{dc}^{cyl}\right) \tag{8.27}$$

where C_V is constant, V_{ref} is the reference value of the dc capacitor voltage and V_{dc}^{cyl} is the cycle average value of the dc capacitor voltage. In order to influence the control rapidly, we have chosen this average to be the running average.

Example 8.14: Let us consider the same system and parameters as given in Examples 8.12 and 8.13. The controller gains are chosen as

$$C_P = C_V = -0.15$$

The system is started from rest, i.e., the compensator is connected at the end of first half cycle after the average load power is obtained. The capacitor is assumed to be precharged to 2.0 per unit and this value is also chosen as the set point for this voltage. The results are shown in Figure 8.29. It can be seen that the dc capacitor voltage settles within about 0.3 s. The average power drawn by the DSTATCOM is also plotted in this figure. It has a steady-state value of -0.035 per unit indicating that the DSTATCOM is drawing power from the supply. The steady state terminal voltage waveforms are also shown in Figure 8.29. It can be seen that these are balanced without any distortion.

ΔΔΔ

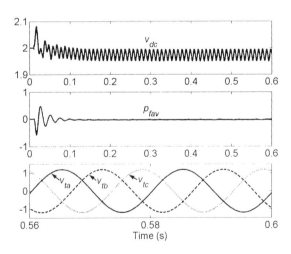

Figure 8.29. Performance of a voltage controlled DSTATCOM with dc capacitor control

8.5.2 Output Feedback Control of DSTATCOM in Voltage Control Mode

The state feedback controller presented above assumes that the system parameters are known. However as we have mentioned before, the feeder parameters are at best the Thevenin equivalent of the upstream distribution system. Since the Thevenin equivalent can change at any time depending on the load, it is desirable that these parameters are not used in the voltage controller design. Below we present a voltage control technique that only requires the timing information from the source v_s for synchronization.

Consider the network shown in Figure 8.30. In this u is the switching variable that can take on values ± 1 corresponding to the states of the inverter. To derive a control law, we assume for the time being that u is equal to a continuous signal u_c. We shall derive the state space equation for the part system enclosed in the dotted path in Figure 8.30. Let us define the following state and input vectors

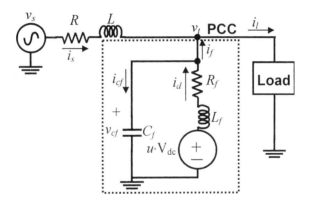

Figure 8.30. Equivalent circuit of voltage compensated distribution system

$$x^T = [v_t \quad i_d] \quad z^T = [u_c \quad i_f]$$

The state space equation of the system is then given by

$$\begin{aligned}\dot{x} &= \begin{bmatrix} 0 & 1/C_f \\ -1/L_f & -R_f/L_f \end{bmatrix} x + \begin{bmatrix} 0 & -1/C_f \\ V_{dc}/L_f & 0 \end{bmatrix} z \\ &= Ax + Bz\end{aligned} \quad (8.28)$$

The continuos-time state equation is then discretized as

$$x(k+1) = \phi\, x(k) + \theta\, z(k) \qquad (8.29)$$

where k is the kth sampling instant and the matrices ϕ and θ, for a sampling time of ΔT, is given as

$$\phi = e^{A\Delta T} \text{ and } \theta = \int_0^{\Delta T} e^{A\tau} B\, d\tau$$

Let us define the elements of the matrices given in (8.29) as follows

$$\phi = \begin{bmatrix} \phi_{11} & \phi_{12} \\ \phi_{21} & \phi_{22} \end{bmatrix} \text{ and } \theta = \begin{bmatrix} \theta_{11} & \theta_{12} \\ \theta_{21} & \theta_{22} \end{bmatrix}$$

We can then write from (8.29)

$$v_t(k+1) = \phi_{11} v_t(k) + \phi_{12} i_d(k) + \theta_{11} u_c(k) + \theta_{12} i_f(k) \qquad (8.30)$$

We shall now design a deadbeat controller discussed in Section 5.7.5. For a reference voltage of v_t^*, the following cost function is chosen

$$J = \{v_t(k+1) - v_t^*(k+1)\}^2 \qquad (8.31)$$

The minimization of this function results in the following control input

$$u_c(k) = \frac{v_t^*(k+1) - \phi_{11} v_t(k) - \phi_{12} i_d(k) - \theta_{12} i_f(k)}{\theta_{11}} \qquad (8.32)$$

Once u_c is obtained the control input u is obtained in the hysteresis band control of (8.15). Let us consider the following example.

Example 8.15: Let us consider the same system as given in Example 8.13 in which it is assumed that the VSIs are lossless and supplied by a fixed dc source instead of the dc storage capacitor. We have chosen $|V_t^*|$ to be 1.2 per unit and assume that the phase of the desired terminal voltage is obtained from (8.25). The control gain (C_P) is chosen as – 0.01. The results are shown in Figure 8.31 in which the compensator is switched on after the first half cycle. It can be seen that the tracking error is forced to zero within one cycle

8. Realization and Control of DSTATCOM

of the compensator being connected. However the phase-b takes longer to stabilize. It can be seen that it takes about 0.2 s to stabilize.

We now investigate how the voltage controller behaves when a voltage sag occurs. The source voltages of the three phases are reduced to 75% of their nominal values used before at 0.21 s. It is desired that DSTATCOM still maintains the voltage to a peak value of 1.2 per unit. The results are shown in Figure 8.32. It can be seen that there is no transient in the terminal voltage magnitude. This implies that the power required by the load is almost unchanged. It is well known that the power flow between two sources that are connected by an inductor with a value of X is given by

$$P = \frac{V_1 V_2}{X} \sin \delta$$

where V_1 and V_2 are the rms voltage magnitudes and δ is the angle between these two voltages. Therefore since the source voltage has reduced, the power supplied to the load can be held constant by introducing additional lag in the terminal voltage with respect to the source voltage. It can be seen from Figure 8.32 (b) that the phase angle controller increases the phase lag such that power supplied to the load is held constant. It is to be noted that since the source is unbalanced, the phase ϕ_t is an equivalent phase angle that is required to maintain the power flow and not the load angle δ used in the equation given above.

ΔΔΔ

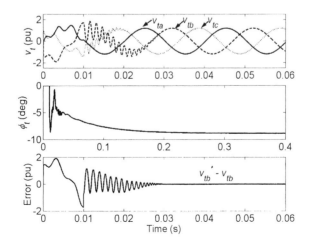

Figure 8.31. Performance of voltage controlled DSTATCOM under output feedback control

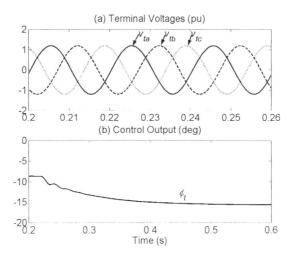

Figure 8.32. Performance of voltage controlled DSTATCOM under voltage sag

The above example assumes that the compensator is supplied by a constant dc source. In practical circuit however the system will be supplied by a dc storage capacitor. This can be done using (8.27) in the same manner as discussed in the previous sub-section. Various simulation and experimental results with voltage controlled DSTATCOM operating under deadbeat control are given in [11].

8.6 Conclusions

In this chapter we have presented practical structures of DSTATCOM that can be used in power distribution systems. We have also discussed how the DSTATCOM can be used in weak supply point connections. In particular these devices can be used both as current compensators to compensate harmonics generated by loads and also as voltage regulators to regulate the voltage of a particular distribution system bus. In general these algorithms are found to perform satisfactorily. Note that we have used only capacitor filters to bypass the harmonics generated by inverter switching. There are however possibilities of using LC filters for this purpose.

8.7 References

[1] A. Ghosh and A. Joshi, "The use of instantaneous symmetrical components for balancing a delta connected load and power factor correction," *Electric Power Systems Research*, Vol. 54, pp. 67-74, 2000.

[2] M. K Mishra, A. Ghosh and A. Joshi, "A new STATCOM topology to compensate loads containing ac and dc components," *IEEE Power Engineering Society Winter Meeting*, Singapore, January, 2000.

[3] M. K. Mishra, A. Joshi and A. Ghosh, "A new closed loop control scheme for capacitor voltage equalization in shunt compensator using neutral current injection," *Proc. IEEE-PES Winter Meeting-2001*, Columbus, Ohio, 2001.

[4] M. K. Mishra, A. Joshi and A. Ghosh, "Control strategies for capacitor voltage equalization in neutral clamped shunt compensator," *Proc. IEEE-PES Winter Meeting-2001*, Columbus, Ohio, 2001.

[5] M. K. Mishra, A. Joshi and A. Ghosh, "Control schemes for equalization of capacitor voltages in neutral clamped shunt compensator," To appear in *IEEE Trans. Power Delivery*.

[6] C. A. Quinn, N. Mohan and H. Mehta, "Active filtering of harmonic currents in three-phase, four-wrire systems with three-phase and single-phase nonlinear loads," *Proc. IEEE Applied Power Electronics Conf. (APEC)*, pp. 829-836, 1992.

[7] C. A. Quinn, N. Mohan, and H. Mehta, "A Four wire current-controlled converter provides harmonic neutralization in three-phase, four-wire systems," *Proc. Applied Power Electronic Conf. (APEC'93)*, pp. 841-846, 1993.

[8] K. Hoffman and G. Ledwich, "Improved power system performance using inverter based resonant switched compensators," *IEEE Power Electronics Specialists Conf. (PESC)*, pp. 205-210, 1994.

[9] A. Ghosh and G. Ledwich, "Load compensating DSTATCOM in weak ac systems," To appear in *IEEE Trans. Power Delivery*.

[10] G. Ledwich and A. Ghosh, "A flexible DSTATCOM operating in voltage and current control mode," *Proc. IEE – Generation, Transmission & Distribution*, Vol. 149, No. 2, pp. 215-224, 2002..

[11] M. K. Mishra, A. Ghosh and A. Joshi, "Operation of a DSTATCOM in voltage control mode," To appear in *IEEE Trans. Power Delivery*, Pre-print No. PE-523PRD (01-2002).

Chapter 9

Series Compensation of Power Distribution System

In the previous two chapters we have discussed the shunt compensation of power system loads. It has been shown that shunt compensator can be very effective in balancing any unbalance in the load currents and also for cleaning up harmonic pollution in the load, provided that the supply is balanced. In this way, a shunt compensator protects the utility system from the ill effects of customer loads. In this chapter we shall show that a series compensator is the dual of a shunt compensator – it protects a sensitive load from the distortion in the supply side voltage. The basic principle of a series compensator is simple: by inserting a voltage of required magnitude and frequency, the series compensator can restore the load side voltage to the desired amplitude and waveform even when the source voltage is unbalanced or distorted. Usually a series compensator is used to protect sensitive loads during faults in the supply system.

A power electronic converter based series compensator that can protect critical loads from all supply side disturbances other than outages is called a dynamic voltage restorer (DVR). This device employs IGBT solid-state power-electronic switches in a pulse-width modulated (PWM) inverter structure. The DVR is capable of generating or absorbing independently controllable real and reactive power at its ac output terminal. Like in a DSTATCOM, the DVR is made of a solid-state dc to ac switching power converter that injects a set of three-phase ac output voltages in series and synchronism with the distribution feeder voltages. The amplitude and phase angle of the injected voltages are variable thereby allowing control of the real and reactive power exchange between the DVR and the distribution system. The dc input terminal of a DVR is connected to an energy source or an energy storage device of appropriate capacity. The reactive power exchanged between the DVR and the distribution system is internally generated by the DVR without ac passive reactive components. The real power exchanged at the DVR output ac terminals is provided by the DVR input dc terminal by an external energy source or energy storage system.

A typical DVR connection is shown in Figure 9.1. It is connected in series with the distribution feeder that supplies a sensitive load. For a fault clearing or switching at point A of the incoming feeder or fault in the distribution feeder-1, the voltage at feeder-2 will sag. Without the presence of the DVR, this will trip the sensitive load causing a loss of production. The DVR can protect the sensitive load by inserting voltages of controllable amplitude, phase angle and frequency (fundamental and harmonic) into the distribution feeder via a series insertion transformer as shown in Figure 9.1. It is however to be mentioned that the rating of a DVR is not unlimited. Thus a DVR can only supply partial power to the load during very large variations (sags or swells) in the source voltage. As we shall demonstrate in this chapter that during steady state operations, the DVR can compensate for the inductive drop in the line by inserting a voltage in quadrature with the feeder current. During this phase, the dc storage need not supply any real power except for the losses in the converter. The DVR can also limit a fault current by injecting a leading voltage in quadrature with the fault current thereby increasing the effective fault impedance of the distribution feeder.

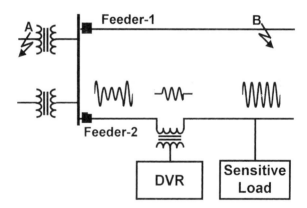

Figure 9.1. DVR connection for voltage sag correction of sensitive loads

In August 1996, Westinghouse Electric Corporation installed world's first DVR in Duke Power Company's 12.47 kV substation in Anderson, South Carolina. This was installed to provide protection to an automated rug manufacturing plant. Prior to this connection, the DVR was first installed at the Waltz Mill test facility near Pittsburgh for full power tests. The test results are discussed in [1]. The next commissioning of a DVR was done by Westinghouse in February 1997 in Powercor's 22 kV distribution system at

Stanhope, Victoria, Australia. This was done to protect a diary milk processing plant. The saving that may result from the installation of this DVR is estimated at over $100,000 per year [2]. In the next phase of development, DVRs that can be mounted on an overhead platform supported by two poles were fabricated. The first ever DVR that is mounted on a platform was installed to protect Northern Lights Community College and several other smaller loads in Dawson Creek, British Columbia, Canada [3].

In this chapter we shall present two different structures of the DVR. These are shown in Figure 9.2. In the structure of Figure 9.2 (a), the dc bus of the VSI realizing the DVR is supplied from the feeder through a rectifier. Therefore the DVR can absorb real power from the feeder through the dc bus. This is not possible for the structure of Figure 9.2 (b) in which the DVR is supplied by a dc storage capacitor. Therefore the DVR in this structure must operate in the mode in which it will have no real power exchange with the ac system in the steady state.

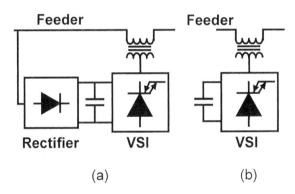

Figure 9.2. Two possible structures of DVR

In addition to voltage compensation through DVR, a hybrid structure of series active and shunt passive filter has been proposed for harmonic neutralization of nonlinear loads. We shall discuss this method at the end of the chapter.

9.1 Rectifier Supported DVR

Let us first consider the DVR structure shown in Figure 9.2 (a). Let us consider the simplified distribution system shown in Figure 9.3. It is assumed that the system is compensated by an ideal series compensator. We can then define the following components of the distribution system.

- Ideal Series Compensator: represented by the instantaneous voltage sources v_{fa}, v_{fb} and v_{fc}.
- Supply Voltages: represented by the instantaneous voltage sources v_{sa}, v_{sb} and v_{sc}.
- Load Voltages: represented by the instantaneous voltages v_{la}, v_{lb} and v_{lc}.
- Terminal (PCC) Voltages: represented by the instantaneous voltages v_{ta}, v_{tb} and v_{tc}.
- Line Currents: represented by i_{sa}, i_{sb} and i_{sc}. Note that these currents will also flow through the load and therefore we might also call them load currents.
- Sensitive Loads: represented by the impedances Z_{la}, Z_{lb} and Z_{lc}. It is assumed that these loads are balanced, i.e, $Z_{la} = Z_{lb} = Z_{lc}$.

The series compensator is connected between a terminal bus on the left and a load bus on the right. The instantaneous voltages of the terminal (PCC) and load buses are denoted by v_t and v_l respectively with subscripts a, b and c denoting the phases with which they are associated. The voltage sources are connected to the series compensator terminals by a feeder with an impedance of $R + jX$. In this study we assume that the balanced loads are of RL type, given by $Z_{la} = Z_{lb} = Z_{lc} = R_l + jX_l$. It is to be noted that the phase angle ϕ_l between the load terminal voltage v_l and the line current i_s depends on the load impedance and is independent of the line impedance or the series compensator voltage.

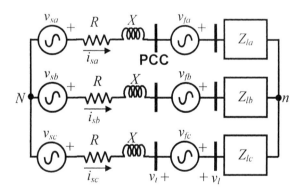

Figure 9.3. Schematic diagram of a series compensator connected power system

From Figure 9.3 we observe that

$$v_l = v_t + v_f \qquad (9.1)$$

9. Series Compensation of Power Distribution System

It is desired that the DVR regulates the load voltage. The reference voltage of the DVR (v_f^*) is then given by [4]

$$v_f^* = v_l^* - v_t \tag{9.2}$$

where v_l^* is the desired load voltage. Theoretically we can choose v_l^* to have any arbitrary magnitude and angle. Let us now consider the following example.

Example 9.1: Let us assume that a load is supplied by a balanced source with a peak voltage of $\sqrt{2}$ per unit. Let the feeder and load impedances in per unit be (see Figure 9.3)

$$R + jX = 0.05 + j0.3 \text{ and } R_l + jX_l = 2 + j1.5$$

The DVR will be connected at the end of the first half cycle (0.01 s). We desire that the peak of the load voltage be $\sqrt{2}$ per unit.

The results are shown in Figure 9.4 in which the reference load voltage has an angle of $-20°$ in Case-1, while it has an angle of $+20°$ in Case-2. These reference angles are measured with respect to the source voltage and all quantities in Figure 9.4 are given in per unit. It can be observed from Figure 9.4 (a) and (c) that the load voltage becomes the desired voltage as soon as the DVR is connected. However to understand the behavior of the instantaneous powers shown in Figure 9.4 (b) and (d), let us for the moment assume that the dc bus of the DVR is supplied by an ideal dc source that injects or absorbs any amount of power. Then for Case-1, part of power demanded by the load (p_l) is supplied by the terminal (p_t) and the other part comes from the compensator (p_f). The behavior in Case-2 however is different. In this case the terminal supplies more power than demanded by the load. Therefore the excess power is returned to the dc source connected to the DVR.

In actual situation when the dc bus of the DVR is supplied by the terminal through the rectifier shown in Figure 9.2 (a), the reverse power flow may damage the rectifier unit. We must therefore always ensure that the phase angle of the desired load bus voltage is less than that of the terminal voltage.

∆∆∆

To alleviate the problem of reverse power flow we must compute the phase angle of the positive sequence of the terminal voltage (V_{t1}) and make the reference load voltage (v_l^*) to lag this voltage by a small amount. To accomplish this we shall use the on-line fundamental symmetrical component extraction algorithm given in Section 3.3.1. The positive sequence extraction is used because a disturbance in the supply side creates a temporary unbalance in the system. The idea is illustrated in the following example.

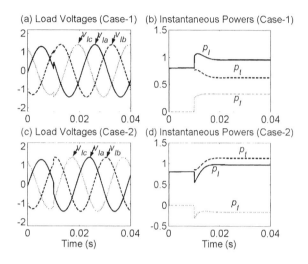

Figure 9.4. Performance of rectifier supported DVR for two different phase angles: Case-1 when load voltage lags the source voltage by 20° and Case-2 when it leads by 20°

Example 9.2: Let us consider the same system as given in Example 9.1. We desire that the reference load voltage has a magnitude of √2 per unit while its phase angle lags that of the phase angle of the positive sequence of the terminal voltage by 5°. The compensator is connected at the end of the first half cycle. Further after two cycles (0.04 s) a voltage disturbance is created in the source voltage that lasts for about one and a half cycle. During the disturbance, the voltages of phases-a and b sag to 75% and 80% of the nominal values respectively, while that of the phase-c rises to 120% of the nominal value. We have not introduced any change in the phase angles of these voltages. The results are shown in Figure 9.5. It can be seen that the load voltages are held constant during the disturbance. Also during the voltage disturbance, the terminal voltages are unbalanced and as a consequence the DVR voltages are also unbalanced. Therefore both the instantaneous powers supplied by the terminal and the DVR oscillate at 100 Hz. However there is no reverse power flow through the DVR. Since the load voltage is held balanced and loads are balanced, the load power remains constant except for minor disturbances at the beginning and end of the voltage sag/swell.

9. Series Compensation of Power Distribution System

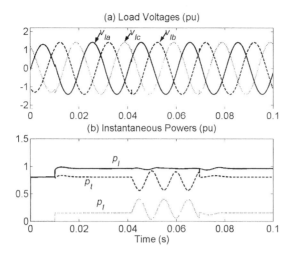

Figure 9.5. Performance of rectifier supported, phase angle controlled DVR during voltage sag/swell

To investigate the performance of the DVR during the load change we have assumed that the load current has reduced suddenly such that load impedance is three times the value given in Example 9.1. The results are shown in Figure 9.6 in which the load reduction takes place when the system is in the steady state. It can be seen that no transient is observed in the load voltages. Further there is a reduction in all the three powers indicating that load demand has reduced. However even during this load reduction, the flow of power through the DVR remains in the positive direction.

△△△

In addition to the power flow consideration, the rating of the DVR also plays an important role in voltage correction. Figure 9.7 depicts the per unit voltage to be injected by the DVR for four different symmetrical sags in the source voltage. Since the sag is symmetric, only phase-a voltage of the DVR is shown in this figure, as the DVR will inject a balanced voltage. It can be seen as the sag magnitude increases, the peak of the injected voltage increases. In fact for a sag in which the peak of the source voltage reduces to 50% its nominal value, the peak voltage to be injected is 0.85 per unit, i.e., the voltage has an rms value 0.6 per unit. This will obviously increase the rating of the device. Therefore the device rating and the maximum available compensation achievable are important issues. These issues are discussed in [5,6]. In addition, the protection issues are discussed in [5].

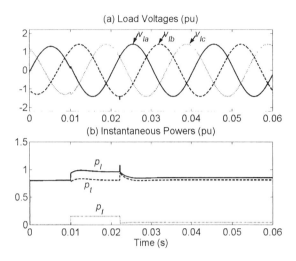

Figure 9.6. Performance of rectifier supported, phase angle controlled DVR during load reduction

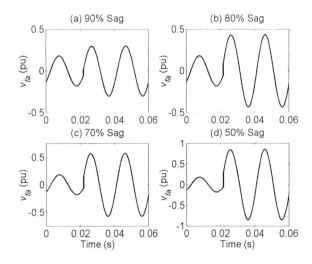

Figure 9.7. DVR voltage injection requirements for different voltage sags

9.2　DC Capacitor Supported DVR

In this section we shall consider the DVR structure shown in Figure 9.2 (b). It will be demonstrated that this series compensator can not only act as a voltage restorer but also as a voltage regulator by pure series reactive injection. This implies that the DVR does not absorb or supply any real

9. Series Compensation of Power Distribution System

power in the steady state. Let us first develop the analytical aspects and illustrate these by examples of the simplified distribution system shown in Figure 9.3.

9.2.1 Fundamental Frequency Series Compensator Characteristics

First we shall present the sinusoidal steady state analysis of a series compensator connected power system. In the analysis presented below we assume that the magnitude of the source voltage is V per unit and we want to regulate the magnitude of the load voltage to V per unit by injecting a voltage from the series compensator. We stipulate the following condition on the compensator [7].

– **Condition-1**: The series compensator need not supply any real power in the steady state. This implies that the phase angle difference between series compensator voltage phasor and line current phasor must be $\pi/2$ in the steady state.

Under this condition, we can divide the operation of the series compensator into three different cases depending on the feeder and load impedances. These are discussed below.

Case 1: When the Line Resistance is Negligible, i.e., $R = 0$: The phasor diagram for this case is shown in Figure 9.8. The only way the load and source voltage magnitudes can be equal is when the series compensator completely compensates for the reactive drop in the feeder. This will force the source and load voltages to be in phase.

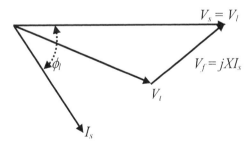

Figure 9.8. Phasor diagram for Case-1

Case-2: When the load is resistive, i.e., $X_l = 0$: The phasor diagram for this case is shown in Figure 9.9. It can be seen that the magnitude of the source and load voltages will never be equal in this case unless the condition that the series compensator must not supply (or absorb) real power is relaxed.

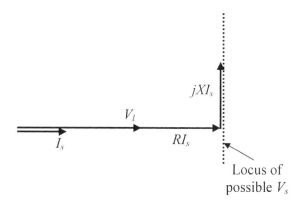

Figure 9.9. Phasor diagram for Case-2

Case-3: This is the most general case in which the load current lags the load voltage and the feeder resistance is not neglected. To draw a phasor diagram, we assume that the load voltage is fixed at V per unit and the source voltage is allowed to vary. Since the primary target is to make the magnitudes of V_l and V_s equal, the locus of desirable V_s is the semicircle as shown in Figures 9.10 and 9.11.

Figure 9.10 shows the limiting behavior. Let us assume that the resistive drop (RI_s) in the feeder is greater than the length *SP*. Since the series compensator must inject voltage in quadrature with the load current, it will not be possible to get the source voltage to be equal to V per unit. Even though the source voltage can be fixed anywhere along the line *NM*, the maximum of $|V_l|/|V_s|$ is obtained when the source voltage is equal to *OM*. On the other hand, if the RI_s drop is exactly equal to *SP*, the load and source voltages can be made equal by aligning the source voltage with the line current. The magnitude of the source voltage is then equal to *OQ*.

Let us again consider the limiting case shown in Figure 9.10 in which we assume that the magnitude of the source voltage (*OM*) is equal to 1.0 per unit and that of the load voltage is V per unit. We then have the distance $OT = V \cos\phi_l$. Hence the distance *MT* will be equal to $1 - V \cos\phi_l$. Therefore we can write $RI_s = 1 - V \cos\phi_l$.

Now suppose the RI_s drop is less than the limiting value (i.e., *SP* in Figure 9.10). The series compensator must then compensate the entire reactive drop in the feeder and provide additional injection such that the magnitude of the source voltage becomes V per unit. It can be seen from Figure 9.11 that there are two possible intersection points with the semicircle – one at A and the other at B. This implies that two possible values of series compensator voltage can be obtained for this case. In the first case the source voltage will be along OA, while in the other case it will be along OB. It is needless to say that the best choice is the A intersection requiring much smaller voltage injection from the series compensator.

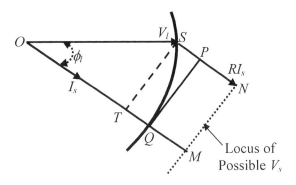

Figure 9.10. Phasor diagram of the limiting condition for Case-3

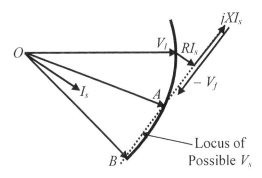

Figure 9.11. Phasor diagram showing multiple solutions for Case-3

To obtain a valid solution we require that

$$RI_s \leq 1 - V\cos\phi_l \quad \Rightarrow \quad I_s \leq \frac{1 - V\cos\phi_l}{R} \tag{9.3}$$

The changes in the source current with the changes in the load power factor angle are shown in Figure 9.12 for different values of the line resistance and $V = 1.0$ per unit. From this it can be seen that as the line resistance increases, the current drawing capacity of load that has to be compensated decreases. This implies that if the load requires more current than is permissible, the series compensator will not be able to regulate the load voltage to 1.0 per unit.

Alternatively, we can also regulate the load voltage to a value that is other than 1.0 per unit. Figure 9.13 shows the system load current characteristics for different values of V. It can be seen that as the requested load voltage V decreases, the maximum current drawing capacity of the load increases. At the same time, a restriction is also put on the minimum current that can be drawn by the load. Similarly, as V increases, the current drawing capacity of the load decreases. Clearly, even for zero load current, a voltage of 1.05 per unit cannot be achieved for low power factor angles.

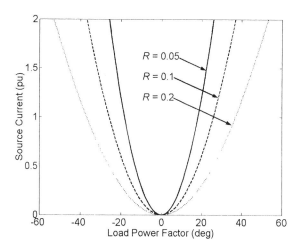

Figure 9.12. Compensatable source current versus power factor, source and load voltages equal: a range of feeder resistance

Example 9.3: In this example, we illustrate the procedure for the steady-state computation of series compensator voltage. Let the feeder and load impedances in per unit be

$$R + jX = 0.05 + j0.3 \text{ and } R_l + jX_l = 2 + j1.5$$

We now connect a series compensator aiming to regulate the load voltage to 1.0 per unit. Let us assume that $V_l = 1\angle 0°$ per unit. The line current is then

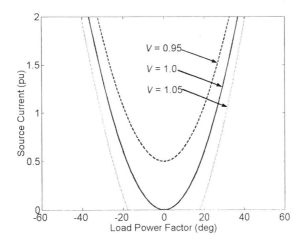

Figure 9.13. Compensatable source current versus load power factor: a range of target voltages

$$I_s = \frac{1\angle 0°}{R_l + jX_l} = 0.4\angle -36.87° \text{ per unit}$$

For zero series compensator real power, its voltage must be in quadrature to the line current. We then have

$$V_f = |V_f| e^{j(\angle I_s + 90°)} = |V_f| \angle 53.13° = |V_f|(a_1 + jb_1) \qquad (9.4)$$

where $a_1 + jb_1$ is a unit phasor at $90°$ to I_s. Again from Figure 9.3 we get

$$V_s + V_f = V_l + (R + jX)I_s = V_l + a_2 + jb_2 \qquad (9.5)$$

where $a_2 + jb_2$ represents the feeder drop.

Substituting (9.4) in (9.5) and rearranging we get

$$V_s = V_l + a_2 + jb_2 - |V_f|(a_1 + jb_1) \qquad (9.6)$$

Assuming that the magnitude of the source voltage $|V_s|$ is known, we get the following quadratic equation from the magnitude condition

$$\left|V_{f}\right|^{2} - 2\{a_{1}(|V_{l}|+a_{2})+b_{1}b_{2}\}|V_{f}|+(|V_{l}|+a_{2})^{2}+b_{2}^{2}-|V_{s}|^{2}=0 \qquad (9.7)$$

Let us assume that the magnitude of the source voltage is 1.0 per unit. Then, as the magnitude of the load voltage is 1.0 per unit, solving the above equation we get

$$|V_{f}| = 0.1476 \, \& \, 1.2924$$

It was mentioned before, these two values of $|V_f|$ correspond to two operating points (A and B) in Figure 9.11. We would obviously choose the lower of the two values because of the lower rating requirements on the compensator.

∆∆∆

9.2.2 Transient Operation of Series Compensator when the Supply is Balanced

The above example demonstrates how a series compensator can be controlled. However, the equations derived above assume sinusoidal steady state operation and full system knowledge. It is thus important to derive the required steady state quantities using half cycle averaging. To extract the sinusoidal steady state quantities, we shall use the derivations presented in Section 3.3.1. Below we present an algorithm that is implemented in the following steps:

1. The system quantities are phase locked to a reference point.
2. Half cycle averaged positive sequence of the line current and source voltage is then extracted with respect to the phase lock.
3. The magnitude of series compensator voltage is calculated based on (9.7).
4. The series compensator voltage is then synthesized with this magnitude and an angle that leads the line current angle by 90°.

We shall call this the Type-1 control. Let us illustrate with the help of the following example.

Example 9.4: Let us consider the same system as in Examples 9.1 and 9.3. The peak of the source voltage is $\sqrt{2}$ per unit and we want to regulate the load voltage such that its peak is also $\sqrt{2}$ per unit. Under this condition, the peak value of the compensator voltage should be equal to $\sqrt{2} \times 0.1476$ as derived in Example 9.3. The system frequency is chosen as 50 Hz. We now use the averaging process given in Section 3.3.1 to obtain the source voltage and load current. It is assumed that the series compensator is realized by three ideal voltage sources and thus any change in the series compensator voltage is reflected in the load voltage instantaneously. It is however not feasible in practical cases where a voltage source inverter is used to implement the series compensator. We shall discuss this aspect later.

The system response is shown in Figure 9.14. It is to be noted that the series compensator is connected to the system after half a cycle (10 ms) once the first average is obtained. We use moving average subsequently such that the value of fundamental voltage and current is available at each sampling instant. The steady state terminal voltage is given by

$$V_t = V_s - (R + jX)I_s = 0.916\angle -5.26°$$

It can be seen from Figure 9.14 (a) that the peak of the terminal is about 1.30 ($\approx \sqrt{2} \times 0.916$) per unit. Figure 9.14 (b) and (c) depict the load voltage and series compensator voltage respectively. The series compensator power is shown in Figure 9.14 (d). It can be seen that it draws power as soon as the series compensator is connected. The load voltage is suddenly changed from 1.30 per unit to $\sqrt{2}$ per unit at the end of first half cycle. Since the load is passive, this results in an increase in the average load power. Since the source is not able to meet this increased power demand instantaneously, the series compensator helps in riding over the transient by supplying the real power temporarily. However, the transient does not last long and the power becomes zero within less than a cycle.

△△△

In the control technique presented above it is assumed that entire circuit parameters are known as well as the source voltage. This however may not be feasible. Alternatively, the series compensator voltage must be synthesized based on local measurements only. To accomplish that we note from Figure 9.3 that $v_t + v_f = v_l$. We then get [7]

$$V_t = V_l - |V_f|(a_1 + jb_1) \qquad (9.8)$$

Note that as in (9.4) the term $a_1 + jb_1$ is a unit phasor that is 90° from the line current. Assuming that $V_l = |V_l|\angle 0°$, the above equation leads to the following quadratic

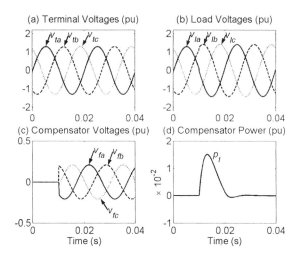

Figure 9.14. System response with Type-1 control

$$\left|V_f\right|^2 - 2a_1\left|V_l\right|\left|V_f\right| + \left|V_l\right|^2 - \left|V_t\right|^2 = 0 \qquad (9.9)$$

This algorithm will be referred to as the Type-2 control. It requires the measurement of the local quantities only. To implement this algorithm we need the fundamental of the series compensator terminal voltage (v_t) along with the line current. The following example illustrates the Type-2 control.

Example 9.5: Let us consider the same system as given in Example 9.4. The system response is shown in Figure 9.15. The system behavior in this case is almost identical to the behavior shown in Figure 9.14.

∆∆∆

9.2.3 Transient Operation when the Supply is Unbalanced or Distorted

The above example illustrates the advantage of using a series compensator when the supply is balanced. However, one of the main reasons for the use of series compensator is to produce clean balanced sinusoidal load voltages even when the supply is unbalanced or distorted. We thus have to modify the above algorithm to accommodate this. To accomplish this, condition-1 must first be modified as

9. Series Compensation of Power Distribution System

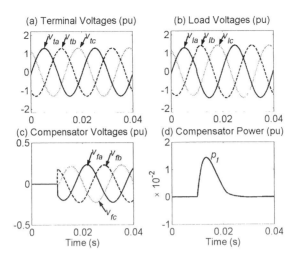

Figure 9.15. System response with Type-2 control

- **Condition-2**: The positive sequence of the series compensator power must remain zero in the steady state. Since the reference load voltages and hence load currents are balanced, this implies that the phase angle difference between the positive sequence of the series compensator voltage phasor and the line current phasor must be $\pi/2$ in the steady state.

In the presence of unbalance or harmonics this condition stipulates that the instantaneous series compensator power will have a zero mean, but will also contain a periodic term. We now divide the series compensator terminal voltage as

$$v_t = v_{t1} + v_t^{rest} \tag{9.10}$$

where v_{t1} is the positive sequence component of v_t and v_t^{rest} is the remaining portion containing the influence of unbalance and harmonics. As per condition-2 we can then write

$$V_{f1} = |V_{f1}| e^{j(\angle I_s + 90°)} = |V_{f1}|(a_1 + jb_1) \tag{9.11}$$

where $a_1 + jb_1$ is a unit phasor at 90° to I_s. We then modify (9.9) to obtain the magnitude of the fundamental frequency component of series compensator voltage from the quadratic

$$|V_{f1}|^2 - 2a_1|V_l||V_{f1}| + |V_l|^2 - |V_{l1}|^2 = 0 \qquad (9.12)$$

The fundamental component of the series compensator voltage is obtained from the above equation. The instantaneous series compensator voltage is then obtained by solving the following equation after phase locking the fundamental component with the reference point

$$v_f = v_{f1} - v_t^{rest} \qquad (9.13)$$

This will provide for the desired correction of the positive sequence term and will cancel all negative or zero sequence components as well as harmonics.

Example 9.6: Let us consider the same system as given in Examples 9.4 and 9.5. Let the supply voltages be unbalanced such that the voltage peaks of phase a, b and c are 1.0, 0.9 and 1.2 per unit respectively. In addition to the unbalance we have also added 5th and 7th harmonics to the source voltage with their magnitudes being inversely proportional to their harmonic number. The source voltages are shown in Figure 9.16 (a). The series compensator is connected at the end of the first half cycle when the extraction of the fundamental quantities starts. The load voltages are shown in Figure 9.16 (b). It can be seen that these voltages become balanced harmonic free with a peak of √2 per unit as soon as the series compensator is pressed in service. The magnitude of the source voltage is maximum in the phase-c. Thus the series compensator voltage correction for this phase is also maximum. This voltage is shown in Figure 9.16 (c). It can be seen that peak voltage requirement is about 0.5 per unit. The instantaneous series compensator power is shown in Figure 9.16 (d). This oscillating power has a zero mean once the initial transients die off at around 0.02s.

△△△

9.2.4   Series Compensator Rating

Since a series compensator is connected in series with the feeder, the full feeder current flows through it. This current mainly depends on the load voltage and impedance. Thus the power rating of a series compensator will mainly depend on the series compensator voltage. So far in our discussion we have assumed that the series compensator is capable of supplying unlimited amount of voltage. This however may not be feasible. Consider the following example.

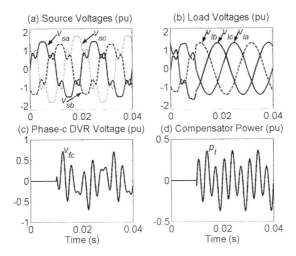

Figure 9.16. Performance of series compensator under unbalanced and distorted supply condition

Example 9.7: Assume that a balanced RL load is supplied by the feeder system shown in Figure 9.3. The feeder and load impedances are given in per unit by

$$R + jX = 0.05 + j0.3 \text{ and } R_l + jX_l = 0.5 + j0.3$$

Assuming that $V_l = 1\angle 0°$ per unit, the line current is

$$I_s = 1.715\angle -30.96° \text{ per unit}$$

We then have

$$V_f = |V_f|(0.5145 + j0.8575)$$

Substituting in (9.7) and assuming $|V_s| = 1.0$ per unit and solving we get

$$|V_f| = 0.6969 \,\&\, 1.3611$$

The lesser of these two values is 0.6969 per unit. Usually, power electronic switches and passive elements are used to realize a series compensator. Thus the capital cost of the installation, maintenance and running cost of the compensator device will increase with the increase in power level rating. If we regulate the load voltage at a lower value instead, the magnitude of the required voltage injection reduces. This is given in Table 9.1 where only the feasible value (lower of the two computed values) is entered.

ΔΔΔ

Table 9.1. Required series voltage injection for desired load voltage

Desired $|V_l|$ in per unit	Required $|V_l|$ in per unit
1.0	0.6969
0.98	0.6269
0.96	0.5635
0.94	0.5048
0.92	0.4497
0.90	0.3976
0.88	0.3478
0.86	0.3001
0.84	0.2543
0.82	0.2099
0.80	0.1670

The above example demonstrates that by dropping the magnitude of the load voltage requirement we can reduce the injection requirement of the series compensator. At the same time it is also possible that a certain target voltage may not be achievable. For example, for the system of Example 9.7, the quadratic (9.7) yields two complex conjugate roots $1.096 \pm j0.096$ if the target load voltage is 1.065 per unit. This mean that there is no feasible solution for the series compensator voltage injection if the target load voltage is greater than 1.06 per unit. The target voltage must be reduced under such a situation. But the question is how one does that on-line?

Fortunately there is a direct relationship between the terminal voltage, power factor of the load and the maximum possible achievable load voltage. Refer to the quadratic given in (9.12). Given a value of $|V_{t1}|$ and a target $|V_l|$, (9.12) will produce two real values of $|V_{l1}|$ provided that the target is feasible. Otherwise, as explained before, it will produce two complex conjugate roots indicating a lack of a feasible solution. We can then surmise that the maximum achievable value of $|V_l|$ is that for which (9.12) yields a single solution. This is given by

$$|V_l| = \frac{|V_{t1}|}{\sqrt{1-a_1^2}} \qquad (9.14)$$

Substituting (9.14) into (9.12) we obtain the series compensator voltage as

9. Series Compensation of Power Distribution System

$$|V_{f1}| = a_1 |V_l| \qquad (9.15)$$

Note that

$$a_1 + jb_1 = \cos(90° - \phi_l) + j(90° - \phi_l) = \sin\phi_l + j\cos\phi_l$$

where ϕ_l is the angle by which i_s lags v_l. Hence we can write

$$1 - a_1^2 = 1 - \sin^2\phi_l = \cos^2\phi_l$$

We can thus modify (9.14) to obtain

$$|V_l| = \frac{|V_{l1}|}{\cos\phi_l} \qquad (9.16)$$

Example 9.8: Let us consider the same system as given in Example 9.5 with Type-2 control. With the system operating in the steady state, the magnitude of the supply voltage suddenly drops to 0.5 per unit in all three phases. This is shown in Figure 9.17 (a). It is needless to say that it is impossible to regulate the load voltage at $\sqrt{2}$ per unit during this time without any real power injection. The best that can be done is to maximize the load voltage. This is done through (9.14) and (9.15) and the results are shown in Figure 9.17 (b-d). The peak of the maximum load voltage that can be supplied during this time is around 0.79 per unit and the peak of the series compensator voltage is around 0.31 per unit. These two voltages are shown in Figure 9.17 along with the compensator power. It can be seen that the steady state value of power remains zero.

∆∆∆

The compensating device itself imposes another limit on the maximum permissible output voltage of the series compensator. We shall investigate this aspect later in this chapter. For the time being, let us assume that the maximum voltage that the compensator can reproduce is denoted by v_f^{max}. Then, no matter however large the voltage demand is, the compensator can only produce up to a maximum of v_f^{max} in any of the three phases. This compensation will clip the load voltage and that in turn will result in distortion in both load voltage and current. The following example illustrates this idea.

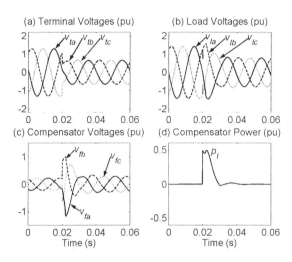

Figure 9.17. Behavior of series compensated system during voltage sag

Example 9.9: Let us consider the same system of Example 9.6. In that example, the peak of the positive sequence series compensator voltage injection required is 0.127 per unit, while the peak of the overall voltage requirements of the three phases are 0.5353, 0.5690 and 0.7142 per unit respectively. If the source was free of harmonics, these peak requirements will reduce to 0.1606, 0.3054 and 0.2061 per unit respectively. We thus see that the major part of the compensation effort is required to cancel the harmonic content of the source. We now choose v_f^{max} of 0.4243 (= $0.3 \times \sqrt{2}$) per unit and run the compensator such that it saturates the voltage injection once it exceeds this requirement.

The system response is shown in Figure 9.18. In these plots, we only display the steady state results from the end of the 1st cycle. Figure 9.18 (a) depicts the actual voltage of phase-c (v_{tc}) and its extracted fundamental voltage (v_{tcf}). It can be seen that the fundamental extraction has stabilized by the time the 2nd cycle starts. The load and injected voltages of the three phases are shown in Figure 9.18 (b)-(d). In these plots, the upper and lower permissible limits of the injected voltage are also indicated. It can be seen that when the series compensator voltage saturates to a maximum or minimum, a distortion occurs in the load voltage. This distortion is maximum in phase-c as it has the maximum peak voltage. Table 9.2 lists the total harmonic distortion (THD) of the source, terminal and load voltages. It can be seen even when the compensator is forced to hit the limit, the THD of the load voltage is considerably lower than the source.

∆∆∆

9. Series Compensation of Power Distribution System

Table 9.2. Total harmonic distortion of various voltages

Voltages		THD in percentage
Source	v_{sa}	24.58
	v_{sb}	24.58
	v_{sc}	24.58
Terminal	v_{ta}	26.65
	v_{tb}	27.04
	v_{tc}	26.11
Load	v_{la}	2.37
	v_{lb}	3.18
	v_{lc}	5.94

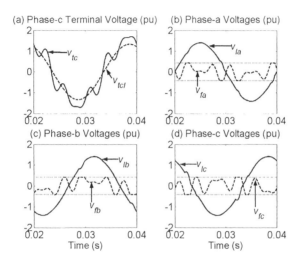

Figure 9.18. System behavior when limits on compensator voltages are imposed

9.2.5 An Alternate Strategy Based on Instantaneous Symmetrical Components

So far we have used the vector formulation for generating DVR reference voltages. We can also use a power based strategy for the generation of these voltages [8]. For this we shall use the local measurements and stipulate that the DVR does not supply or absorb any real power in the steady state. We shall assume that the loads are balanced and we would like to generate a set of balanced voltages at the load terminal. Since the desired load voltages are balanced sinusoids and the loads are balanced, the instantaneous load power is constant. Again as the DVR does not consume or supply any real power, the constant load power must be equal to the average power entering the PCC (terminal bus).

Based on the above considerations and from the instantaneous power equation given in (3.31) of Chapter 3 we get

$$p_l = 2\operatorname{Re}(v_{la1} i_{sa1}^*) = 2|v_{la1}||i_{sa1}|\cos(\phi_{vla1} - \phi_{isa1}) = p_{lav} \qquad (9.17)$$

where ϕ_{vla1} and ϕ_{isa1} are the angles of the vectors v_{la1} and i_{sa1} respectively and p_{lav} is the average power entering the terminal bus. Note that in the above equation the zero sequence components of (3.31) is zero as v_{la0} is zero. From (9.17) we get

$$\phi_{vla1} = \cos^{-1}\left(\frac{p_{lav}}{2|v_{la1}||i_{sa1}|}\right) + \phi_{isa1} \qquad (9.18)$$

We now define the desired instantaneous symmetrical components of the load voltages as

$$\begin{bmatrix} v_{la0} \\ v_{la1} \\ v_{la2} \end{bmatrix} = \sqrt{\frac{3}{2}} \begin{bmatrix} 0 \\ V_m e^{j\phi_{vla1}} \\ V_m e^{-j\phi_{vla1}} \end{bmatrix} \qquad (9.19)$$

where V_m is the desired magnitude of the rms load voltage. Once the instantaneous symmetrical components are obtained from (9.19), we can construct the desired instantaneous voltages using inverse transform given in (3.29). These desired load voltages can then be used to obtain the reference DVR voltages using (9.2).

Example 9.10: Let us consider the same system as given in Example 9.4 in which we desire to regulate the load voltage at $\sqrt{2}$ per unit. The system response when the DVR reference voltages are generated using (9.19) is shown in Figure 9.19. In this the average power (p_{lav}) is obtained through a moving average filter. The DVR is connected at the end of the first half cycle once the first power average is obtained. It can be seen from Figure 9.19 (a) that the load voltages are nearly balanced and have a peak of $\sqrt{2}$ per unit. However the line currents are not balanced as shown in Figure 9.19 (b) and the DVR voltages are distorted as shown in Figure 9.19 (c). The DVR power is oscillating even though the mean of this oscillation is zero. This is shown in Figure 9.19 (d)

ΔΔΔ

9. Series Compensation of Power Distribution System

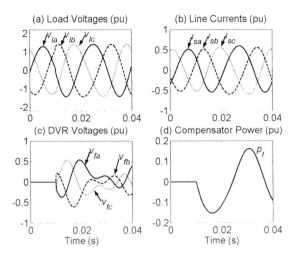

Figure 9.19. System performance on direct application of (9.19)

Since the source and load voltages are balanced and so are the feeder and load impedances, the DVR inserted voltages must be balanced and the compensator power must decay to zero after the initial transients. The reason for this behavioral pattern not being followed in Figure 9.19 is that the sudden connection of the DVR at the end of the first half cycle temporarily unbalances the line currents. This will force the load power to oscillate. Even though p_{lav} is equal to the average real load power, (9.17) is violated during this transient. Without any corrective mechanism, the DVR cannot be forced to correct this violation. This will lead to steady state distortion in its voltage and unbalance in currents.

To alleviate this problem we must keep in mind that as the load is balanced, the load currents will contain the fundamental positive sequence only. The real power at the load terminal then depends only on the positive sequence of the load voltage and line current as required by (9.17). Thus at the point of the insertion of the DVR, the fundamental positive sequence of the line current must be used. Let us denote the fundamental of the instantaneous positive sequence current as i_{sa1f}. This can be directly computed from the samples as

$$i_{sa1f} = e^{j\omega t}\left[\frac{1}{T_0}\int_0^{T_0} i_{sa1} e^{-j\omega t}\, dt\right] = |i_{sa1f}|\angle\phi_{isa1f} \qquad (9.20)$$

The integral in (9.20) is evaluated by using a half cycle averaging through an MA filter. The term in the square brackets containing the integral in (9.20) corresponds to the computation of the fundamental component using the complex Fourier transform. Therefore (9.20) will also reject harmonics in i_s. Note that the algorithm (9.20) is the instantaneous symmetrical component version of (3.56). We can then modify (9.18) as

$$\phi_{vla1} = \cos^{-1}\left(\frac{P_{lav}}{2|v_{la1}||i_{sa1f}|}\right) + \phi_{isa1f} \qquad (9.21)$$

We can then obtain the DVR algorithm using (9.20), (9.21), (9.19) and (9.2).

Example 9.11: Let us consider the same system as given in Example 9.10. The system response with the modified algorithm is shown in Figure 9.20. It can be seen that the load voltages, line currents and DVR inserted voltages are also balanced. Also the DVR power becomes zero after the initial transient dies out.

△△△

Example 9.12: Let us now consider the same unbalanced and distorted supply voltage as given in Example 9.6. The source voltages for this case are shown in Figure 9.16 (a). The DVR is connected to the system at the end of the first half cycle. The results are shown in Figure 9.21. It can be seen that both the load voltages and the line currents become balanced sinusoids as soon as the DVR is connected. The DVR power, shown in Figure 9.21 (d), exhibits the similar zero mean oscillation as given in Figure 9.16 (d).

△△△

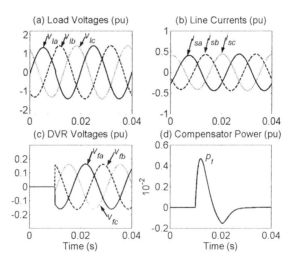

Figure 9.20. System performance with modified reference generation scheme

9. Series Compensation of Power Distribution System

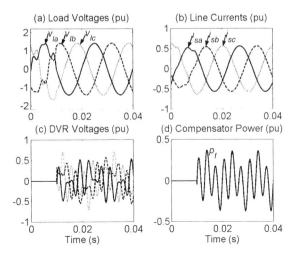

Figure 9.21. System performance when the supply voltage is distorted

The instantaneous symmetrical component based algorithm can also be applied for DVR voltage insertion during a voltage sag and swell. These results are given in [8].

9.3 DVR Structure

The examples in the previous sections assume that the series compensator is realized by ideal voltage sources. In this section we shall develop a DVR structure in which the voltage sources are realized by three voltage source inverters (VSIs). This structure is similar to the DSTATCOM structure of Figure 8.1. In this section we shall study two different filter realizations. One of these two is shown in Figure 9.22. It can be seen just as in Figure 8.1, the three VSIs are connected to a common dc storage capacitor. In this figure each switch represents a power semiconductor device and an anti-parallel diode combination. Each VSI is connected to the network though a transformer and a capacitor filter (C_f). The transformers not only reduce the voltage requirement of the inverters but also provide isolation between the inverters. This prevents the dc storage capacitor from being shorted through switches in different inverters.

In the DVR structure of Figure 9.22, the capacitor filter is connected across the secondary of the transformer. This prevents switching frequency harmonics from entering the system. The main drawback of the system of Figure 9.22 is that the direct connection of VSI to the transformer primary results in losses in the transformer. The high frequency flux variation causes significant increase in transformer iron losses. To avoid this, a switch frequency LC filter (L_f–C_f) is placed in the transformer primary as shown in Figure 9.23 [9]. The secondary of the transformer is directly connected to the feeder. This will constrain the switch frequency harmonics to mainly in the primary side of the transformer. Either of the DVR realization can be controlled through output feedback [7] or state feedback [11]. We discuss these two realizations below.

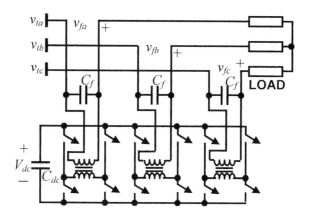

Figure 9.22. DVR structure with capacitor filter

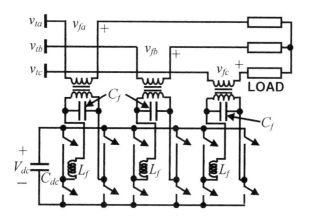

Figure 9.23. DVR structure with LC filter

9.3.1 Output Feedback Control of DVR

In this type of control, the square of the error between reference DVR voltage and actual DVR voltage is minimized to obtain a deadbeat control action. The single-phase equivalent circuit of the DVR with capacitor filter is shown in Figure 9.24. Here $u \cdot V_{dc}$ denotes the switched voltage generated at the inverter output terminals, the inductance L_T represents the leakage inductance of each transformer. The switching losses of the inverter and the copper loss of the connecting transformer are modeled by a resistance R_T.

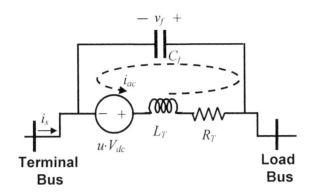

Figure 9.24. Single phase equivalent circuit of DVR with capacitor filter

We can construct the state space model of the system from the equivalent circuit shown in Figure 9.24. Let us define a state vector as $x^T = [v_f \ i_{ac}]$. We then get the following state space model from Figure 9.24

$$\dot{x} = \begin{bmatrix} 0 & 1/C_f \\ -1/L_T & -R_T/L_T \end{bmatrix} x + \begin{bmatrix} 0 & -1/C_f \\ V_{dc}/L_T & 0 \end{bmatrix} \begin{bmatrix} u_c \\ i_s \end{bmatrix} \qquad (9.22)$$

The single-phase equivalent circuit of the DVR with the LC filter is shown in Figure 9.25. In this figure L_T denotes the leakage inductance of each of the transformers and R_{in} denotes the switching losses of each inverter. The copper losses of the transformers are neglected here for simplicity. Note that these losses can be incorporated by a resistor in series with the inductor L_T. Defining a state vector as $x^T = [v_{cf} \ i_{ac}]$, we get the following state space model from Figure 9.25

$$\dot{x} = \begin{bmatrix} 0 & 1/C_f \\ -1/L_f & -R_{in}/L_f \end{bmatrix} x + \begin{bmatrix} 0 & -1/C_f \\ V_{dc}/L_f & 0 \end{bmatrix} \begin{bmatrix} u_c \\ i_s \end{bmatrix} \qquad (9.23)$$

The DVR voltage is given by

$$v_f = v_{cf} - L_T \frac{di_s}{dt} \qquad (9.24)$$

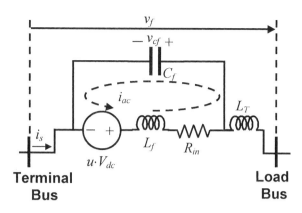

Figure 9.25. Single phase equivalent circuit of DVR with LC filter

The state equation (9.22 or 9.23) is discretized to obtain

$$x(k+1) = Fx(k) + Gu(k) \tag{9.25}$$

Let us define the output by v_y. It is equal to v_f for (9.22) and equal to v_{cf} for (9.23). Defining the elements of matrix F as f_{ij} and the elements of matrix G as g_{ij} we can write

$$v_y(k+1) = f_{11} v_y(k) + f_{12} i_{ac}(k) + g_{11} u_c(k) + g_{12} i_s(k) \tag{9.26}$$

The minimization of the deadbeat performance index results in the following control law

$$u_c(k) = \frac{v_{ref}(k+1) - f_{11} v_y(k) - f_{12} i_{ac}(k) - g_{12} i_s(k)}{g_{11}} \tag{9.27}$$

where v_{ref} is the instantaneous reference value of the DVR. Once u_c is obtained the control input u is obtained in the hysteresis band control discussed before.

The switching band control scheme can be used only when the control variable $u(k)$ is available. Again obtaining $u(k)$ is dependent on the availability of the reference voltage v_{ref}. The reference voltage for the DVR with the capacitor filter (Figure 9.24) is available in a straightforward manner from (9.13) since the output for this case is v_f. This however is not true for the DVR with LC filter (Figure 9.25). We have to generate a reference voltage for v_{cf}.

From Figure 9.25 we can write the phasor fundamental positive sequence component of terminal voltage as

$$V_{t1} = V_l - V_{cf1} + j\omega L_T I_s \qquad (9.28)$$

Since the DVR must inject zero power, the voltage across the filter capacitor can be written as

$$V_{cf} = |V_{cf1}|(a_1 + jb_1) \qquad (9.29)$$

where $a_1 + jb_1$ is a unit vector that leads the line current by 90°. We can then use the following expression for the line current

$$jI_s = |I_s|(a_1 + jb_1) \qquad (9.30)$$

Now defining $X_T = \omega L_T$ (9.28) can be written as

$$V_{t1} = V_l - (|V_{cf1}| - X_T|I_s|)(a_1 + jb_1) \qquad (9.31)$$

The above equation results in the following quadratic

$$|V_{cf1}|^2 - 2|V_{cf1}|(X_T|I_s| + a_1|V_l|) + |V_l|^2 - |V_{t1}|^2 + X_T^2|I_s|^2 \\ + 2a_1 X_T |V_l||I_s| = 0 \qquad (9.32)$$

The DVR voltage must not only supply the instantaneous positive sequence component of v_{cf} but also the harmonic component of v_t. Therefore

$$v_{cf} = v_{cf1} - v_t^{rest} \qquad (9.33)$$

Example 9.13: Let us consider the same system of Example 9.6 in which the source voltages are unbalanced and distorted. We assume that the DVR is supplied by a constant dc source with the value of V_{dc} being 0.6 per unit. The DVR parameters are given in Table 9.3. A hysteresis band of 0.05 per unit is chosen for the switching controller. The results with capacitor and LC filters are shown in Figures 9.26 and 9.27 respectively. In both these figures the terminal and load voltages, compensator power and tracking error of phase-a are shown. It can be seen that both these structures perform satisfactorily with the compensator power being zero mean in either case.

Table 9.3. DVR parameters

With capacitor filter		With LC filter	
$X_f = 1/\omega C_f$	4.0 per unit	$X_f = 1/\omega C_f$	4.0 per unit
$X_T = \omega L_T$	0.1 per unit	$X_T = \omega L_T$	0.1 per unit
R_T	0.05 per unit	$X_f = \omega L_f$	0.025 per unit
		R_{in}	0.05 per unit

ΔΔΔ

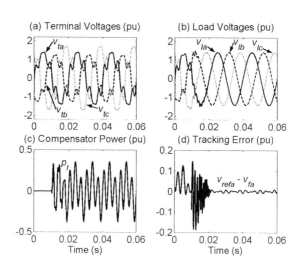

Figure 9.26. DVR under deadbeat control with capacitor filter

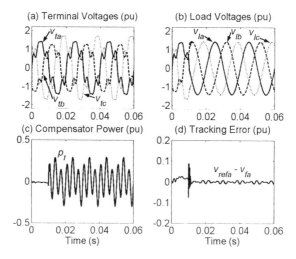

Figure 9.27. DVR under deadbeat control with LC filter

9.3.2 State Feedback Control of DVR

Let us first develop the state feedback controller with the system with capacitor filter. The single-phase equivalent circuit of the compensated system is shown in Figure 9.28. With respect to this figure, let us define the following state and input vectors

$$x^T = \begin{bmatrix} i_1 & i_2 & v_f \end{bmatrix}, \quad z^T = \begin{bmatrix} v_s & u_c \end{bmatrix}$$

The state space model of the system is then

$$\dot{x} = \begin{bmatrix} -\dfrac{R+R_l}{L+L_l} & 0 & \dfrac{1}{L+L_l} \\ 0 & -\dfrac{R_T}{L_T} & -\dfrac{1}{L_T} \\ -\dfrac{1}{C_f} & \dfrac{1}{C_f} & 0 \end{bmatrix} x + \begin{bmatrix} \dfrac{1}{L+L_l} & 0 \\ 0 & \dfrac{V_{dc}}{L_T} \\ 0 & 0 \end{bmatrix} z \quad (9.34)$$

We shall now use a reduced order feedback in which we feedback only i_2 and v_f. The feedback equation is then given by

$$u_c = -K\begin{bmatrix} i_2 - i_{2ref} & v_f - v_{ref} \end{bmatrix} \quad (9.35)$$

where i_{2ref} and v_{ref} are the reference values of i_2 and v_f respectively.

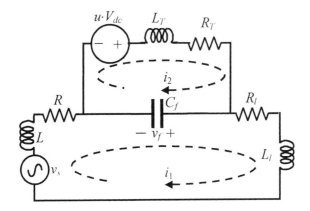

Figure 9.28. Distribution system equivalent circuit of DVR with capacitor filter

Let us define the following upper and lower levels of switching band [7]

$$V_{up} = u_c + h$$
$$V_{dn} = u_c - h \qquad (9.36)$$

In (9.36) the hysteresis band (h) constrains the switching frequency. An H-bridge inverter is capable of supplying V_{dc}, 0, $-V_{dc}$ across its output. We refer to these as switch states $+1$, 0, -1 respectively. This three-level switching control is selected according to Table 9.4. For the section where the desired DVR voltage v_{ref} is positive, the inverter is switched between 0 and $+1$, while it is switched between 0 and -1 when v_{ref} is negative.

Table 9.4. Three-level switching selection

Conditions		Switch Values (u)
$v_{ref} \geq 0$	$v_f > V_{up}$	0
$v_{ref} \geq 0$	$v_f < V_{dn}$	$+1$
$v_{ref} < 0$	$v_f > V_{up}$	-1
$v_{ref} < 0$	$v_f < V_{dn}$	0

In addition to the switching band control loop, an additional loop is required to correct the voltage in the dc storage capacitor against the losses in the inverter and transformer. Furthermore, as we have discussed before, the DVR has to supply real power during transients. All this may cause the capacitor voltage to fall. To correct these deviations, a small amount of real power must be drawn from the source to replenish the losses. To accomplish this we introduce a simple proportional-plus-integral (PI) controller to regulate the dc capacitor voltage around a reference value of V_{dcref}. The PI controller is of the form

$$z_c = K_p e + K_I \int e\, dt \qquad (9.37)$$

where $e = V_{dcref} - V_{dc}^{av}$, V_{dc}^{av} being the average voltage of the capacitor over a complete cycle. The unit of z_c is radian and in steady state is indicative of the losses in the converter.

We now modify (9.11) such that

$$V_{f1} = |V_{f1}| e^{j(\angle I_s + 90° - z_c)} = |V_{f1}|(\tilde{a}_1 + j\tilde{b}_1) \qquad (9.38)$$

As a result the quadratic (9.12) is modified to

9. Series Compensation of Power Distribution System   367

$$|V_{f1}|^2 - 2\tilde{a}_1|V_l||V_{f1}| + |V_l|^2 - |V_{t1}|^2 = 0 \qquad (9.39)$$

Under this condition, the phase difference between the line current and the DVR voltage differs slightly from 90°. The following example demonstrates the theoretical limits of the DVR operation.

Example 9.14: Consider the load and feeder impedances are same as given in Example 9.1. They are supplied by a set of balanced voltages with a peak value of √2 per unit. It is assumed that all system parameters are referred to the feeder side of the transformers. The system parameters are the same as given in Example 9.13 (Table 9.3). The LQR control design used in Chapter 8 is used here. The LQR gain matrix is generated using $Q = diag(0\ 1\ 20)$ and $r = 0.1$, where $diag$ denotes a diagonal matrix. For the above system parameters we get $K = [-4.07\ 4.47\ 13.04]$. We now use a reduced order feedback in which the source current is neglected. It is to be noted that the DVR reference voltage v_f^* is obtained from the reference voltage generation scheme. However, there exists no such mechanism for the generation of the current reference i_2^*. It is therefore taken as 0. The capacitor and the capacitor control loop parameters are given in per unit by

$$V_{dcref} = 0.6,\ X_{dc} = 1/\omega C_{dc} = 0.1326,\ K_P = 1\ \text{and}\ K_I = 10$$

The dc capacitor is precharged to 0.6 per unit and the DVR is connected at the end of the first half cycle. A hysteresis band (9.36) of 0.05 per unit is chosen. The PI controller is started from rest. The results are shown in Figure 9.29. It can be seen that the controller settles in 10 cycles (0.2 s). The excursion in the capacitor voltage is within 0.05 per unit and it also settles within 0.1 s. The load voltages are balanced sinusoids.

Let us now assume that the system is operating in steady state for one cycle (20 ms) with capacitor voltage feedback when the peak of the source voltages is suddenly reduced to 0.4×√2 per unit. As a result of this, the terminal voltage drops to a level such that it is impossible to regulate the load voltage at 1.0 per unit. To operate the DVR under this condition, we must modify the conditions given in (9.14) and (9.15) since the limiting value is now obtained from the quadratic (9.39). This limiting value is

$$|V_l| = \frac{|V_{t1}|}{\sqrt{1-\tilde{a}_1^2}} \qquad (9.40)$$

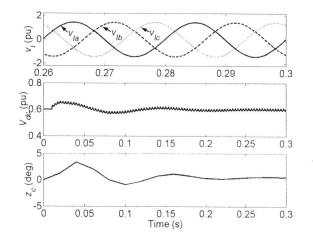

Figure 9.29. DVR performance with state feedback and dc capacitor control loop

and the DVR voltage under the limiting condition is

$$|V_{f1}| = \tilde{a}_1 |V_l| \qquad (9.41)$$

The system response is shown in Figure 9.30 in which the terminal voltages and the load voltages are shown in (a) and (b) respectively. The maximum possible load voltage is calculated based on the Fourier extraction of the line current and terminal voltage. As mentioned before this extraction depends on the half cycle averaging and this averaging process is continuously running. The load voltage limit, shown in Figure 9.30 (c), thus varies continuously till the system oscillations die out. It has a value of 1.0 per unit before the transient. The dc capacitor response voltage is shown in Figure 9.30 (d).

∆∆∆

Let us now consider the following example that demonstrates the practical limit. As we have seen before that the practical limit on the target load voltage is usually put by the DVR itself. It will be clear from the following example that the storage device limits the operation of the DVR.

Example 9.15: Consider the system parameters as in Example 9.14. With the system in steady state, the peak of the phase-c of the source voltage suddenly sags to 0.2 per unit at the end of the 1st cycle. It however is restored to √2 per unit at the end of the 3rd cycle through clearing of the fault. The

9. Series Compensation of Power Distribution System

system response is shown in Figure 9.31. The terminal and load voltages are shown in Figure 9.31 (a) and (b) respectively. It can be seen that the DVR fails to maintain the load voltage at $\sqrt{2}$ per unit during the transient.

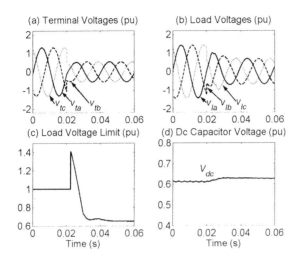

Figure 9.30. Performance of DVR when all phases dip

There are two reasons behind this behavior. The first one is due to the reduction in the target load voltage as per (9.40). The second one is the saturation of the inverter voltage due to the inability of the dc storage capacitor to supply the required voltage. This is shown in Figure 9.31 (c) where the desired DVR voltage of phase-c (v_{fc}^*) is given by dashed line and the actual DVR voltage (v_{fc}) is shown by solid line. The dc capacitor behavior is shown in Figure 9.31 (d). It can be seen that capacitor charges momentarily before discharging to supply the drop in the faulted phase. If the drop is maintained, the capacitor loop PI control action will restore the capacitor voltage to its nominal value with a time constant of 0.15 sec.

ΔΔΔ

For the DVR with LC filter, the transformer inductance (L_T) is placed in series with the load inductance (L_l). Therefore the state space representation remains the same except that the parameters of the state matrices are changed. Therefore the state feedback control equation will also remain the same except that in this case the reference is the voltage across the capacitor C_f rather than the DVR reference voltage itself. This reference voltage can be calculated using (9.32) and (9.33). Additionally the capacitor voltage feedback loop can also be added to the above equations using the same procedure discussed above.

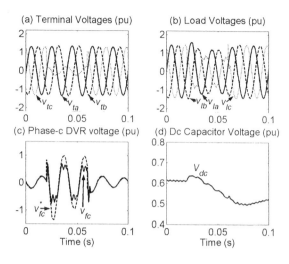

Figure 9.31. Performance of DVR when one phase sags

9.4 Voltage Restoration

So far our discussion was focused on voltage regulation and restoration using a series compensator that ideally requires no real power in the steady state. In this configuration the series compensator is kept on-line all the time to maintain voltage at the load terminals. It has been shown that the series compensator, which is supplied from a dc storage capacitor, ordinarily needs real power to replenish any losses in the converter circuit. It also needs real power to ride over any transient. However, as we have demonstrated, this power can be drawn from the source through feedback control of capacitor voltage.

The series compensator can also be used in an alternate form in which it comes on line only when there is a voltage sag. Otherwise it stays inactive. Consider the phasor diagram of Figure 9.32 (a). This represents the steady state operation of the circuit when the series compensator acts as a voltage restorer only. The supply voltage during the steady state operation is V_s^{old} and it leads the load voltage by an angle δ_{old}. Now suppose a fault reduces the supply voltage to V_s^{new} that leads the load voltage by an angle δ_{new}. The series compensator then must inject a voltage such that the vector sum of load voltage and line drop remains unchanged and equal to V_s^{old}. This is shown in Figure 9.32 (b).

The phasor diagram shown in Figure 9.32 (a) is only to illustrate the restoration behavior for transient control of sudden voltage dip. Since we can only use the local measurements, neither the source voltage nor the feeder impedance can be used for series compensator control. The voltage

9. Series Compensation of Power Distribution System

restoration function however is very straightforward. The series compensator voltage is obtained from the following equation

$$V_f = V_l^{pf} - V_t \tag{9.42}$$

where V_l^{pf} is the measured prefaulted voltage at the load terminal. It is to be noted that voltage restoration using (9.42) implies real power exchange during any transient. A series compensator that is supplied by a dc storage rather than a dc capacitor can easily accomplish that. Dynamic voltage compensation using high speed flywheel energy storage system (FESS) has been reported in [11]. Alternatively, as we have discussed before, the dc link capacitor can be supplied from a rectifier.

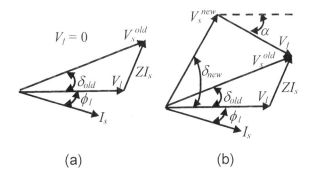

Figure 9.32. Phasor diagram of series operation: (a) steady state operation and (b) transient voltage restoration

Example 9.16: Let us consider the same system as Example 9.1. When the system operates in the nominal steady state, the peak of the terminal voltage and load voltage is 1.2964 per unit. A fault occurs that reduces the peak of the supply voltage to 0.6 per unit from √2 per unit at the end of the 1st cycle. The supply voltage is restored to its nominal steady state value at the end of 3rd cycle. During the voltage sag the series compensator must inject a voltage such that the load voltage is held at the prefaulted level without any jump in the phase angle. The magnitude and phase of the prefaulted voltage are obtained through the fundamental extraction algorithm discussed before. The system response is shown in Figure 9.33. It can be seen that apart from spikes at the beginning and at the end of the transient the load voltage remains at constant peak throughout. The real power absorbed by the compensator is shown in Figure 9.33 (d). Since, for operation in the nominal steady state, the voltage injection requirement is zero, the real power requirement is also zero. However, during the transient the compensator consumes power to maintain the load voltage.

ΔΔΔ

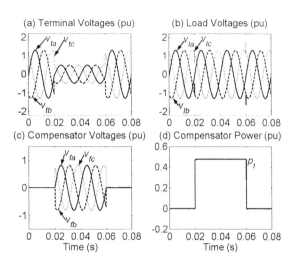

Figure 9.33. Transient voltage restoration using DVR

9.5 Series Active Filter

A series compensator, which injects a voltage in series, can also act as a series active filter to isolate the source from harmonics generated by loads. Consider the distribution system shown in Figure 9.3. If the load is unbalanced, then by injecting a voltage in series we shall be able to correct this unbalance at the PCC (terminal bus).

Example 9.17: Consider a three-phase, four-wire distribution system that is supplied by a balanced voltage with a peak value of √2 per unit. It supplies an unbalanced load at the end of a balanced feeder. The feeder impedance is $0.05 + j0.3$ per unit and the load parameters in per unit are

$$Z_{la} = 2 + j1.5, \; Z_{lb} = 2.55 + j1.25 \text{ and } Z_{lc} = 1 + j1.2$$

We shall correct this unbalance through a series compensator. To accomplish this we assume that a bidirectional converter that can inject or absorb power supplies the series compensator. We use a very simplified algorithm to correct the voltage. We first compute the fundamental positive

9. Series Compensation of Power Distribution System

sequence of the load voltage and then stipulate that the series compensator injects voltages such that the terminal bus voltage is equal to this voltage.

The results are shown in Figure 9.34. In this we have connected the converter at the end of the first cycle (0.02 s). It can be seen that both the terminal voltages and source currents become balanced, but the load voltage remains unbalanced. Also the power entering the terminal becomes constant.

ΔΔΔ

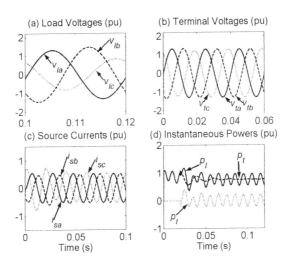

Figure 9.34. Load balancing using series compensator

Now suppose in addition to the unbalance, the load also draws harmonic currents. We must then provide a path for this harmonic current to flow. This can be accomplished by connecting a shunt capacitor at the load terminal. This will provide a path for the load harmonic current to flow such that the nonlinear current does not flow through the source. The series compensator then will be able to balance the voltages of the terminal bus and, as a consequence, the source currents will also be balanced. Let us consider the following example.

Example 9.18: Let us consider the same system as given in Example 9.17. In this case we have assumed that the load also contains a converter that draws a squarewave current with a peak of 0.35 per unit. The reactance value of the shunt capacitor is 7.0 per unit. We follow the same principle for the series compensator control, i.e., find the fundamental positive sequence of the load voltage first and then make the compensator inject a voltage such that the terminal voltage is equal to this voltage. The results are shown in Figure 9.35 in which the compensator is connected at the end of the first cycle. Note that before the compensator is connected at 0.02 s, the terminal voltage is equal to the load voltage. It can be seen that form Figure 9.35 (b) that the load voltage is badly distorted before the connection of the compensator. However, once the compensator is connected, the terminal voltage and load current get balanced. Also the power entering the terminal also becomes constant. However the compensator power does not have mean of zero. This means that the compensator must be capable of either supplying or absorbing power. The load voltage however remains unbalanced and distorted. Note that in this figure a uniform time axis for all the plots are not chosen as all quantities have different settling time.

ΔΔΔ

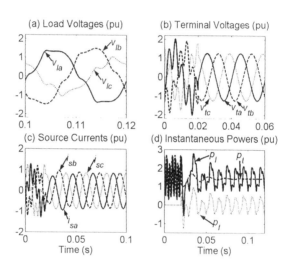

Figure 9.35. Load balancing and harmonic compensation using a series device and a shunt capacitor filter

The above example shows that the series compensator, when used with a shunt capacitor, can clean up the source current and the terminal voltage. The load voltage however still remains unbalanced and distorted with the distortion level being unacceptable. To alleviate this problem, we can connect tuned filters at the load terminals. The per phase equivalent circuit of the series active and shunt passive filter is shown in Figure 9.36. In this figure, the feeder impedance is denoted by $R + jX$. The shunt capacitor filter is denoted by C_f and the load impedance is denoted by Z_l. We can use a set of tuned filter banks to eliminate the lower order harmonics.

Example 9.19: Let us consider the same system as given in Example 9.18. We now compensate the system with both shunt filter capacitor and

9. Series Compensation of Power Distribution System

tuned filters as shown in Figure 9.36. Three different tuned filters are used to eliminate the 3rd, 5th and 7th harmonics. The results are shown in Figure 9.37. It can be seen from Figure 9.37 (b) that prior to the connection of the series active filter at 0.02 s, the tuned filters have sufficiently managed to eliminate the harmonic components of the load voltage. However with the connection of the series compensator both terminal voltage and source currents get balanced. The load voltage now contains significantly fewer amounts of harmonics. However this voltage now has both magnitude and phase unbalance.

ΔΔΔ

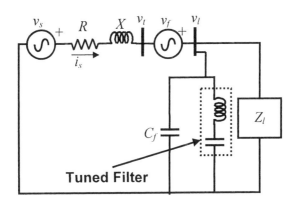

Figure 9.36. Schematic diagram of series active and shunt passive filter compensated system

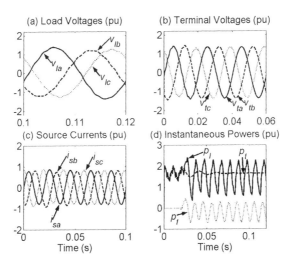

Figure 9.37. Compensation characteristics series active and shunt passive filter combination

The concept of using a combination of series active and shunt passive filters has been proposed in [12]. In [13] its transient characteristics and system stability are discussed. This scheme has been refined further to put a set of shunt passive filters and a series filter that is connected in series with the shunt filters [14,15]. The passive filter improves the harmonic characteristics of the load voltage and the active filter improves the filtering characteristics of the tuned filters. The advantage with this scheme is that the size of the active filter used is much smaller as compared to the active filter that is connected in series with the feeder. A review of the various filtering schemes is presented in [16]. In [17] a series compensator is discussed for the correction of both supply side unbalance and voltage regulation.

9.6 Conclusions

In this chapter we have presented a systematic study of a dynamic voltage restorer that can tightly regulate voltage at the load terminals against any variation in the supply side voltage. It has been shown that the DVR can get a dc voltage support externally or it can operate through a dc capacitor in which it consumes no real power in the steady state. The capability of the device is clearly demonstrated through steady state analysis. Based on this analysis, a number of options to obtain the time varying DVR reference voltages are formulated. Finally a suitable topology to realize the DVR by voltage source inverters (VSIs) is also discussed. All discussions are supplemented by simulation results using MATLAB. As shown in this chapter that a DVR is a voltage regulator, voltage restorer and voltage conditioner – all put into one.

In addition it has been shown that through a combination of tuned filter and series voltage injection, supply side harmonics can also be eliminated. We have however only demonstrated the operating principle of this combination. The topics such as the control of this combination and the reference series compensator voltage generation are not discussed here. Some of these topics are discussed in [12,13].

9.7 References

[1] N. H. Woodley, L. Morgan and A. Sundaram, "Experience with an inverter-based dynamic voltage restorer," *IEEE Trans. Power Delivery*, Vol. 14, No. 3, pp.1181-1186, 1999.

[2] N. H. Woodley, A. Sundaram, B. Coulter and D. Morris, "Dynamic voltage restorer demonstration project experience," *12th Conf. Electric Power Supply Industry (CEPSI)*, Pattaya, Thailand, 1998.

[3] N. H. Woodley, K. S. Berton, C. W. Edwards, B. Coulter, B. Ward, T. Einarson and A. Sundaram, "Platform-mounted DVR demonstrated project experience," 5th *International Transmission & Distribution Conf, Distribution 2000*, Brisbane, 2000.

[4] Y. H. Zhang, D. M. Vilathgamuwa and S. S. Choi, "An experimental investigation of dynamic voltage restorer," *IEEE Power Engineering. Society Winter Meeting*, Singapore, 2000.

[5] S. W. Middlekauff and E. R. Collins, "System and customer impact: considerations for series custom power devices," *IEEE Trans. Power Delivery*, Vol. 13, No. 1, pp. 278-282, 1998.

[6] G. T. Heydt, W. Tan, T. LaRose and M. Negley, "Simulation and analysis of series voltage boost technology for power quality enhancement," *IEEE Trans. Power Delivery*, Vol. 13, No. 4, pp. 1335-1341, 1998.

[7] A. Ghosh and G. Ledwich, "Compensation of distribution system voltage using dynamic voltage restorer (DVR)," To appear in *IEEE Trans. Power Delivery*.

[8] A. Ghosh and A. Joshi, "A new algorithm for the generation of reference voltages of a DVR using the method of instantaneous symmetrical components," *IEEE PE Letters in IEEE Power Engg. Review*, Vol. 22, No. 1, pp. 63-65, 2002.

[9] C. G. Hochgraf, "Power inverter apparatus using a transformer with its primary winding connected the source end and a secondary winding connected to load end of an ac power line to insert series compensation," *United States Patent*, Number US005905367A, May 1998.

[10] A. Ghosh and G. Ledwich "Structures and control of a dynamic voltage regulator (DVR)," *IEEE Power Engineering Society Winter Meeting*, Columbus, Ohio, 2001.

[11] R. S. Weissbach, G. G. Karady and R. G. Farmer, "Dynamic voltage compensation on distribution feeders using flywheel energy storage," *IEEE Trans. Power Delivery*, Vol. 14, No. 2, pp.465-471, 1999.

[12] F. Z. Peng, H. Akagi and A. Nabae, "A new approach to harmonic compensation in power systems – a combined system of shunt passive and series active filters," *IEEE Trans. Industry Applications*, Vol. 26, No. 6, pp. 983-990, 1990.

[13] F. Z. Peng, H. Akagi and A. Nabae, "Compensation characteristics of the combined system of shunt passive and series active filters," *IEEE Trans. Industry Applications*, Vol. 29, No. 1, pp. 144-152, 1993.

[14] H. Fujita and H. Akagi, "A practical approach to harmonic compensation in power systems – series connection of passive and active filters," *IEEE Trans. Industry Applications*, Vol. 27, No. 6, pp. 1020-1025, 1991.

[15] H. Fujita, T. Yamasaki and H. Akagi, "A hybrid active filter for damping of harmonic resonance in industrial power systems," *IEEE Trans. Power Electronics*, Vol. 15, No. 2, pp. 215-222, 2000.

[16] H. Akagi, "New trends in active filters for power conditioning," *IEEE Trans. Industry Applications*, Vol. 32, No. 6, pp. 1312-1322, 1996.

[17] A. Campos, G. Joos, P. D. Ziogas and J. F. Lindsay, "Analysis and design of a series voltage unbalance compensator based on a three-phase VSI operating with unbalanced switching functions," *IEEE Trans. Power Electronics*, Vol. 9, No. 3, pp. 269-274, 1994.

Chapter 10

Unified Power Quality Conditioner

A unified power quality conditioner (UPQC) is a device that is similar in construction to a unified power flow conditioner (UPFC) [1]. The UPQC, like a UPFC, employs two voltage source inverters (VSIs) that are connected to a common dc energy storage capacitor. One of these two VSIs is connected in series with the ac line while the other is connected in shunt with the same line. A UPFC is employed in a power transmission system to perform shunt and series compensation at the same time. Similarly a UPQC can also perform both the tasks in a power distribution system. However, at this point the similarities in the operating principles of these two devices end. Since a power transmission line generally operates in a balanced, distortion (harmonic) free environment, a UPFC must only provide balanced shunt or series compensation. A power distribution system, on the other hand, may contain unbalance, distortion and even dc components. Therefore a UPQC must operate under this environment while providing shunt or series compensation.

The UPQC is a relatively new device and not much work has been reported on it yet. It has been viewed as a combination of series and shunt active filters in [2,3]. In [3] it has been shown that it can be used to attenuate current harmonics by inserting a series voltage proportional to the line current. Alternatively, the inserted series voltage is added to the voltage at the point of common coupling such that the device can provide a buffer to eliminate any voltage dip or flicker. It is also possible to operate it as a combination of these two modes. In either case, the shunt device is used for providing a path for the real power to flow to aid the operation of the series connected VSI. Also included in this structure is a shunt passive filter to which all the relatively low frequency harmonics are directed. Experimental results with a relatively stiff voltage source are also provided in [3].

In this chapter we shall demonstrate that a UPQC combines the operations of a DSTATCOM and a DVR together. The series component of the UPQC inserts voltage so as to maintain the voltage at the load terminals balanced and free of distortion. Simultaneously, the shunt component of the UPQC injects current in the ac system such that the currents entering the bus to which the UPQC is connected are balanced sinusoids. Both these objectives must be met irrespective of unbalance or distortion in either source or load side. In addition, we shall prefer to operate both the series and shunt compensators such that they do not supply or absorb any real power from the ac system during the steady state.

10.1 UPQC Configurations

Let us first assume that the combination of an ideal series voltage source and an ideal shunt current source represents the UPQC. There are two possible ways of connecting this device at the point of common coupling (PCC). The single-line diagrams of these two schemes are shown in Figures 10.1 and 10.2. In these figures the voltage at the PCC is referred to as the terminal voltage v_t. The load voltage, load current and source current are denoted by v_l, i_l and i_s respectively. The voltage and current injected by the UPQC are denoted by v_d and i_f respectively. The source voltage is denoted by v_s, while R and L constitute the feeder impedance. We shall restrict our discussions to three-phase, four-wire systems only.

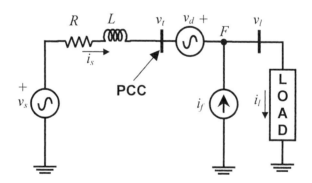

Figure 10.1. The right-shunt UPQC compensation configuration

Let us assume that both the source voltages and load currents are unbalanced and distorted. We stipulate that the UPQC shall perform the following two functions

- Convert the feeder (source) current (i_s) to balanced sinusoids through the shunt compensator.
- Convert the load voltage (v_l) to balanced sinusoids through the series compensator and also regulate it to a desired value.

This can be achieved by employing either of the two configurations shown in Figures 10.1 and 10.2. They are termed as right-shunt and left-shunt respectively depending on the placement of the shunt compensator vis-à-vis the series compensator. We shall discuss the characteristics of these two configurations separately.

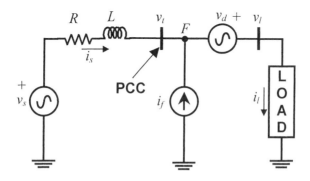

Figure 10.2. The left-shunt UPQC compensation configuration

10.2 Right-Shunt UPQC Characteristics

In this section we shall present the operating characteristics of the right-shunt UPQC. For all the subsequent discussions, the following notations will be used.

- The instantaneous quantities will be denoted by the lower case letters (e.g. v) and the upper case letters (e.g. V) will denote the phasor quantities.
- The subscripts a, b and c will denote the three phases.
- The subscripts 0, 1 and 2 will denote the zero, positive and negative sequences respectively.

We shall now characterize the operation of the UPQC through the voltage and current that it injects into the ac system. Let us first assume that the load and source are unbalanced but not distorted (harmonic-free).

In the right-shunt structure of Figure 10.1, we note that $v_t + v_d = v_l$. Now if the source voltage is unbalanced, the terminal voltage will also be unbalanced. Therefore to obtain a balanced voltage at the load terminal, v_d must cancel out the unbalance in the terminal voltage. From this condition we get

$$V_{d0} = -V_{t0} \quad V_{d2} = -V_{t2} \tag{10.1}$$

The above condition stipulates that V_l is strictly positive sequence. Let us assume that the magnitude of the positive sequence voltage is $|V_{l1}|$. Note that the angle of this voltage will depend on the power factor of the load.

We shall now generate the injected positive sequence voltage such that the series compensator does not require any positive sequence power. Observe that the current flowing through the series compensator is the source current i_s. We then must inject a positive sequence voltage that has phase difference of 90° with I_{s1}. This yields the following equation

$$V_{t1} + |V_{d1}|(a_1 + jb_1) = V_{l1} \tag{10.2}$$

where $a_1 + jb_1$ is a unit vector that is 90° to I_{s1}. Assuming $V_{l1} = |V_{l1}|\angle 0°$, (10.2) results in the following quadratic that is similar to the one derived for the DVR in Chapter 9

$$|V_{d1}|^2 - 2a_1|V_{l1}||V_{d1}| + |V_{l1}|^2 - |V_{t1}|^2 = 0 \tag{10.3}$$

As in the case of the DVR, if the desired voltage level $|V_{l1}|$ is achievable, the quadratic of (10.3) has two positive real solutions of $|V_{d1}|$. It is needless to say that the minimum of these two solutions will be chosen, as this will require less effort by the UPQC.

For shunt current injection, we can see from Figure 10.1 that by applying KCL at point F we get $i_s + i_f = i_l$. Therefore the shunt compensator must inject current to cancel the zero and negative sequences of the load current. Thus we get

$$I_{f0} = I_{l0} \quad I_{f1} = 0 \quad I_{f2} = I_{l2} \tag{10.4}$$

This will ensure that a purely positive sequence current equal to the positive sequence load current I_{l1} flows from the source. Therefore through (10.1) to (10.4), both unbalances in the source voltages and load currents are cancelled by the right-shunt UPQC.

10. Unified Power Quality Conditioner

Note from Figure 10.1 that if the compensation is perfect, a purely positive sequence current (I_{s1}) will flow through the series compensator. Since the positive sequence injected voltage (V_{d1}) is in quadrature with this current as per (10.2), the positive sequence power consumed by the series compensator will be zero. Moreover since the current through the series compensator is strictly positive sequence, the negative and zero sequence power consumed by it will also be zero. Therefore the average power consumed by the series compensator is zero. Again since the load voltage is strictly positive sequence and shunt current source does not inject any positive sequence current, the average power injected by the shunt compensator is also zero. We can therefore conclude that the right-shunt UPQC does not inject or absorb any power.

It is to be noted that the algorithm given in (10.1) to (10.4) requires the phasor measurements. For this we shall use the on-line fundamental sequence component extraction algorithm given in Chapter 3. The following example illustrates the working principles of the right-shunt UPQC.

Example 10.1: Let us consider a three-phase, four-wire distribution system that is supplied by an unbalanced source. The instantaneous source voltages are given in per unit by

$$v_{sa} = \sqrt{2}\sin(100\pi t)$$
$$v_{sb} = \sqrt{2} \times 1.2\sin\left(100\pi t - 120°\right)$$
$$v_{sc} = \sqrt{2} \times 0.85\sin\left(100\pi t + 120°\right)$$

It supplies an unbalanced Y-connected load through a balanced feeder with an impedance of $0.05 + j0.3$ per unit per phase. The load is a parallel combination of a balanced Y-connected resistive load and an unbalanced Y-connected RL load. The value of resistive load is 10 per unit per phase and the impedances of the unbalanced load are $1.999 + j1.5001$ per unit, $2.5479 + j1.252$ per unit and $0.9991 + j2.99901$ per unit for phases a, b and c respectively.

The UPQC is connected at the end of the first half cycle (10 ms) after the fundamental quantities are obtained. An ideal voltage source and an ideal current source as shown in Figure 10.1 represent the UPQC. The results are shown in Figure 10.3. The terminal voltages and the load currents are shown in Figure 10.3 (a) and (b) respectively. It can be seen that they are unbalanced before and after the UPQC is connected. The load voltages and source currents are shown in Figure 10.3 (c) and (d) respectively. It can be seen that they become balanced within a cycle of the connection of the UPQC to the ac system.

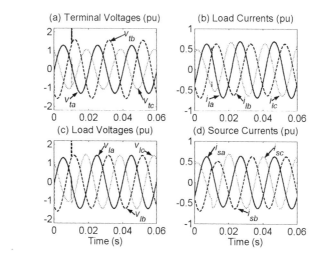

Figure 10.3. System voltages and currents with right-shunt UPQC

The instantaneous powers are shown in Figure 10.4. The terminal power (p_t) and the load power (p_l) are shown in Figure 10.4 (a) and (b) respectively. The power through the series compensator (p_d) and the shunt compensator (p_f) are shown in Figure 10.4 (c). To investigate the behavior of these powers closely, their running averages with a window of 10 ms (half a cycle) are calculated. These are shown in Figure 10.4 (d). In these running averages, the first values are obtained after the first half cycle. It can be seen that the averages of the load and terminal powers become equal in the steady state and the average powers through the series and shunt compensators become zero. We can therefore conclude that the algorithm based on (10.1) to (10.4) makes the UPQC operate in the zero power injection/absorption mode.

Figure 10.5 shows the instantaneous reactive powers in various parts of the circuit. The reactive powers are computed using the relations (3.21) and (3.23) given in Chapter 3. It can be seen that the sum of the terminal reactive power (q_t) and the series compensator reactive power (q_d) becomes constant in the steady state with a mean equal to the mean of the reactive power requirement of the load (q_l). The mean of the shunt compensator reactive power (q_f) is zero. Since the shunt compensator only cancels the unbalance in the load current, it only provides the oscillating component of the load reactive power, while the mean of the load reactive power is supplied by the combination of the terminal and series compensator.

ΔΔΔ

10. Unified Power Quality Conditioner

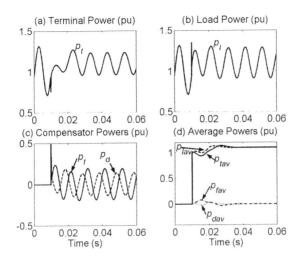

Figure 10.4. Instantaneous and average powers with right-shunt UPQC

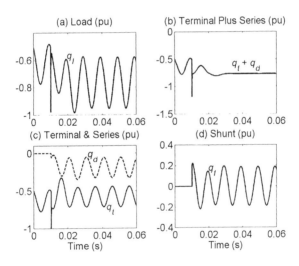

Figure 10.5. Instantaneous reactive powers with right-shunt UPQC

To make the shunt compensator to share a part of the load reactive power, we have to modify (10.4) while keeping in mind that the shunt compensator also must cancel the zero and negative sequence of the load current. We therefore get

$$I_{f0} = I_{l0} \quad I_{f2} = I_{l2} \tag{10.5}$$

Equation (10.5) ensures that the shunt compensator supplies the oscillating reactive component of the load reactive power. We can therefore distribute the average component of the load reactive power between the shunt compensator and that supplied by the terminal and series compensator combined. However whatever may be the reactive power injection by the shunt compensator, it must be ensured that the power through the shunt compensator is zero. Since the average values of both the active and reactive power through the shunt compensator depends only on the positive sequence current, we get

$$|V_{l1}||I_{f1}|\cos\theta = 0$$
$$|V_{l1}||I_{f1}|\sin\theta = \beta \times q_{lav} \qquad (10.6)$$

where θ is the angle between V_{l1} and I_{f1}, β is a scalar that defines how much reactive power must be supplied by the shunt compensator and q_{lav} is the average of the instantaneous load reactive power. Note that the first part of (10.6) is satisfied for $\theta = \pm 90°$ as $|V_{l1}|$ and $|I_{f1}|$ are non-zero. Then for inductive loads, i.e., when q_{lav} is negative, the angle of the positive sequence current I_{f1} must lag V_{l1} by 90° to satisfy the second part of (10.6). The magnitude of I_{f1} will then be given by

$$|I_{f1}| = \beta \frac{|q_{lav}|}{|V_{l1}|} \qquad (10.7)$$

The following example demonstrates the reactive power sharing.

Example 10.2: Let us consider the same system as given in Example 10.1. The results for $\beta = 0$ are shown in Figures 10.3 to 10.5. Let us now choose β to be equal to 1. The results are shown in Figure 10.6 in which the compensator is connected after 10 ms. It can be seen that both the load voltages and the source currents become balanced sinusoids within about one and a half cycle. The load and terminal average powers become equal in the steady state and, as a consequence, the power injected by the UPQC becomes zero. The reactive powers are shown in Figure 10.6 (d). Since β is equal to one, the shunt compensator must supply the entire reactive power (average plus mandatory oscillating components) of the load. This is evident from Figure 10.6 (d) where the traces of q_l and q_f coincide. It can be seen that the shunt compensator forces power factor of the source current to be unity at the load terminal as the sum of the terminal reactive power and the series compensator reactive power is zero. Note that these two components

10. Unified Power Quality Conditioner

(q_t and q_d) are not individually zero but they cancel each other to arrive at the zero reactive power for $\beta = 1$. In Figure 10.6 (b) along with the source currents, scaled version of the phase-a load voltage ($v'_{la} = v_{la}/1.5$) is plotted to show that the source currents are in phase with the load voltage. This also indicates a unity power factor operation.

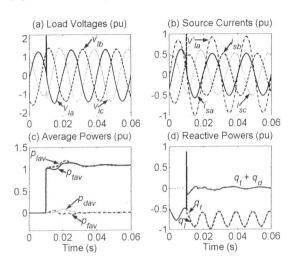

Figure 10.6. System response with right-shunt UPQC with shunt reactive injection for $\beta = 1$

Figure 10.7 depicts the system reactive powers for two different values of β, namely $\beta = 0.5$ and $\beta = -0.5$. From Figure 10.7 (a), it can be seen that the reactive power at the load terminal ($q_t + q_d$) is equal to the average of the reactive power supplied by the shunt compensator (q_f) for $\beta = 0.5$. This means that the average reactive power required by the load is shared equally by the shunt compensator and the terminal and series compensator combination. The reactive powers entering the terminal and series compensator are shown in Figure 10.7 (b). It can be seen that their oscillating components cancel each other. The results for $\beta = -0.5$ are shown in Figure 10.7 (c and d). The shunt compensator now supplies positive reactive power. To offset this the reactive power at the load terminal (i.e., combination of terminal and series, $q_t + q_d$) becomes more negative than the load reactive power. This means that the source current now leads the load voltage. Obviously this will force the source to supply more reactive power than required by the load, the balance being absorbed by the shunt compensator.

△△△

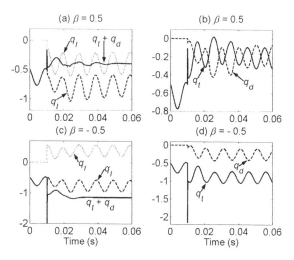

Figure 10.7. Instantaneous reactive powers for two different values of β

10.3 Left-Shunt UPQC Characteristics

Let us now consider the left-shunt UPQC shown in Figure 10.2. Like the right-shunt structure, this must also cancel the unbalance in the terminal voltage. Therefore we get the following equation that guarantees V_l is strictly positive sequence.

$$V_{d0} = -V_{t0} \quad V_{d2} = -V_{t2} \tag{10.8}$$

To generate the injected positive sequence voltage, we note that the current flowing through the series compensator is the load current (i_l) and not the source current (i_s) like in the case of the right-shunt structure. We shall then inject a positive sequence voltage that has phase difference of 90° with I_{l1}. This yields the following equation

$$V_{t1} + |V_{d1}|(a_2 + jb_2) = V_{l1} \tag{10.9}$$

where $a_2 + jb_2$ is a unit vector that is 90° to I_{l1}. Assuming $V_{l1} = |V_{l1}|\angle 0°$, (10.9) results in a quadratic that is similar to the one given in (10.3). This is given by

$$|V_{d1}|^2 - 2a_2|V_{l1}||V_{d1}| + |V_{l1}|^2 - |V_{t1}|^2 = 0 \tag{10.10}$$

10. Unified Power Quality Conditioner

To generate reference current through the shunt compensator, it must be ensured that the positive sequence power through the compensator is zero. Furthermore, if a scalar multiple γ defines the fraction of the load reactive power that can be supplied by the positive sequence reactive power of the shunt compensator, we get the following relations

$$|V_{t1}||I_{f1}|\cos\phi = 0$$
$$|V_{t1}||I_{f1}|\sin\phi = \gamma \times q_{lav} \qquad (10.11)$$

where ϕ is the angle between V_{t1} and I_{f1} and should be $\pm 90°$. The reference sequence currents of the shunt compensator are then given by

$$I_{f0} = I_{l0} \quad I_{f2} = I_{l2} \quad I_{f1} = \gamma \frac{|q_{lav}|}{|V_{t1}|} \qquad (10.12)$$

The following example illustrates the working principles of the left-shunt UPQC.

Example 10.3: Let us consider the same system of Examples 10.1 and 10.2. The UPQC is connected at the end of the first half cycle (10 ms). For this we have chosen a γ of 1. The results are shown in Figures 10.8 and 10.9. Figure 10.8 shows the load voltages and the source currents. It can be seen that they are balanced sinusoids. The average active powers and the instantaneous reactive powers are shown in Figure 10.9. It can be seen that the average terminal active power becomes equal to the average load active power (Figure 10.9 a). However the average powers through shunt and series compensators are not zero. As can be seen from Figure 10.9 (b) that p_{dav} has a negative steady state value while p_{fav} has a positive steady state value. But their net effect is zero. This implies that there is a real power exchange between these two compensators. Since the value of γ is chosen as 1, the mean of q_f is equal to the mean of q_l, but they may not be identical in the steady state. This is evident from Figure 10.9 (c). The sum $q_t + q_d$ is neither constant nor has a mean of zero.

△△△

We have thus seen that unlike the right-shunt UPQC, the left-shunt UPQC does not operate in the zero power injection/absorption mode. Equations (10.9) and (10.11) only guarantee that the positive sequence powers through the series and shunt compensators are zero. However since i_l is unbalanced and so is v_d, there will be negative and zero sequence powers through the series compensator. Similarly due to the presence of negative and zero sequence components in the terminal voltage v_t and shunt compensator current i_f, the power through the shunt compensator is not zero. However as seen from Figure 10.9 (a and b), the averages of these two powers cancel each other such that the average load power is equal to that supplied by the terminal.

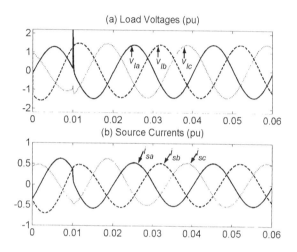

Figure 10.8. Load voltages and source currents with left-shunt UPQC

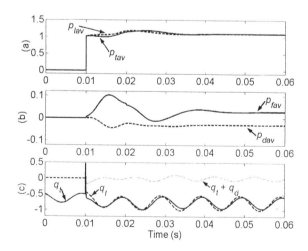

Figure 10.9. Average active and instantaneous reactive powers with left-shunt UPQC

Comparing the characteristics of these two structures we can conclude the following.

- The right-shunt UPQC operates in a zero power injection/absorption mode, while the left-shunt UPQC cannot operate in this mode.
- The right-shunt UPQC can make the power factor unity at the load terminal, while the power factor at the load terminal depends on the load for the left-shunt UPQC.
- The shunt compensator in the right-shunt UPQC can supply the entire requirement of the reactive power by the load whereas the shunt compensator in the left-shunt UPQC can only supply the mean of the load reactive power.

Overall the characteristics of the right-shunt UPQC are superior to those of the left-shunt UPQC.

10.4 Structure and Control of Right-Shunt UPQC

So far we have discussed the characteristics of the two UPQCs using ideal models. In this section we shall present the structure and control of the right-shunt UPQC.

10.4.1 Right-shunt UPQC Structure

As mentioned earlier this device is built around two voltage source inverters that are supplied by a common dc storage capacitor. Also the VSIs must be able to inject unbalanced voltages and currents. Therefore we have chosen a structure that consists of six H-bridge inverters and six single-phase isolating transformers. The schematic diagram of the right shunt structure is shown in Figure 10.10.

In the right-shunt UPQC structure, the outputs of the VSIs realizing the shunt compensator are directly connected to three single-phase transformers. The secondary terminals of these transformers are connected in Y, with the neutral point being connected to the load neutral. The phase sides of the secondary terminals are connected in shunt with the distribution feeder. Three filter capacitors, one for each phase, are also connected in shunt to provide paths for the switching frequency harmonics generated by the three VSIs. Three LC filters are connected at the three output terminals of the three VSIs realizing the series compensator. The secondary terminals of the three single-phase transformers are then connected to inject voltages in series with the distribution feeder. The LC filters are used to bypass the switching frequency harmonics.

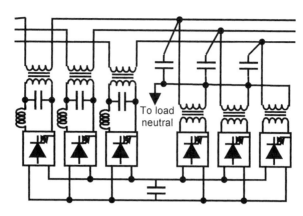

Figure 10.10. Schematic diagram of a right-shunt UPQC

10.4.2 Right-Shunt UPQC Control

The single-phase equivalent circuit of the right-shunt UPQC is shown in Figure 10.11. In this figure R and L indicate the feeder parameters. The load is denoted by R_l and L_l. The LC filter across the series inverter is represented by L_d and C_d, while the resistance R_d represents the inverter losses. The inductance L_T denotes the leakage inductance of the transformers connected in series. The leakage inductance of the shunt transformer is denoted by L_f while the resistance R_f represents losses. The shunt filter capacitor is denoted by C_f. Note that the voltage at the load terminal is the voltage across this filter capacitor. The series injected voltage v_d is as indicated in Figure 10.11, while the voltage v_{sd} is the voltage across the capacitor C_d. Note that both the series and shunt inverters are supplied by a common capacitor. The voltage across this capacitor is denoted by V_{dc}. The switched voltages across the series and shunt inverter output terminals are then denoted by $V_{dc}.u_1$ and $V_{dc}.u_2$ respectively.

In right-shunt UPQC structure of Figure 10.10, the tracking control stability problem is avoided through the choice of LC filter for the series inverter and capacitor filter for the shunt inverter. Consider the equivalent circuit of Figure 10.11. Instead of the LC filter, had we used a capacitor filter, the inductance L_T will be zero as the transformer leakage inductance will then be in series with the series inverter. However a feedback control is required to track the voltages across both the capacitors. Note from Figure 10.11 that in the absence of L_T tracking voltage v_l by the shunt compensator will make the tracking voltage v_d across the series compensator redundant and vice versa. If we now force the two inverters to track the voltages across their respective capacitors, the stability problem will arise as one controller will interfere with the tracking performance of the other. Therefore to avoid the tracking problem the LC filter structure for the series inverter is more suitable for the right-shunt UPQC. Also the LC filter restricts the switching frequency harmonics in the primary of the transformer connected to it.

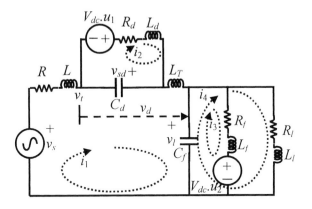

Figure 10.11. Single-phase equivalent circuit of right-shunt UPQC

To derive a state space model of the system of Figure 10.11, notice that the circuit of contains six state variables – four loop currents and two capacitor voltages. We can then define the following state vector

$$x^T = \begin{bmatrix} i_1 & i_2 & i_3 & i_4 & v_{sd} & v_l \end{bmatrix} \qquad (10.13)$$

The circuit of Figure 10.11 also contains three forcing functions – the source voltage v_s and switching variables u_1 and u_2. Let us replace u_1 and u_2 by the continuous-time variables u_{c1} and u_{c2} respectively and define the following control vector

$$u^T = \begin{bmatrix} u_{c1} & u_{c2} \end{bmatrix} \qquad (10.14)$$

The state space equation of the system is then given by

$$\dot{x} = Ax + B_1 u + B_2 v_s \qquad (10.15)$$

where the matrices A and B_1 and B_2 are given by

$$A = \begin{bmatrix} -\dfrac{R}{L+L_T} & 0 & 0 & 0 & \dfrac{1}{L+L_T} & -\dfrac{1}{L+L_T} \\ 0 & -\dfrac{R_d}{L_d} & 0 & 0 & -\dfrac{1}{L_d} & 0 \\ 0 & 0 & -\dfrac{R_f}{L_f} & 0 & 0 & \dfrac{1}{L_f} \\ 0 & 0 & 0 & -\dfrac{R_l}{L_l} & 0 & \dfrac{1}{L_l} \\ -\dfrac{1}{C_d} & \dfrac{1}{C_d} & 0 & 0 & 0 & 0 \\ \dfrac{1}{C_f} & 0 & -\dfrac{1}{C_f} & -\dfrac{1}{C_f} & 0 & 0 \end{bmatrix}$$

$$B_1 = \begin{bmatrix} 0 & 0 \\ \dfrac{V_{dc}}{L_d} & 0 \\ 0 & -\dfrac{V_{dc}}{L_f} \\ 0 & 0 \\ 0 & 0 \\ 0 & 0 \end{bmatrix} \quad B_2 = \begin{bmatrix} \dfrac{1}{L+L_T} \\ 0 \\ 0 \\ 0 \\ 0 \\ 0 \end{bmatrix}$$

We now have to formulate a state feedback control system to obtain the switching variables. Note that the state variables can be written in terms of the network parameters as follows

$$\begin{aligned} i_s &= i_1 \\ i_l &= i_4 \\ i_f &= i_4 - i_1 \\ i_{cf} &= i_1 - i_3 - i_4 \\ i_{cd} &= i_2 - i_1 \end{aligned} \qquad (10.16)$$

where i_{cf} and i_{cd} are the charging currents through the capacitors C_f and C_d respectively. We now define a transformed state vector z and relate with the state vector x using (10.16) as

$$z = \begin{bmatrix} i_f \\ i_{cf} \\ v_l \\ i_{cd} \\ v_{sd} \\ i_l \end{bmatrix} = \begin{bmatrix} -1 & 0 & 0 & 1 & 0 & 0 \\ 1 & 0 & -1 & -1 & 0 & 0 \\ 0 & 0 & 0 & 0 & 0 & 1 \\ -1 & 1 & 0 & 0 & 0 & 0 \\ 0 & 0 & 0 & 0 & 1 & 0 \\ 0 & 0 & 0 & 1 & 0 & 0 \end{bmatrix} x = Px \qquad (10.17)$$

The state space equation (10.15) is then transformed as

$$\dot{z} = PAP^{-1}z + PB_1 u + PB_2 v_s \qquad (10.18)$$

The control input is then given by

$$u = -K(z - z_{ref}) \qquad (10.19)$$

where K is the LQR gain matrix that is computed using the matrices PAP^{-1} and PB_1. Once u_{c1} and u_{c2} are computed using (10.19), the switching functions u_1 and u_2 are obtained using the hysteresis band control described in Chapter 8.

For the state feedback controller to perform satisfactorily, the reference z_{ref} for the transformed state vector z must be chosen judiciously obeying the network laws. Of the six elements of the vector z, it is rather difficult to form a reference for the load current i_l. This element is therefore eliminated from the state feedback and a reduced order feedback is used instead of the full state feedback of (10.19). As we have seen in Chapter 8 that this does not destabilize the system. Of the remaining five elements, the reference for i_f is obtained from the reference generation equations (10.5) and (10.7). Therefore we have to generate reference for the four remaining variables.

Note that the desired load voltage is balanced and therefore has positive sequence only. Therefore the reference load voltage and the current through the filter capacitor of the shunt compensator are given by

$$\begin{aligned} V_{lref} &= V_{t1} + V_{d1ref} \\ I_{cfref} &= \omega C_f V_{lref} e^{j90°} \end{aligned} \qquad (10.20)$$

where ω is the fundamental frequency. The magnitude of V_{d1ref} is the solution of the quadratic (10.3), while its angle is defined by the unit vector $a_1 + jb_1$. Note that the balanced sinusoidal desired source current is equal to

$$I_{sref} = I_{l1} - I_{f1ref} \qquad (10.21)$$

where the magnitude of I_{f1ref} is obtained from (10.7), while its angle lags that of V_{lref} by 90°.

Also note from Figure 10.11 that the voltage v_{sd} is given by

$$v_{sd} = v_d + L_T \frac{di_1}{dt} = v_d + L_T \frac{di_s}{dt} \qquad (10.22)$$

Therefore the positive sequence of the reference voltage across the capacitor C_d is given by

$$V_{sd1ref} = V_{d1ref} + \omega L_T I_{sref} e^{j90°} \qquad (10.23)$$

The zero and negative sequence reference voltages are then

$$\begin{aligned} V_{sd0ref} &= V_{d0ref} = -V_{t0} \\ V_{sd2ref} &= V_{d2ref} = -V_{t2} \end{aligned} \qquad (10.24)$$

Once these sequence components are obtained we use the inverse symmetrical component transform to obtain the three-phase voltage reference V_{sdref}. Consequently the current through this capacitor is given by

$$I_{cdref} = \omega C_d V_{sdref} e^{j90°} \qquad (10.25)$$

Note that the reference quantities are obtained in the phasor domain. These are then converted into instantaneous domain and tracked using the state feedback law of (10.19). The following example shows the effects of the state feedback control law.

Example 10.4: Let us consider a system in which the source voltages are unbalanced and are given in per unit by

$$\begin{aligned} v_{sa} &= \sqrt{2} \sin(100\pi t) \\ v_{sb} &= \sqrt{2} \times 1.2 \sin(100\pi t - 120°) \\ v_{sc} &= \sqrt{2} \times 0.85 \sin(100\pi t + 120°) \end{aligned}$$

10. Unified Power Quality Conditioner

The balanced feeder impedance is given by $0.5 + j0.3$ per unit per phase and the unbalanced load impedances are given in per unit by

$$Z_{la} = 2 + j1.5 \quad Z_{lb} = 2.55 + j1.25 \quad Z_{lc} = 1 + j2.3$$

The UPQC parameters in per unit are

$$X_{cf} = \frac{1}{\omega C_f} = 7.02 \quad X_f = \omega L_f = 0.2 \quad R_f = 0$$

$$X_{cd} = \frac{1}{\omega C_d} = 4.0 \quad X_d = \omega L_d = 0.01 \quad X_T = \omega L_T = 0.1 \quad R_d = 0$$

It is assumed that a dc source rather than a dc capacitor supply all the inverters. The dc source voltage is chosen as 2.5 per unit.

The system is started with the VSIs being blocked and the system steady state is obtained. The UPQC is switched on half a cycle after the system reaches the steady state, once the phasor variables are obtained. The LQR is designed with the following weighting matrices

$$Q = \begin{bmatrix} 20 & & & & \\ & 1 & & & \\ & & 10 & & \\ & & & 1 & \\ & & & & 10 \\ & & & & & 0 \end{bmatrix} \quad R = \begin{bmatrix} 0.1 & \\ & 0.1 \end{bmatrix}$$

A hysteresis band of 0.1 per unit is chosen for the control of both the shunt and series inverters. The results are shown in Figure 10.12 for which it is assumed that β in (10.7) is zero, i.e., the shunt compensator only compensates for the unbalance in the load currents. It can be seen from Figure 10.12 (a) and (b) that the load voltages and source currents get balanced within 2 cycles of the UPQC being switched on. Figure 10.12 (c) shows the average powers. It can be seen that the terminal average power becomes equal to the load average power in the steady state. From Figure 10.12 (d) it can be seen that the reactive power supplied by the shunt compensator is oscillating with a mean of zero. As a consequence the terminal and the series compensator supplies the average of the reactive power required by the load. This is because β is chosen as zero.

ΔΔΔ

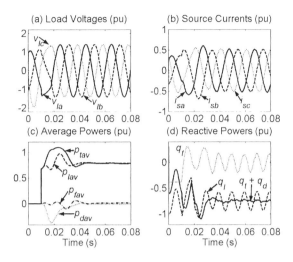

Figure 10.12. Correcting unbalance using a right-shunt UPQC

10.4.3 Harmonic Elimination using Right-Shunt UPQC

So far in our discussions we have restricted our attention to canceling unbalance only. A UPQC can also act as a harmonic isolator. In case both the load and source contain harmonic, the UPQC can prevent the harmonic in the source voltage from appearing at the load terminal and, at the same time, can prevent the load current harmonics from flowing into the source. Below we present an algorithm through which it can be achieved. This algorithm uses a combination of state feedback control for shunt compensator and deadbeat output feedback control for series compensator.

The state feedback control is achieved through the control law

$$u_{c2} = -K \begin{bmatrix} i_f - i_{fref} \\ i_{cf} - i_{cfref} \\ v_l - v_{lref} \end{bmatrix} \qquad (10.26)$$

where the gain matrix K is obtained through LQR design. The series compensator is controlled by the deadbeat control law that is discussed in Section 9.3.1. The control law is given by

$$u_{c1}(k) = \frac{v_{sdref}(k+1) - f_{11} v_{sd}(k) - f_{12} i_2(k) - g_{12} i_s(k)}{g_{11}} \qquad (10.27)$$

10. Unified Power Quality Conditioner

As mentioned earlier that for the control law to perform satisfactorily, a set of consistent references is required. To obtain the reference set we note that both the load current and terminal voltage now have fundamental and harmonic components. We can therefore write

$$i_l = i_l^{fund} + i_l^{har}$$
$$v_t = v_t^{fund} + v_t^{har} \qquad (10.28)$$

where the superscripts *fund* and *har* denote the fundamental and harmonic components respectively. In Section 10.4.2 we have discussed the reference generation procedure for the fundamental components. In addition, the shunt filter must cancel the harmonic contents of the load current. We therefore get

$$i_{fref} = i_f^{fund} + i_l^{har} \qquad (10.29)$$

where i_f^{fund} is generated using the equations (10.5) and (10.7). Since the voltages at the load terminal must be balanced sinusoids, (10.20) provides valid references for v_l and i_{cf}.

In a similar way the reference for v_{sd} is given by

$$v_{sdref} = v_{sd}^{fund} - v_t^{har} \qquad (10.30)$$

where v_{sd}^{fund} is the solution of (10.23) and (10.24). It is to be noted that the desired current through the inductance L_T is balanced. Therefore the harmonic voltage that must appear across v_d must also appear across v_{sd} itself. This results in the above equation. The following example illustrates the harmonic elimination property of the right-shunt UPQC.

Example 10.5: Let us now consider the same system as given in Example 10.4 except that the source voltages now contain the 5th and 7th harmonic components with their magnitudes being inversely proportional to their harmonic numbers and the load also draws a squarewave current with a peak of 0.35 per unit. The source voltages and load currents are shown in Figure 10.13 (a) and (b) respectively. The weighting for the state feedback controller and the hysteresis bands are the same as given in Example 10.4. Further it is also assumed that the VSIs are supplied by a dc source with a value of 2.5 per unit.

The UPQC is connected at 0.01 s (i.e., at the end of the first half cycle) with the value β being set to zero. The load voltages and source currents are shown in Figure 10.13 (c) and (d) respectively. It can be seen both of them become balanced within about one and half cycle. The average powers and instantaneous reactive powers are shown in Figure 10.14 (a) and (b) respectively. It can be seen that they behave as expected. The series and shunt tracking errors, i.e., $(v_{dref} - v_d)$ and $(i_{fref} - i_f)$ are shown in Figure 10.14 (c) and (d) respectively. It can be seen that they rapidly converge to zero. This proves the efficacy of the UPQC control.

ΔΔΔ

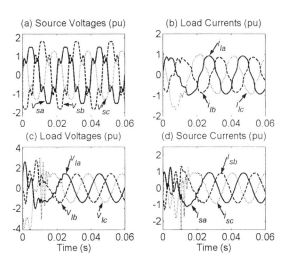

Figure 10.13. Canceling load and source unbalance and harmonics using right-shunt UPQC

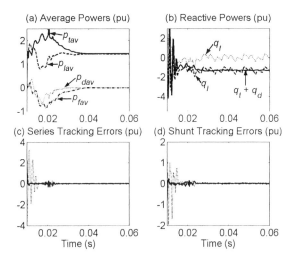

Figure 10.14. Average power, reactive power and tracking errors using right-shunt UPQC when both load and source are unbalanced and distorted

10. Unified Power Quality Conditioner

The above examples assume that the UPQC is supplied by dc source. In practice however the dc source is replaced by a dc capacitor. Note that the right-shunt UPQC operates in zero power injection/absorption mode. Therefore the dc capacitor, once charged, will have a constant average voltage in an ideal lossless case. In practical situations however the dc capacitor voltage must be regulated against the losses in the inverter and transformer. As we have shown in Chapters 8 and 9 that the dc capacitor control involves drawing additional power from the ac system to replenish the losses in the circuit.

We can use the shunt compensator for the dc capacitor voltage regulation. Since the shunt compensator should cancel the negative and zero sequence currents of the load, we can only use the positive sequence component to draw additional power. We can then modify (10.6) as

$$|V_{l1}||I_{f1}|\cos\theta = p_{loss}$$
$$|V_{l1}||I_{f1}|\sin\theta = \beta \times q_{lav}$$
(10.31)

where p_{loss} is the output of the dc capacitor control loop. This term can be derived from a proportional-plus-integral (PI) control output of the negative feedback of the average dc capacitor voltage in the same manner as discussed in Chapters 8 and 9. Note that for a small amount of p_{loss}, θ will be close to 90°.

10.5   Structure and Control of Left-Shunt UPQC

In this section we shall discuss the structure and control of the left-shunt UPQC. There are many similarities between the two UPQC structures and controls. We shall however highlight their differences in this section.

10.5.1   Left-Shunt UPQC Structure

The schematic diagram of a left-shunt UPQC is shown in Figure 10.15. In this we have used only capacitor filter both for the series and the shunt inverters. The tracking problem associated with the right-shunt UPQC will not arise in the left-shunt UPQC even if we use capacitor filters with both shunt and series inverters. Consider the equivalent circuit of the left-shunt UPQC shown in Figure 10.16. In this case the shunt inverter tracks the terminal voltage v_t and the series inverter maintains the voltage v_l across the load by tracking the voltage v_d. Therefore the two inverters track two different quantities even if they are interdependent. It is therefore expected that the stability problem will not arise for the left-shunt UPQC structure of Figure 10.15.

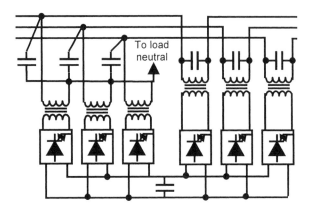

Figure 10.15. Schematic diagram of a left-shunt UPQC

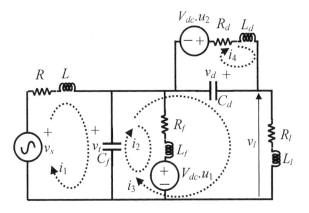

Figure 10.16. Equivalent circuit of a left-shunt UPQC

10.5.2 Left-Shunt UPQC Control

In the single-phase equivalent circuit of the left-shunt UPQC, shown in Figure 10.16, the leakage inductance of the series transformer is denoted by L_d. The switched voltage across the shunt and series inverter output terminals are then denoted by $V_{dc}.u_1$ and $V_{dc}.u_2$ respectively. The other parameters are the same as those given for the right-shunt UPQC of Figure 10.11. The state space equation of this circuit can also be written with six state variables and two input variables. The state and input vectors are

10. Unified Power Quality Conditioner

$$x^T = \begin{bmatrix} i_1 & i_2 & i_3 & i_4 & v_t & v_d \end{bmatrix} \quad u^T = \begin{bmatrix} u_{c1} & u_{c2} \end{bmatrix}$$

and the state space equation of the system is then given by

$$\dot{x} = Ax + B_1 u + B_2 v_s \tag{10.32}$$

where

$$A = \begin{bmatrix} -\dfrac{R}{L} & 0 & 0 & 0 & -\dfrac{1}{L} & 0 \\ 0 & -\dfrac{R_f}{L_f} & 0 & 0 & \dfrac{1}{L_f} & 0 \\ 0 & 0 & -\dfrac{R_l}{L_l} & 0 & \dfrac{1}{L_l} & \dfrac{1}{L_l} \\ 0 & 0 & 0 & -\dfrac{R_d}{L_d} & 0 & -\dfrac{1}{L_d} \\ \dfrac{1}{C_f} & -\dfrac{1}{C_f} & -\dfrac{1}{C_f} & 0 & 0 & 0 \\ 0 & 0 & -\dfrac{1}{C_d} & \dfrac{1}{C_d} & 0 & 0 \end{bmatrix}$$

$$B_1 = \begin{bmatrix} 0 & 0 \\ -\dfrac{V_{dc}}{L_f} & 0 \\ 0 & 0 \\ 0 & \dfrac{V_{dc}}{L_d} \\ 0 & 0 \\ 0 & 0 \end{bmatrix} \quad B_2 = \begin{bmatrix} \dfrac{1}{L} \\ 0 \\ 0 \\ 0 \\ 0 \\ 0 \end{bmatrix}$$

Note that the state variables can be written in terms of the network parameters as follows

$$\begin{aligned} i_s &= i_1, \quad i_l = i_3, \quad i_f = i_3 - i_1 \\ i_{cf} &= i_1 - i_2 - i_3 \\ i_{cd} &= i_4 - i_3 \end{aligned} \tag{10.33}$$

We can then perform a suitable state transformation such that all the pertinent signals appear as state variables. The transformation is given by

$$z = \begin{bmatrix} i_f \\ i_{cf} \\ v_t \\ i_l \\ i_{cd} \\ v_d \end{bmatrix} = \begin{bmatrix} -1 & 0 & 1 & 0 & 0 & 0 \\ 1 & -1 & -1 & 0 & 0 & 0 \\ 0 & 0 & 0 & 0 & 1 & 0 \\ 0 & 0 & 1 & 0 & 0 & 0 \\ 0 & 0 & -1 & 1 & 0 & 0 \\ 0 & 0 & 0 & 0 & 0 & 1 \end{bmatrix} x = Px \qquad (10.34)$$

The state equation (10.32) then can be transformed into

$$\dot{z} = PAP^{-1}z + PB_1u + PB_2v_s = \Lambda z + \Gamma_1 u + \Gamma_2 v_s \qquad (10.35)$$

We can then design a state feedback controller of the form $u = -K(z - z_{ref})$ and use hysteresis band tracking control. The feedback gain matrix can be calculated using the LQR design.

We can then use the reduced order feedback by neglecting the load current. The harmonic components of the load current and the terminal voltages are cancelled using the same method discussed in the previous section. Let us consider the following example.

Example 10.6: For this example we shall consider a system in which the source voltages are unbalanced while the load currents are both unbalanced and distorted. The system parameters are the same as used in Examples 10.4 and 10.5 except that the value of L_d is chosen as 0.2 per unit.

The UPQC is assumed to be supplied by a 2.5 per unit dc source. The UPQC is connected at the end of the first half cycle after the fundamental quantities are obtained. At this instant the series compensator is connected first. After another one and half cycle the shunt compensator is connected. This is done to prevent excessive current to flow through the circuit when both compensators are connected simultaneously. This problem was not encountered with the right-shunt UPQC. It is to be noted that both the filter capacitors are precharged at the point of connection of the compensators.

The results are shown in Figure 10.17. It can be seen that both the load voltages and source currents become balanced sinusoids. The terminal voltages remain unbalanced. However it becomes free of harmonics once the UPQC action takes place.

ΔΔΔ

10. Unified Power Quality Conditioner

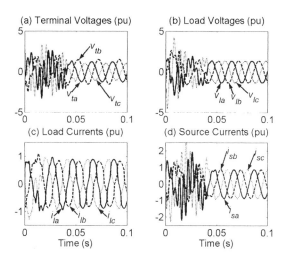

Figure 10.17. System quantities with a left-shunt UPQC

The voltage source in the above example is assumed to be unbalanced but distortion-free. Consider the equivalent circuit of Figure 10.16. If the voltage source is distorted then, for the source current to be distortion-free, the terminal voltage must also contain all the harmonic components of the source voltage [4]. This can be accomplished through the back projection algorithm discussed in Chapter 8, which will complicate the control system. This problem however does not exist with the right shunt structure. The dc capacitor control of the left-shunt structure is discussed in [4].

10.6 Conclusions

In this chapter we have discussed an operation of UPQC in which it can simultaneously correct for the unbalance and distortion in the source voltage and load current. It has been shown that the right-shunt UPQC structure can operate in the zero power injection/absorption mode thereby making the dc capacitor control simpler. However the size of the dc capacitor must be chosen such that it has the ride through capability during any transient. The issues involved with size and control of the dc capacitor need to be explored further.

A UPQC is an expensive device, as it requires two sets of inverters. However it is much more flexible than any single inverter based device. Thus its full capability must be investigated further. Also new applications of this device must be explored to justify its cost.

10.7 References

[1] L. Gyugyi, "Unified power flow control concept for flexible ac transmission systems," *Proc. IEE*, Pt. C, Vol. 139, pp. 323-331, 1992.
[2] F. Kamran, T. G. Habelter, "Combined deadbeat Control of a series-parallel converter combination used as universal power filter," *Proc. IEEE Power Electronic Specialist Conf. (PESC)*, pp. 196-201, 1995.
[3] H. Fujita and H. Akagi, "The unified power quality conditioner: the integration of series- and shunt-active filters," *IEEE Trans. Power Electronics*, Vol. 13, No. 2, pp. 315-322, 1998.
[4] A. Ghosh and G. Ledwich, "A unified power quality conditioner (UPQC) for simultaneous voltage and current compensation," *Electric Power Systems Research*, Vol. 59, pp. 55-63, 2001.

Chapter 11

Distributed Generation and Grid Interconnection

Thus far we have considered point compensation and the correction of the voltage or current at a particular location in the network. This chapter considers the voltage profile of lines with distributed loads and the impact of real or reactive power from one or more injection points. The consideration of real power injection is due to a developing trend for distributed generation (DG) and the need to consider their impacts on the distribution system.

11.1 Distributed Generation – Connection Requirements and Impacts on the Network

Throughout the 20th century economies of scale have driven generation plant sizes larger with single stations exceeding 3000 MW. A parallel trend is now developing with significantly smaller sized generation units connected at the distribution level. Several factors contribute to this trend. Greenhouse gas issues are behind a push in many countries for encouraging the use of renewable energy. Many of the technologies draw upon dispersed energy such as solar, wind and wave which require many smaller sized units.

The economies of local generation using gas in units such as microturbines is approaching that of a central generation when the transmission and distribution overheads are taken into account. Fuel cells are still more expensive but are showing great promise for low cost reliable small size generation units. Solar cells are inherently modular in small panels with large installations using more of the same panels. This provides limited opportunities for economies of scale.

The move to open competitive markets in electricity has increased the uncertainties of supply. A notable example is the Californian market in 2000/01 where customers saw increased cost of energy and rolling blackouts. In response to this uncertainty of central supply there was a massive increase in demand for back-up generation with the possibility of generation back into the grid when conditions were well suited. In medium sized industrial plants or large buildings with a significant heating load, co-generation is becoming more attractive. The local generation of electricity also provides waste heat that can supply much of the heating needs of the local process.

This expansion of distributed generation has the potential to significantly change the nature of the distribution system and the associated power quality issues. Many of the technologies require an inverter stage for grid connection. The inverter is required for fuel cells and for solar cells since the source is inherently dc and needs to be converted to ac supply. Microturbines operate at very high speeds. The use of direct connected generators that thus produce alternating voltage at a high frequency requires inverters to convert to the supply frequency. Many of the wind turbine systems are induction generators but the efficiencies of variable speed turbines and the controllability of connection to weak grid points are encouraging a market in inverter connected wind generation systems.

11.1.1 Standards for Grid Connection

This development and its potential for rapid expansion has prompted a lot of work on the development of standards for inverter connection to the grid. This prospective explosion of inverters connected to the grid has the potential for significant deterioration of the quality of supply for all users. The standards are aimed at limiting the adverse impact in the areas of harmonic injection from inverters and flicker generated by inverters. In addition, inverters should stop generation after the power utility has opened the main supply. Continued generation would constitute a hazard for the utility workers. This is referred to as anti-islanding protection. The various standards that are available for grid interconnections are given in [1-4].

11.1.2 Key Requirements in Standards

The draft Australian standard requires passive means of detection of islanding which includes under and over voltage protection and under and over frequency protection. Typically if an inverter is isolated with some load, then there will be a gross mismatch of real power leading the synchronization process to generate a lower and lower frequency. Once the frequency is outside the specified limits the inverter will trip. If there is a reactive power mismatch, the voltage will tend to stabilize outside the required voltage limits leading to an inverter trip.

11. Distributed Generation and Grid Interconnection

The system is also required to trip in less than 2 sec even when there is real and reactive balance with the load. This requires an active method of detection of loss of supply. Examples of anti-islanding include

- Shifting the frequency of the inverter away from nominal conditions in the absence of a reference frequency (frequency shift).
- Allowing the frequency of the inverter to be inherently unstable in the absence of a reference (frequency instability).
- Periodically varying the output power of an inverter (power variation).
- Monitoring for sudden changes in the impedance of the grid by periodically injecting a current pulse (current injection).

11.1.3 Grid Friendly Inverters

In all the above issues, the controllability of inverters offers the potential for doing much more than limiting the worst effects on power quality. Any inverter operating at less than the rated real power can provide reactive power at comparatively low cost. These units have the potential for

- Distribution system voltage support
- Harmonic absorption
- Voltage dip reduction
- Flicker suppression
- Reduction of outages

These aspects are of particular importance in rural areas where the quality of supply is expected to be much lower than that of the urban areas. The voltage support can be provided using the additional capacity of the inverter. If an inverter rated at 5 kVA were being used to export 3 kW then 4 kVAr would be available to assist with line voltage support with a small increase in losses. There are many modes in which an inverter can operate. One mode could include aiming to have zero harmonic voltage at the point of connection. Another would be to create a harmonic current absorption proportional to the harmonic voltage. In this mode the inverter would emulate a shunt resistor to the harmonic frequencies. The dip suppression and flicker reduction would occur in the same way by supplying reactive power or short periods of real power to keep the terminal voltage close to its reference sinewave.

The reduction of outages requires a more explicit agreement with the utility company such that the distributed generation becomes a source of supply for a block of load that would otherwise be isolated from supply due to an upstream fault. This explicit islanding would require coordinated control of the isolating breaker such that when supply was restored upstream the local generation could be adjusted to bring the voltages in-phase prior to re-synchronization. In an urban environment with low probability of isolation and many points of potential synchronization there is low justification for this explicit islanding. For rural loads with one line of supply suffering high fault rates there is a much greater case for the development of islands fed from inverter or synchronous machine sources provided the available generation is sufficient for the expected load of the proposed island.

11.1.4 Angle Stability for Inverters

While connecting a synchronous machine to supply, the stability of the connection is a significant issue. Inverters do not have the problem of rotor inertia while changing the angle of the injected voltage. In most cases the desirable solution is to synthesize the inverter current to be in phase with the mains voltage. The difficulty is in the process of synchronization of the reference to the actual voltage. When there is a sudden step of load downstream, the angle of the voltage at the connection point takes a step. For correct synthesis the usual solution of a phase locked loop to learn the phase of the voltage connection point often has a lag of several cycles with a corresponding distortion of the injection. This can be avoided by having the injected current to be a scaled version of the line voltage. This avoids any synchronization issue but will have a distorted injection of current whenever the voltage is distorted rather than the injection acting to correct for the distortion of the voltage.

11.1.5 Issues for Distributed Generation

There are several aspects that need to be addressed to ensure quality of supply as the penetration of generation expands. The main issues at low levels of penetration are

- impact on protection systems,
- dynamic Interaction between generators and
- conflict in voltage control for generators in proximity.

One aspect that has shown a potential for difficulties for some utilities is the use of fixed tap transformers which are preset based on expected loading patterns to achieve the best voltage profile. With significant DG on the feeder the reversal of the power flow can lead to transformers which used to provide a downstream boost is now providing an upstream voltage boost.

11. Distributed Generation and Grid Interconnection

At high levels of DG penetration on a feeder the additional issue of scheduling of generation becomes important to avoid line overload and severe voltage rise problems. In some circumstances particularly for rural communities with poor supply reliability islanding may be an attractive proposition on failure of the main supply. When several generators are present in the island, generation control becomes essential for voltage and frequency stabilization. These requirements can be met without central control, but reconnection to the mains on restoration of supply needs a more precise frequency control if a smooth resynchronization is to be achieved.

11.2 Interaction and Optimal Location of DG

The dynamics of distribution networks with distributed generation for different penetration scenarios have been discussed in [5]. In this paper the impact of DG is assessed using both eigenanalysis and individual channel analysis and design (ICAD) for small-signal analysis, and non-linear time domain simulations for transient stability analysis. In this continuous time analysis of multi-machine scenarios, the participation of the generator subsystems: rotor dynamics, synchronous machine, AVR/exciter and governor/turbine in the oscillation modes have been examined through modal analysis to detect the degree of interaction. However, this report investigates longer-term dynamic interaction of voltage controller assuming the angle has reached its steady state. The following analysis has been performed in discrete time domain (rather than continuous time domain) assuming angle transients have been settled. Therefore it only gives an indication of interaction. Short-term angle stability interaction can be performed following the techniques discussed in [5].

11.2.1 EigenAnalysis and Voltage Interaction

Let us assume that a single line system consists of N number of load buses and two distributed generators (DG_1 and DG_2) are connected at two different locations (for example, $N-2$ and $N-6$) with their internal buses $N+1$ and $N+2$ respectively. Bus voltage and current of this system are related as

$$Y_{BUS}V_{BUS} = I_{BUS} \qquad (11.1)$$

Equation (11.1) can be expanded for the whole system as follows

$$\begin{bmatrix} y_{1,1} & y_{1,2} & \cdots & y_{1,N} & y_{1,N+1} & y_{N+2} \\ y_{2,1} & y_{2,2} & \cdots & y_{2,N} & y_{2,N+1} & y_{2,N+2} \\ \vdots & \vdots & \vdots & \vdots & \vdots & \vdots \\ y_{N,1} & y_{N,2} & \cdots & y_{N,N} & y_{N,N+1} & y_{N,N+2} \\ y_{N+1,1} & y_{N+1,2} & \cdots & y_{N+1,N} & y_{N+1,N+1} & y_{N+1,N+2} \\ y_{N+2,1} & y_{N+2,2} & \cdots & y_{N+2,N} & y_{N+2,N+1} & y_{N+2,N+2} \end{bmatrix} \begin{bmatrix} V_s \\ V_2 \\ \vdots \\ V_N \\ V_{DG1} \\ V_{DG2} \end{bmatrix} = \begin{bmatrix} I_s \\ 0 \\ \vdots \\ 0 \\ I_{DG1} \\ I_{DG2} \end{bmatrix}$$
(11.2)

Noting that the Y_{bus} is a symmetric matrix, (11.2) can be partitioned into sub-matrices as

$$\begin{bmatrix} Y_1 & Y_2 & Y_4 \\ Y_2^T & Y_3 & Y_5 \\ Y_4^T & Y_5^T & Y_6 \end{bmatrix} \begin{bmatrix} V_s \\ V_x \\ V_{DG} \end{bmatrix} = \begin{bmatrix} I_s \\ 0 \\ I_{DG} \end{bmatrix}$$
(11.3)

where

$$Y_1 = [y_{1,1}], \quad Y_2 = [y_{1,2} \ y_{1,3} \ \cdots \ y_{1,N}], \quad Y_4 = [y_{1,N+1} \ y_{1,N+2}],$$

$$Y_3 = \begin{bmatrix} y_{2,2} & y_{2,2} & \cdots & y_{2,N} \\ \vdots & \vdots & \vdots & \vdots \\ y_{N,2} & y_{N,3} & \cdots & y_{N,N} \end{bmatrix}, \quad Y_5 = \begin{bmatrix} y_{2,N+1} & y_{2,N+2} \\ \vdots & \vdots \\ y_{N,N+1} & y_{N,N+2} \end{bmatrix},$$

$$Y_6 = \begin{bmatrix} y_{N+1,N+1} & y_{N+1,N+2} \\ y_{N+2,N+1} & y_{N+2,N+2} \end{bmatrix}, \quad V_x = \begin{bmatrix} V_2 \\ \vdots \\ V_N \end{bmatrix}, \quad V_{DG} = \begin{bmatrix} V_{DG1} \\ V_{DG2} \end{bmatrix}, \quad I_{DG} = \begin{bmatrix} I_{DG1} \\ I_{DG2} \end{bmatrix}$$

From (11.3) we obtain

$$Y_2^T V_S + Y_3 V_X + Y_5 V_{DG} = 0 \tag{11.4}$$

Rearranging the above equation we get

$$V_X = -Y_3^{-1}\left(Y_2^T V_S + Y_5 V_{DG}\right) \tag{11.5}$$

By applying the principle of superposition, the relative changes of voltages due to DG injection can be obtained. By substituting $V_S = 0$ in (11.4) and

11. Distributed Generation and Grid Interconnection

examining the response of the system to 1.0 per unit voltage at DG_1 and DG_2 individually, we get

$$V'_{X_{DG1}} = -Y_3^{-1} Y_5 \begin{bmatrix} 1 \\ 0 \end{bmatrix} \quad \text{and} \quad V'_{X_{DG2}} = -Y_3^{-1} Y_5 \begin{bmatrix} 0 \\ 1 \end{bmatrix} \tag{11.6}$$

The absolute values of relative changes of voltages at DG connection points can be extracted from the matrix in equation (11.6) as

$$A_{DG} = \begin{bmatrix} V_{DG1_{N-2-1}}^{Relative} & V_{DG1_{N-6-1}}^{Relative} \\ V_{DG2_{N-2-1}}^{Relative} & V_{DG2_{N-6-1}}^{Relative} \end{bmatrix} \tag{11.7}$$

where $N-2-1$ and $N-6-1$ are the locations of relative changes of voltages for DG1 and DG2 in the matrix of (11.6). Therefore, the measured voltages at DG connection points in this example system are

$$V_m = \begin{bmatrix} V_{m1} \\ V_{m2} \end{bmatrix} = A_{DG} V_{DG} \tag{11.8}$$

where V_{m1} and V_{m2} are the measured voltages for DG1 and DG2 respectively. For a close-loop control system with proportional control gain K_1 for DG_1 and K_2 for DG_2, the DG voltages can be obtained as

$$V_{DG1} = -K_1 (V_{m1} - V_{ref}) \quad \text{and} \quad V_{DG2} = -K_2 (V_{m2} - V_{ref}) \tag{11.9}$$

where, V_{ref} is the reference voltage.
DG voltages at the next stage are

$$V_{DG}^{i+1} = V_{DG}^{i} - KV_m - KV_{ref} \tag{11.10}$$

where

$$K = \begin{bmatrix} K_1 & 0 \\ 0 & K_2 \end{bmatrix}$$

If V_{ref} is assumed to be zero, (11.9) can be rearranged as

$$V_{DG}^{i+1} = (I_2 - KA_{DG}) V_{DG}^{i} \tag{11.11}$$

where I_2 is a (2×2) identity matrix, V_{DG}^{l+1} is the DG voltage at the next stage, $k+1$ and V_{DG}^{l} is the DG voltage at current stage k. Equation (11.11) is in the form of $x(k+1) = Fx(k)$ and therefore eigenvalue analysis of coefficient matrix A (where in this case $F = I_2 - KA_{DG}$) can predict the level of interaction and system instability contributed by the DG.

Eigenanalysis is useful for the analysis of small-signal stability of low frequency oscillations and for the design of corrective controls. The modes of oscillation can be clearly identified by eigenvalues of the system matrix (A) at which the damping and frequency of each mode change with different operating conditions. The examination of eigenvectors of individual modes helps to determine the characteristics of modes and assists in developing mitigating measures. Eigenanalysis of different DG locations in the network and different network loadings has been carried out in this study to predict interaction and system instability caused by DGs.

From (11.7), the indication of voltage interaction between two DGs in this example system can be observed and an *Interaction Index* can be introduced to predict the contribution of interaction by each DG. The diagonal elements of the matrix in (11.7) are the relative changes of self-voltages of DGs located at a particular location. The off-diagonal elements in each row are the relative changes of voltages at other locations contributed by DG located in the position that produces relative change of self-voltage in that row. Therefore, the ratio of off-diagonal and diagonal elements in the column can be defined as an Interaction Index that will indicate the interaction of DGs. For this example system, Interaction Index for DG1 interacting with DG2 is

$$Index_{DG1|2}^{Interaction} = \frac{V_{DG2_{N-2-1}}^{Relative}}{V_{DG1_{N-2-1}}^{Relative}} \qquad (11.12)$$

and the *Interaction Index* for DG2 interacting with DG1 is

$$Index_{DG2|1}^{Interaction} = \frac{V_{DG1_{N-6-1}}^{Relative}}{V_{DG2_{N-6-1}}^{Relative}} \qquad (11.13)$$

Therefore, Interaction Index between two DGs can be generalized for a DG_l at a location L interacting with DG_j as,

$$Index_{DGi|J}^{Interaction} = \frac{V_{DGj_{L-1}}^{Relative}}{V_{DGi_{L-1}}^{Relative}} \qquad (11.14)$$

11. Distributed Generation and Grid Interconnection

The *Interaction Index* among multiple DGs can be generalized for a DG_l at a location L interacting with DG_j, DG_{j+1}, DG_{j+2}, ..., DG_{j+n} as

$$Index^{Interaction}_{DGi|J,J+1,J+2,...,J+n} = \begin{bmatrix} \dfrac{V^{Relative}_{DGj_{L-1}}}{V^{Relative}_{DGi_{L-1}}} & \dfrac{V^{Relative}_{DGj+1_{L-1}}}{V^{Relative}_{DGi_{L-1}}} & \dfrac{V^{Relative}_{DGj+2_{L-1}}}{V^{Relative}_{DGi_{L-1}}} & \cdots & \dfrac{V^{Relative}_{DGj+n_{L-1}}}{V^{Relative}_{DGi_{L-1}}} \end{bmatrix}$$

(11.15)

The individual element of the matrix in equation (11.15) will indicate the degree of interaction for a DG with other DGs in different locations. The index-elements in the above equation will give an indication of interaction. It to be is noted that for a stable system, the value of Interaction Index is very much less than 1.0 and close to 0. However, if the value of Index approaches 1.0, it would indicate that the system is close to instability. There is a critical limit for every network which depends on system parameters and loading. The network instability will occur if the network is loaded beyond this critical limit. This may be justified from eigenvalue analysis by comparing the changes of magnitudes of eigenvalues of the network.

11.2.2 Simulation Results of EigenAnalysis and Voltage Interaction

A prototype single line system with two DGs is considered for this analysis [6]. A 120 km SWER (Single Wire Earth Return) network has been modeled with line impedance $Z_l = 1.828 + j0.876$ Ω/km. Source voltage is assumed to be $V_s = 19.1$ kV, while the source Thevenin impedance is $Z_s = 70.53 + j57.73$ Ω. A total of 19 symmetrical line sections have been considered in the SWER feeder to analyze the interaction between the DGs. A line section is defined as a length of feeder of two consecutive buses. An LTC transformer with voltage regulation facility is connected at the beginning of the SWER system. The optimum location of the regulator has been determined by trial and error. For this test system, the regulator is connected at a distance of 10.5% of the entire line length from the source. The high source impedance of the SWER system requires the regulation to be closer to the source. Two DGs of 50 kVA each, of rotary type (Synchronous generator) will be installed on the SWER backbone to investigate the interaction. DG_1 has been kept at position $N - 2$, (where N is the bus number at the far end) and the position of DG_2 is moved from buses 4 to $N - 3$ (closest bus of DG_1), one by one, to observe the interaction of DG-DG. The relative changes of voltages at DG connection points have been examined. Figures 11.1 to 11.3 show the relative changes of voltages at DG connection points for DG_1 at position 18 and DG_2 at 5, 14 and 17 respectively. It is observed that when DG_2 comes in the proximity of DG_1, the relative changes of voltages become closer. The Interaction Indices for DG_1 and DG_2 are calculated and graphically represented in Fig.11.2.

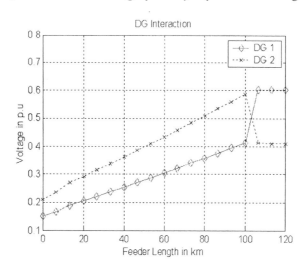

Figure 11.1. Indication of voltage interaction with DG1 at position 90% and DG2 at 85%

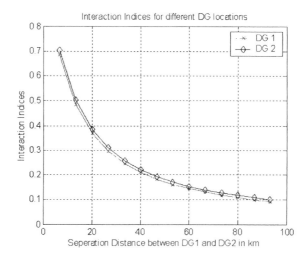

Figure 11.2. Interaction Indices for DG1 and DG2 with different locations of DG2

Clearly as these generators move closer, the potential for adverse interaction rises significantly. From the network viewpoint having high gain voltage control of the line from DG is generally a strong positive. When the potential for generators to be installed in close proximity exists, the network

11. Distributed Generation and Grid Interconnection

owner would be well advised that voltage control with a droop characteristic be installed. As discussed in [7], the use of a droop proportional to rating can avoid these interactions in voltage control while still providing a positive benefit for line voltage profile. If an inverter were connected to this generation with fast voltage control capability, the droop capability would be even more important.

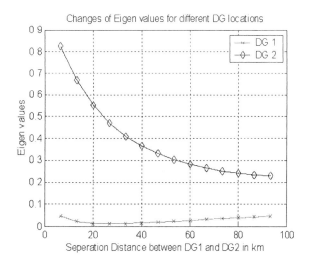

Figure 11.3. Changes of eigenvalues for different DG locations

11.3 Power Quality in DG

The following study has investigated the power quality issues related to DG integration in distribution networks. Voltage dip transients, harmonics and voltage flicker of distribution systems with DG installed are analyzed to ascertain the DG impact on power quality.

11.3.1 Mitigation of Voltage Dip during Motor Start

One of the main power quality problems is voltage dip. Voltage dip is usually caused by sudden load demand, motor start, network fault etc. An induction motor can draw a starting current typically six times the normal current. But the network is designed based on the normal load current and to maintain voltage levels during all possible simultaneous motor starts is unjustified. Therefore, a minimum allowable voltage dip is defined in standards which is considered during the planning and design of networks including motor loads. However, an expansion of network is sometimes necessary due to an increase of load demand. It is difficult to restructure and resize the whole network and the utility seeks to defer the expansion to limit the financial consequences. Unfortunately in rural areas, many loads incorporate induction motors such as pumps, refrigerators, air conditioners and fans. The sizes of motors can be very large in hotels, farms, factories etc. and they can create large voltage dips during starting due to the higher supply impedance of rural networks. The dip becomes worse if more than one motor start at the same time. Motor start transients can cause a large voltage dip at the motor terminal which can last for a few seconds with the worst case occurring at the motor terminals. This dip affects the surrounding customer loads and may cause sensitive loads to malfunction. As the reaction of tap changer is very slow, it takes up to a minute for it to respond which might not be helpful for transient restoration. Also, there is a limit for the number of step sizes and maximum voltage available to raise the voltage level. Thus the customers at the far end of rural lines experience significant voltage variations during motor starts and other transient loading.

DG with voltage control capability in a distribution network can reduce the voltage dips caused by sudden load demand, load change, motor start etc. and hence improve quality of supply. During a motor start, the DG voltage controller responds immediately and reduces the voltage dip. The DG controller needs to be designed in such a way that it is capable of handling the transient situation. The voltage improvement by DG depends on the size of DG and the current demand by the motor at the time of its start. An inverter based DG is capable of fast voltage control and mitigates the motor transient. Even rotary type of DG can help the motor transient without any special control. The DG can also help to speed up the transient restoration. In the proximity of DG, the depth and width of voltage dip are reduced dramatically.

Figure 11.4 shows the voltage dip caused by starting a 35 kW motor on the line modeled as a balanced disturbance. A DG of rotary type is capable of reducing the motor transient effects, since the DG can respond at every instance of time and thus correct for the transients. The speed of response is limited by the time constants on the motor field and will thus take nearly a second to fully respond. If the DG is nearer to the motor, the effects of motor start are reduced. Figure 11.5 shows the improvement of motor terminal voltage by DG with a synchronous generator. It is seen in Figure 11.5 that the depth and width of voltage dip have been reduced by the presence of DG. It helps to speed up the transient restoration. From Figures 11.4 and 11.5 it is seen that the dip lasts 12.5 seconds at 0.92 per unit without DG and reduces to 0.5 second in the presence of rotary DG. For inverter based DG interfacing, the reactive correction from the inverter can respond in milliseconds. The rating of a line support inverter DG may not be sufficient to fully correct for all motor starts but the inverter can significantly stiffen the line voltage to such transients and reduce the depth of dip even close to the connection point of the motor being started.

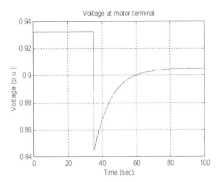

Figure 11.4. Voltage dip at 35 kW motor terminal with 135 kW load and fixed nominal tap (Lowest voltage = 0.871 per unit, 0.9 per unit for 5.5 s., 0.92 per unit for 12.5 s)

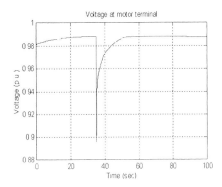

Figure 11.5. Voltage dip at 35 kW motor terminal with 135 kW load, fixed nominal tap and 100 kVA DG-PQ (Lowest voltage = 0.8825 p.u., 0.9 p.u. for 0.4 s., 0.92 p.u. for 0.5 s)

11.3.2 Harmonic Effects with DG

Voltage and current waveforms of the network are distorted due to loads such as adjustable speed drives, arc furnaces, electronic converters, rectifiers etc. As a result, harmonic currents and voltages are generated in the network. Distributed generators may introduce or reduce the harmonics in the network, depending upon the design of generator and selection of technology. Proper selection and design of generator and its control can avoid harmonic injection by the generator into the network. For some cases the DG control can alleviate harmonic effects by reducing harmonic voltages and thus reducing voltage distortion. PWM inverter DG is the best choice if a key objective of DG installation is to reduce harmonics effects.

Pure rotary type generators, with accurate design, can generate voltage waveforms that are free of harmonics. However, depending on the design of the generator windings (pitch of the coils), core non-linearity, grounding and other factors, they may become a source of harmonics and generate distorted voltage with a significant amount of harmonics.

Line commutated inverters or SCR type inverters produce high levels of harmonic current and make an adverse harmonic contribution to the utility system. This poor harmonic performance largely led to the disappearance of the SCR as a mean of connection for DC energy. In current practice, inverters are designed based on IGBTs/MOSFETs that use pulse width modulation to generate the injected sine wave. These inverters are capable of generating a very clean output [8], which is almost free of low-order harmonics. In addition, PWM inverter based connection is capable of very fast control of the synthesized waveform. In some modes, this can be used to keep the voltage at the point of connection totally harmonic-free.

An investigation on harmonic effects is made by simulating a distribution network with different types of DGs. Distributed load totaling 200 kW, 0.8 power factor (lagging) is connected to the system and is modeled as generating a significant amount of fifth-harmonic current. In this study an extreme case of harmonic injection is examined where each customer injects fifth harmonic current at a level of one-fifth of the fundamental current in phase with one another. Two DGs are installed at $N-1$ and $N-4$ (where N is the bus number at the far end) to generate voltage at fundamental frequency and mitigate fifth harmonic effects in the network at steady state. Figure 11.6 shows the voltage distortions by the 5th harmonic current and with and without DG. It is found that without DG the maximum distortion level by this harmonic current is 7.10%. Rotary DGs, with 3% voltage distortion at 5th harmonic can reduce this number to 3.30%. Rotary DGs with zero harmonic generation reduce the peak distortion to 0.93% and PWM inverter based DGs reduce the distortion to 0.65%. The inverter DGs are designed in such a way that the connection point voltages of the DGs at the 5th harmonic become zero. Pure rotary DGs refer to those generating zero harmonic voltage internally. PWM inverter based DG gives the best solution for resolving harmonic problems and provides the best reduction in voltage distortion.

Voltage flicker can be described as dynamic variations of network voltage magnitude and can be sufficient in duration and frequency to allow visual observations of a change in the intensity of electric light source. Human beings are especially sensitive to luminance fluctuations around 8.8 Hz [9]. The International Electromechanical Commission (IEC) has published a norm for a flicker meter, IEC 61000-4-15 [10], which may be used to measure voltage flicker. A DG installation may increase the flicker

11. Distributed Generation and Grid Interconnection

level especially during start/stop or if it has continuous variations in input power because of a fluctuating energy source. That is, if a DG starts or its output fluctuates frequently enough, flicker of lighting loads may be noticeable to customers. Squirrel cage induction generators have a high possibility to make flicker level worse because of an inability to actively control terminal voltage. In the case of an individual wind and solar energy generator, the output will fluctuate significantly as the wind and sun intensity changes. However diversity between the fluctuations of individual generators will reduce the relative size of fluctuations for a group of generators.

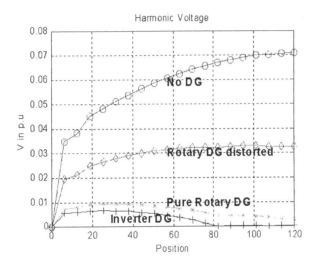

Figure 11.6. Harmonic voltages with and without DG

11.3.3 Voltage Flicker and Voltage Fluctuation

Flickering lamps are disturbing if the total flicker level on the grid is 0.7 to 1 per unit [9]. Voltage flicker caused by variations in real power output due to fluctuations in renewable energy sources depends on the resource, the characteristics of the generator and the impedance of the network. However, voltage flicker and voltage fluctuation can be mitigated by DG through active mitigation policies. Mitigation approaches [8] for induction generators include reduced voltage starts, as well as speed matching and active or passive reactive power compensation. Synchronous generators might require tighter synchronization and voltage matching. Inverters might be controlled to limit inrush currents and changes in output levels. Energy storage devices such as flywheels, variable speed drives for wind generators or batteries might be used to compensate for fluctuating power levels.

11.4 Islanding Issues

DG units operate in parallel with power distribution system. This parallel operation of a DG unit with the utility grid system may result in islanding operation of DG due to a fault in the utility system. Islanding can be defined as the operating condition where a DG maintains the electric supply on a local section of distribution system, which is disconnected from the utility source following an opening of a circuit breaker on a distribution line [11]. Loss of main supply may initiate an independent power island, which is usually caused by tripping of a circuit breaker. Depending upon the capabilities, DG will continuously support the independent power island as long as the system frequency and voltage are maintained within the required operation limits. A power-island can be created without the knowledge of the utility personnel and public and may complicate the issues like safety hazard, grounding system, synchronization, orderly restoration of utility supply and supply quality. However, proper protection with communication can avoid all the above problems and DG can be used for dedicated islanding operation. Islanding operation can support the supply continuity in rural/remote areas [11].

11.4.1 Anti-Islanding Protection

Fault clearance in utility distribution network is performed by relays located close to the fault, which isolates the faulty part from the utility network or causes the main supply breaker trip off. Also, the mains can fail to supply continuously due to the unpredictable troubles in operation and control or abnormalities. In these situations, DG may become islanded from the normal grid supplies, which, without proper arrangements, is undesirable. Islanding operation may leave a section of the network without an earth and the fault level may not be sufficient to operate protection relays. Also, the synchronization facilities of a DG system may not be able to re-synchronize correctly with the main system when it returns. Protection applied to the inter-tie between the DG and the utility network must be capable of detecting the loss of grid supply and tripping the circuit breaker on the inter-tie, if appropriate.

If a small DG unit dedicated for local supply only becomes islanded, power will flow from DG to the utility circuit of island through inter-tie. If the facility containing the DG did not normally export power, islanding could be detected by a reverse power relay monitoring at the inter-tie power-flow. If the DG normally exports power to the grid system, loss of the main supply could cause a severe overloading of the DG unit, which could be detected by under/over voltage and under/over frequency relays.

11. Distributed Generation and Grid Interconnection

If the DG unit were large enough to maintain the system voltage and frequency within specified limits following an island, special relaying techniques would be required to detect islanding. The most efficient and effective method is SCADA that monitors all circuit breakers, switches and isolators of the system between utility system and DG units and produces a transfer trip signal to open the relevant breaker during any abnormality [11].

However, it may not be possible to cover the whole system with SCADA due to the high cost, especially for rural or remote networks. Other islanding detection techniques are in practice, which measure network conditions and detect islanding. These can be categorized into two: active and passive techniques. Some of the active methods are real and reactive power export, error detection methods, system fault level monitoring method etc. [11]. They are not widely used due to the interaction with power system operation. Main passive techniques are under/over voltage and under/over frequency relays, rate of change of frequency (ROCOF) and phase displacement monitoring or vector shift. Among them rate of change of frequency and vector shift are well-known techniques to detect loss of mains, which are discussed in the following subsections.

11.4.2 Vector Shift

During normal operation the terminal voltage of a synchronous generator DG will lag the emf by the rotor displacement angle. If the grid supply is suddenly disconnected from the DG network, the load on the DG will increase and this will cause a shift in the rotor displacement angle, as the terminal voltage will jump to a new value and the phase position will change. Vector shift relays continuously monitor the duration of each cycle and initiate instantaneous tripping if the duration of a cycle changes as compared to the previous cycle by an angle greater than the vector shift setting. The vector shift settings vary between 6° and 12°. The recommended setting is 6°, but on weak network it could be 12° to prevent mal-operation during switching of a large load [12].

11.4.3 Dedicated Islanding Operation

For a variety of reasons, both technical and administrative, the prolonged operation of a power island fed from the distributed generator but not connected to the main distribution network is generally considered to be unacceptable. However, islanding operation offers opportunities for improvement in reliability and supply continuity. If DG is owned by utilities, then number of issues will be reduced and the utility may decide to offer islanding operation. Also, DG may be located at a remote area where utility supply is of poor reliability and power failures or blackouts occur frequently and are sustained for a long time. In this situation, islanding operation of DG may be acceptable [8].

For an islanding operation, the simplest case is when the DG is installed at customer premises near the islanding breaker and when the DG is capable of supplying the maximum customers' loads (or has a local load shedding scheme). During islanding condition, the DG should supply power at the same frequency and voltage levels that existed without islanding. When the main is reenergized, the DG system needs to be re-synchronized with the main supply. A communication scheme is required to resynchronize the DG with the utility system. The communication system may be set up through a power line carrier, dial-up or networking cables between the DG control panel and main breaker and only a few bytes of control data will be transmitted through this system. If the DG is nearer to the main breaker, the cost of communication set up will be less. Through the communication system, the control commands (may be in digital form of 1 or 0) will be sent to the DG governor to step up or step down the frequency level to adjust the main frequency. The frequency of the DG can be adjusted using ripple control. *Check synchronization* will check the phase angle and frequency of both networks and will lock automatically when they have reached the acceptable limits. In the case of main system failure, the main breaker should be opened immediately and a control command should be sent to the DG panel to let it know that the DG has become islanded and it needs to adjust its generation level to meet the customers' demands. The DG control panel should receive a signal when the main supply returns. For the case of multiple DGs in the network, it is possible to design the control and communication schemes for each of them to operate in a common islanded mode and to resynchronize with the main system through ripple control scheme when the main returns. Careful design for control and communication is required for the complex systems with multiple islanded DG.

11.4.4 Rate of Change of Frequency (ROCOF)

ROCOF works on the assumption that following a loss of grid supply DG is required suddenly to supply an increased amount of load. For a synchronous or induction generator or an inverter DG with an artificial frequency droop, the resulting generation deficit will cause a rate of change of frequency. The operating threshold for a ROCOF relay is adjusted between 0.1 Hz/s and 10 Hz/s and the operating time is defined by the number of power frequency measuring periods over which the rate of change of frequency is calculated which can be adjusted between 2 (40ms, minimum) and 100 (2s, maximum) at 50Hz, typically. This technique is usually considered an appropriate one for detecting loss of mains on a distribution network but not ideal for the transmission or subtransmission levels due to the nuisance trips [12].

11.5 Distribution Line Compensation

Real and reactive injection into distribution lines can affect the voltage profile on lines and the power quality in proximity of the injection point. Generators may be sited to make the best use of renewable resources such as wind or solar energy. These points may not be close to connection to high voltage lines and the connection may be required to lines in the distribution system. The generation may be located in customer premises and the real power export be determined by environmental factors or by the attraction of a particularly high pool price. The variability of this loading and the effect of reverse power flow have the potential to adversely affect the power quality in proximity of these generators.

Conversely long distribution lines can be aided in control of voltage profile by the installation and control of distributed generation geared to the needs of the network. This section looks at controlling distributed generation to affect the voltage profile and suppress dips particularly on long rural lines.

Rural distribution lines are typically long with a significant resistive component of line impedance. They are exposed to high levels of weather and environmental faults. Customers on rural lines are more exposed to voltage dips from faults or equipment of other customers. Shunt reactive compensation is an effective means of voltage control for transmission lines but is much less effective when supporting voltage on high resistance lines. This change of effectiveness is illustrated in a study of a line with uniformly distributed load.

A distribution line can be modeled as a source with a series impedance to represent the line itself and equal shunt impedances to represent the loads. The sample system given in Figure 11.7 shows two tap changers to control the line voltage.

11.5.1 Line Voltage Sensitivity

Sensitivity analysis is performed on an example distribution system with real and reactive generation at the remote end to investigate the relation between the line characteristics and network voltage compensation. To demonstrate the relation, three cases of line parameters are considered. They are (a) $R < C$, (b) $R = X$ and (c) $R > X$, where R and X are the line resistance and reactance respectively. Real and reactive injections by DG are varied from zero to three-times the total load in the network. The solution for the voltage on the feeder with distributed load (modeled as admittances) is obtained from the admittance matrix. The results are reported in Figure 11.8, which shows the voltage sensitivity. It is found that if $R \leq X$, reactive power (rather than real power) injection will be most effective to improve voltage profile. Whereas, for high R (i.e., $R = 3X$), capacitive support will be least effective and real power injection will help appreciably to support voltage and improve voltage profile. It is also found that the voltage improvement is better for real/reactive power injection if the networks serve distributed load rather than a concentrated load.

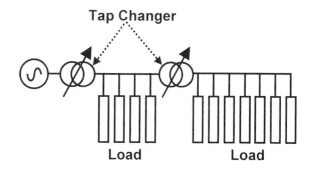

Figure 11.7. Line with distributed load and distributed tap changers

Because of this low effectiveness of shunt reactive control for low voltage distribution lines, this chapter places most emphasis on three different forms of voltage support for distribution lines – series reactive compensation, series real power injection (UPQC, Tap changers) and shunt real power injection (Distributed Generation). As a basis for comparison this section considers a range of real and reactive compensation of lines with high resistive impedance and with distributed load. Initially the study will focus on the effect of heavy load on the line voltage profile. The following section considers the compensation needs of lines under heavy load.

11.5.2 Case-1: Heavy Load

In this we consider various types of compensations. For these studies, the following parameters are used

- Source voltage of 1.0 per unit with negligible source impedance.

11. Distributed Generation and Grid Interconnection

- Line resistance = 0.3 per unit, line reactance = 0.5 per unit and line reactive admittance = 0.2 per unit.
- Total load admittance = 2.0 per unit.

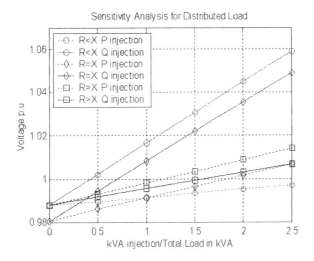

Figure 11.8. Line Voltage Sensitivity for distributed load

Shunt Reactive Compensation: The shunt capacitor is able to support the voltage at its point of connection. The line impedance and loading level determine the curvature of the voltage profile. This level of compensation however required a voltage-ampere (VA) of 1.4 per unit to achieve the profiles shown in Figure 11.9. This is a significant fraction of the total line rating and would not appear to be a cost effective voltage control solution.

Series Reactive Compensation: The series compensation of high voltage transmission lines to reduce reactive impedance and thus increase transfer capacity has been discussed in several references [13] to [17]. The major concern for series compensation has been to

- avoid subsynchronous resonance [13,14,17],
- be able to adjust the level of compensation and
- provide protection for the capacitive element.

For distribution lines, some series capacitive compensation can be beneficial but the last two aspects mentioned above remain. At peak load there is a high reactive demand and voltage control is achieved through use of a large admittance of shunt capacitive compensation. For series capacitance the rise of current with load will intrinsically cause the required reactive power variation with loading. This use of series capacitance to cancel a portion of the line reactance is effective for HV lines but when the line resistance is a significant component of the line voltage drop, the line drop correction will be strongly load power factor dependant. For long lines at light load the line capacitance can be sufficient to cause a significant voltage rise which is known as Ferranti effect. To correct for this voltage rise, the line requires additional inductive reactance. For some lines, this may be supplied by switched shunt reactors. This variation in reactive requirements means that fixed series capacitors will rarely be a complete voltage control solution.

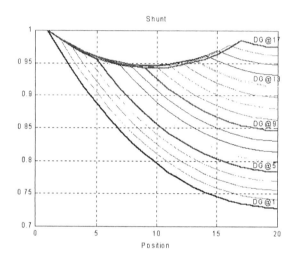

Figure 11.9. Voltage profiles for varying location of shunt reactive compensation

The line compensation variation could be achieved with switched series elements. For example a circuit breaker could be used to bypass the series capacitor. The main difficulty with a circuit breaker solution is that the regular switching to follow the daily load cycle can be an onerous duty on the circuit breaker which may lead to reduced life. Another issue is that unless precise point-on-wave switching is used there can be significant switching transient, which require a series reactance to limit the peak switching current.

The issue of protection also needs to be addressed. At full load operation (i.e., 1.0 per unit current) the voltage across a series capacitor may be less than 5%. For a solid fault on the line the current can easily reach 20 per unit. In this case the voltage across the capacitor can be close to 1.0 per unit and may reach higher under resonant conditions. Thus a capacitor designed to compensate a feeder load with a VA of 1.0 per unit would normally have a

11. Distributed Generation and Grid Interconnection

VA rating of 0.05 per unit but would experience 20 times that during a fault. It would not be economic to provide this level of voltage rating for the capacitor so a voltage limiting protection is required. This can be achieved with a combination of surge arrestors, spark gaps providing high-speed protection and a slower acting bypass switch to more permanently remove the series element from service.

Inverter Based Series Compensation: Inverters can be used to create a variable reactive element in shunt or with transformer coupling can be used in series with the line. The structure of this compensation is the same as for the DVR discussed in Chapter 9. While the inverter can have fast control of the reactive value and does not necessarily suffer from inrush current, issues of external protection are almost the same as for fixed capacitors. The voltage sensitivity of power electronic switches is typically higher than that of passive capacitors and may place tighter requirements on the design of the surge arrestor/spark gap protection design. The voltage reduction for the inverter can be easily achieved under fault conditions by placing the inverter into a zero voltage condition but the main switches are will carry fault current until external protection operates.

Line Compensation Performance: The reactive series compensator would not be expected to be as beneficial as in transmission applications because there is a much higher resistive voltage drop of the distribution line to compensate.

There are 20 nodes in the model line each sharing the load of 2.0 per unit. The series capacitor was inserted at each node in turn, yielding the line voltage profile in Figures 11.10 and 11.11. The reactive support of the line is clear in that the droop up to the injection point is reduced. At the capacitor itself there is a voltage step which depends strongly on the power factor of the load. The VA of the element reduces along the line length as the current reduces. In this case the capacitor impedance was 0.5 per unit. This rating applied at the best profile point at 40% of the line required a VA rating for this series compensation of 0.35 but still did not provide a satisfactory solution to the line voltage profile for this single compensator.

If a series inverter were used to implement this series compensation the control based on the downstream voltage could respond rapidly to compensate for load changes which would otherwise tend to create voltage dips but the effectiveness of maintaining the profile with pure reactive series compensation is questionable.

Series Real Injection: Another option is the use of series real power injection. Consider the series insertion of a voltage in phase with current as shown in Figure 11.12. The power must come from somewhere and the simplest way is to provide it from a shunt element which draws current in phase with the line to neutral voltage.

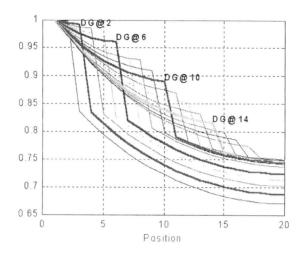

Figure 11.10. Series Capacitor at varied locations bc = 2 load pf = 0.8 lag

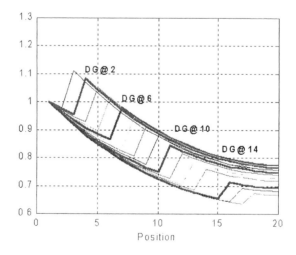

Figure 11.11. Series Capacitor at varied locations bc = 2 load pf = 1.0

With respect to Figure 11.12, the output (side-B) voltage satisfies

$$V_B = V_A + \Delta V \qquad (11.16)$$

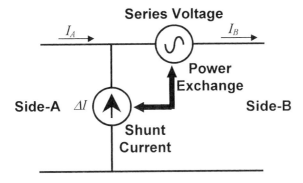

Figure 11.12. Series compensation with real power

where V_A is the side-A voltage and ΔV is the injected voltage. Similarly the currents satisfy

$$I_A = I_B + \Delta I \tag{11.17}$$

and for power balance we have

$$\begin{aligned} V_A I_A &= V_B I_B \\ V_A(I_B + \Delta I) &= (V_A + \Delta V) I_B \end{aligned} \tag{11.18}$$

If we require $\Delta V = \alpha V_A$, then

$$V_A I_A = (1 + \alpha) V_A I_B \tag{11.19}$$

The above equation yields

$$I_A = (1 + \alpha) I_B \tag{11.20}$$

This becomes the standard transformer equation. For other power factors the shunt and series elements can generate the required reactive injections but the fundamental relations remain that the series real power injection with shunt power draw to compensate is equivalent to a transformer without isolation. Controlled series real power injection is equivalent to tap changer operation. For this analysis it is assumed that inverters with shunt and series elements are equivalent to tap changers apart from the speed of response and the lack of automatic control of reactive power. For a high resistance line under heavy load, a set of tap changers has the potential to provide a good voltage profile.

The tap changers can operate independently to step towards making the downstream voltage close to the target of 1.0 per unit voltage. The curve in Figure 11.13 shows the initial voltage changing in a series of steps to make the downstream voltages close to 1.0 per unit. For this load however the cost is a significant suppression of upstream voltage. Part of the difficulty can be the transmission of reactive power so we next consider a shunt reactive element in combination with the tap changers.

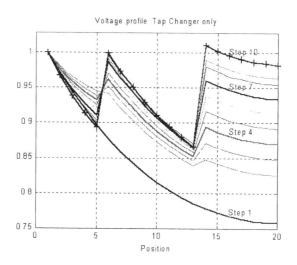

Figure 11.13. Two Tap Changers

Combined Tap Changer and Shunt Reactive Compensation: The tap changer can be combined with a shunt capacitance. Here the shunt capacitor is only switched when the tap changer is above nominal tap. The profile in Figure 11.14 shows the excursion limited to 0.05 per unit voltage once the tapping has finished converging. The line profile without compensation is seen drooping to a voltage of 0.77 per unit. The VA required for this performance is 1.4 per unit.

The maximum deviation from 1.0 per unit is now just above 5% but the shunt reactive component is excessive. The other reactive option is series reactive compensation in combination with a tap changer. A more precise compensation is possible if the level of shunt reactive injection is proportional to an amount above nominal tap. The performance is shown in Figure 11.15. The combined VA for this compensations is 1.2 per unit.

11. Distributed Generation and Grid Interconnection

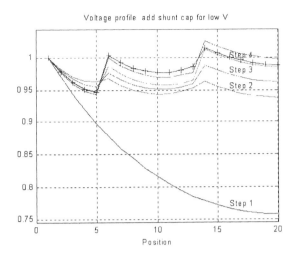

Figure 11.14. Fixed shunt capacitors at tap changers

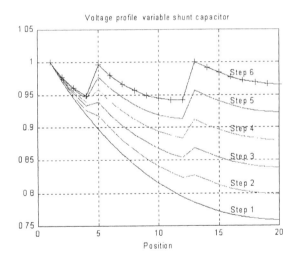

Figure 11.15. Proportional shunt reactive at Tap changers

Combined Tap Changer and Series Reactive Compensation: The inclusion of capacitance with the tap changer itself is easier and requires lower rating when used in series. The profile in Figure 11.16 was obtained with a total VA of 0.316 per unit. The inserted impedances were 0.2 and 0.135 per unit while the currents during the tapping convergence can be seen in Figure 11.17. The current magnitudes increase as the load voltage is raised at the tap changers in Figure 11.17. The voltage step here was 0.15 per unit and converged in 12 steps of iteration.

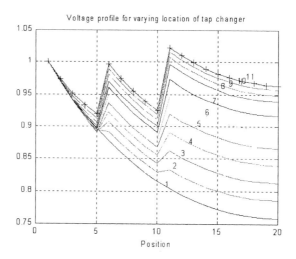

Figure 11.16. Variable series reactive combined with Tap Changers

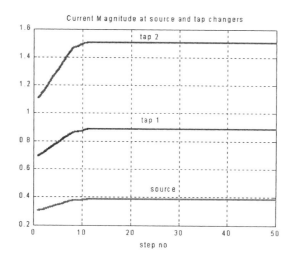

Figure 11.17. Current at tap changer with series reactance

In summary the tap changer with variable series reactive compensation, which increased with the tap position, provides a satisfactory profile with a much lower investment in reactive elements. The heavy load case is not the

11. Distributed Generation and Grid Interconnection

only way that voltage problems arise on distribution lines particularly those of 100 km or more.

11.5.3 Case-2: Light Load

The lightly loaded line can have significant rises in voltage due to the Ferranti effect. When the line capacitive admittance was kept at 0.2 per unit total and the load admittance was kept at 0.002 per unit, the tap changer control achieved the voltage profile shown in Figure 11.18.

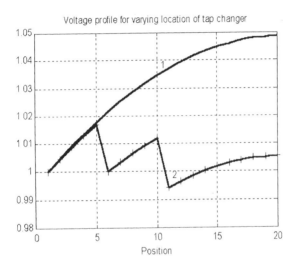

Figure 11.18. Tap changer for light load

11.6 Real Generation

The examples to date required no net injection of energy. For a rural line a single shunt generator towards the end of the line can create an acceptable voltage profile. This indicates that shunt reactive power or Distributed Generation can be a very effective means of voltage control on heavily loaded distribution lines requiring injection at a single point for compensation while for the best alternative three tap changer locations with reactive capacity were required. Fig 11.19 shows the case of total load admittance of 2.0 per unit with generation at 30 degrees leading.

11.7 Protection Issues for Distributed Generation

One of the objections to the inclusion of generation within the distribution network is the potential for adverse impact on the line protection. For generation added to the transmission network, distance protection schemes or unit protection may be used. Distance protection observes the apparent impedance along the line and infers the location of any fault from this value knowing the line impedance per unit distance; if the fault is observed in the line section to be protected then the distance protection unit initiates the opening of the circuit breakers. The distribution network has loads connected along the line sections and the inference of fault location from apparent impedance is more prone to error. Unit protection measures the current into and out of a protected element and the difference in current is an indication of fault current. For distribution lines the difference of currents in a line section is due to loads as well as potential faults. However the main issue is that unit protection requires a secure communication scheme to permit line current differencing. This additional cost of communication generally makes it unattractive for distribution use.

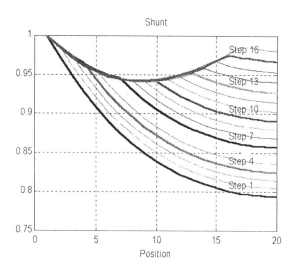

Figure 11.19. Shunt real generation

The DG itself will monitor its current and trips on overcurrent. If the fault is close to a low impedance DG, the line protection may not detect the fault until the DG has tripped since most of the fault current may be supplied from it. This hiding of the effect of faults can delay the response of the main protection and interfere with the provision of backup protection.

Another aspect of concern is the detectability of high impedance faults. When part of the fault current is supplied by the DG and part by the main line the presence of the DG will raise the fault current that can go undetected. This may not always cause particular problems depending on the nature of high impedance faults that are of concern. For many rural lines, the

11. Distributed Generation and Grid Interconnection 437

detection of fault current of a downed conduction is a significant concern. For sandy soils the fault current can be quite low. Fortunately it has been found that the arc voltage often limits the voltage at the faulted point. A constant voltage characteristic of this type of high impedance fault implies that low rated DG does not interfere with fault detection capability [18]

11.8 Technologies for Distributed Generation

There are many possibilities for distributed generation which tends to be defined in terms of the size of the generator being less than 1 MW or in terms of its connection to the distribution system. The main demand is for

– Small photovoltaic or wind power for residential or small office applications.
– Micro turbines and fuel cells tend to be a little larger than required for single residential applications but can serve small industrial sites or apartment buildings.
– Larger size generation can come from microhydro or larger wind.

The technologies, which inherently produce dc such as photovoltaics and fuel cells, need an inverter interface. Microturbines operate at a much higher frequency than mains and require a power electronic interface. Micro hydro and wind have been generally interfaced using induction generators or synchronous machines. With larger wind machines the difficulties of operating an induction generator into a weak grid as well as improved efficiency have led to a rise in inverter interfaced wind generation.

11.9 Power Quality Impact from Different DG Types

In general inverters can be used to strengthen the voltage of a weak supply in the manner discussed in earlier chapters. Induction and synchronous machines can act as points or voltage supply and at least for short transients can strengthen the supply and reduce the effects of voltage dips or harmonics. The induction generator however can be defluxed by a short circuit near its terminals and this results in increasing of the duration of a fault.

Another impact on power quality for the use of distributed generation can come from fluctuations in the real power supplied from wind or the sudden changes of output from a solar cell. For an induction generator interface to a wind generator, gusty wind can produce significant power fluctuations and corresponding variations in the local voltage. When inverters are used in the interface, the reactive power control can help to reduce the effect of real power fluctuations and help in maintaining the local voltage. For a long feeder the voltage magnitude of the two ends may be well controlled but significant power fluctuations will give rise to voltage magnitude and angle fluctuations for customers along the feeder. Thus power quality can be significantly enhanced through the use of inverter interfaces but the effects of wind fluctuations on the flicker for some situations cannot be fully corrected.

Example 11.1: Let us consider a wind generator at the end of a line of impedance $j0.2$ per unit with random variation in wind power around 1.0 per unit.

Induction Generator: The induction generator form of connection can use the rotor inertia to somewhat smooth the variations of power in a gusty wind. The motor parameters used for this example in per unit are $r_p = 0.02$, $x_p = 0.1$ as primary and $r_s = 0.02$ and $x_s = 0.1$ as secondary, while the magnetizing admittance is assumed to be zero. Because the power depends on the cube of wind speed, the peak of the power can vary significantly. The lagging power factor means that as the power level increases, the rotor accelerates and then provides a new level of output power. The two curves in Figure 11.20 below show that the exported power can still have significant variations.

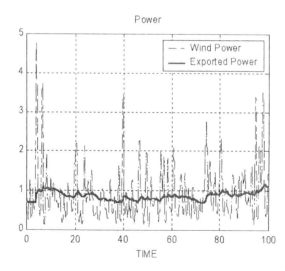

Figure 11.20. Variation in Power from induction generation

The terminal voltage falls as the export power rises and this can give flicker to the customers connected in proximity to this wind generator. This

11. Distributed Generation and Grid Interconnection

is shown in Figure 11.21. Overall there will be a significant range of loading on the feeder giving a substantial variation in voltage level. The histogram of the voltage level emphasises this degree of variation in voltage. The histogram is shown in Figure 11.22.

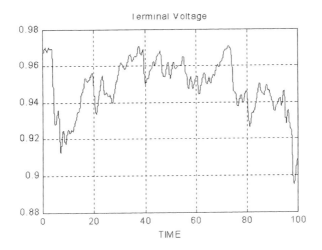

Figure 11.21. Variation in terminal voltage using induction generator

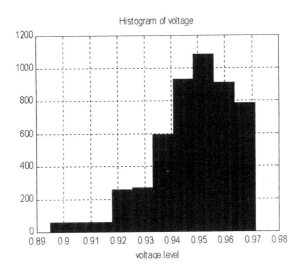

Figure 11.22. Histogram of connection point voltage for induction generator

Inverter at Unity Power Factor: An inverter does not have the luxury of significant energy storage so there is a significant variation in the exported power to maintain the DC bus voltage in reasonable limits. In

Figure 11.23 the variation in export power is almost as large as the variation in wind power. The inverter is current controlled such that the power export is unity at the point of connection. The export level rises an this also gives rise to a voltage drop. As seen in Figure 11.24 (a), this gives a significant variation in terminal voltage.

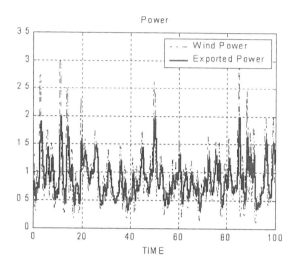

Figure 11.23. Variation in export power from inverter

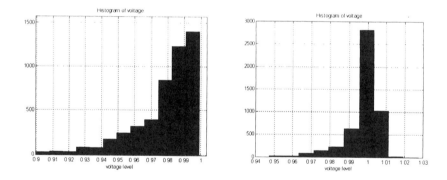

Figure 11.24. Histogram of connection point voltage for current injection (a) unity power factor (b) controlled power factor

Inverter with PI Controlled Power Factor: The best means to correct for voltage variations are to a have fast correction of the reactive power to match the changes in real power. This control uses a PI controller on reactive power with a feedforward term based on wind power levels. This is shown in Figure 11.24(b)

11. Distributed Generation and Grid Interconnection

11.10 Conclusions

The benefits of inverter connected DG on feeder performance can be significant in reducing the voltage variations on the feeder. It can also be used for peak lopping so that line upgrades to meet a rising peak demand can be met. This mode of operation requires the DG to operate by delivering real power at times of peak loading for that particular line. For the remainder of the time, reactive control of the inverter can be used to control voltage and reduce voltage dips on the line.

A market-oriented connection of DG will seek to maximize generation for periods where the pool price may be high. The local network is just a highway to deliver the energy to customers somewhere who are willing to pay a higher price for the energy.

A combined position may be desirable where payment to the generator can come from benefits supplied to the network operator as well as payments from the energy market. Generation must be able to be guaranteed during times of heavy feeder loading if there is to be a benefit from deferral of line capacity upgrades. At other times the reactive capacity may be of benefit to the network but real generation levels will be in response to market prices for energy. This combination of benefit can lead to a greater number of sites where distributed generation is economically attractive.

The use of inverters can be of distinct benefit to network operation. When there is a strong benefit, methods need to be found such that distributed generation is encouraged to establish in regions of the network of greatest benefit. Voltage control, dip reduction and harmonic reduction are the power quality benefits that can be gained without excessive compromise of generation that is primarily built to supply an energy market.

11.11 References

[1] DR012012-4 Draft Australian Standard for Grid Connection of Energy Systems via Inverters.
[2] IEEE P1547/D07 Draft Standard for Interconnecting Distributed Resources with Electric Power Systems.
[3] Distributed Generation Canada Technical Guide 2001.
[4] Draft Interim Guideline for Generator Interconnection to the Wires Owner Distribution System, www.eal.ab.ca.
[5] F. V. Edwards, G. J. W. Dudgeon, J. R. McDonald and W.E. Leithead, "Dynamics of Distribution Networks with Distributed Generation," *IEEE Power Engineering Society Summer Meeting*, Vol. 2, pp. 1032-1037, 2000.
[6] N. Chapman, "Australia's Rural Consumers Benefit from Single-Wire Earth Return Systems," *Transmission & Distribution*, pp. 56-61, April 2001.
[7] H. Sharma and G. Ledwich, "Reactive support from photovoltaic inverters," *Int. Power Engineering Conf. (IPEC)*, pp. 52-57, Singapore, 1997.

[8] P. P. Barker, R. W. de Mello, "Determining the Impact of Distributed Generation on Power Systems: Part 1 – Radial Distribution Systems," *IEEE Power Engineering Society Summer Meeting*, Vol. 3, pp. 1645-1656, 2000.

[9] S. Lundberg, T. Petru, T. Thiringer, "Electrical limiting factors for wind energy installations in weak grids", *Int. J. of Renewable Energy Engineering*, Vol. 3, No. 2, pp. 305-310, 2001.

[10] IEC 61000-4-15: 1997, Electromagnetic Compatibility (EMC), Part 4: Limits, Section 15: Flickermeter – Functional and Design Specifications.

[11] J.E. Kim, J.S. Hwang, "Islanding detection method of distributed generation units connected to power distribution system," *Power System Technology in the Proceedings of Int. Conf. POWERCON*, Perth, pp. 643-647, 2000.

[12] N. Jenkins, R. Allan, P. Crossley, D. Kirschen and G. Strbac, *Embedded Generation*, IEE Power and Energy Series 31, 2000.

[13] K. R Padiyar, *Analysis of Subsynchronous Resonance in Power Systems*, Kluwer Academic Publishers Boston, 1999

[14] P.M. Anderson, B.L. Agrawal and J.E. Van Ness, *Subsynchronous Resonance in Power Systems*, IEEE Press, New York, 1990.

[15] J Seo, S. Moon, J. Park and J. Choe "Design of a robust SSSC supplementary controller to suppress the SSR in the series-compensated," *IEEE Power Engineering Society Winter Meeting*, Vol. 3, pp. 1283-1288, 2001.

[16] G. N. Pillai, A. Ghosh, and A. Joshi, "A robust control of SSSC to improve torsional damping," *IEEE Power Engineering Society Winter Meeting*, Vol. 3, pp. 1115-1120, 2001.

[17] L. Sunil Kumar and A. Ghosh, "Modeling and control design of a static synchronous series compensator," *IEEE Transactions on Power Delivery*, Vol. 14, No. 4, pp. 1448-1453, 1999.

[18] M. Kashem and G. Ledwich "Distributed generation" *Distribution 2001*, Brisbane, November 2001.

Chapter 12

Future Directions and Opportunities for Power Quality Enhancement

There are many developing aspects to the production and delivery of energy. The increased role for power quality will drive an increasing use of compensating devices either as stand alone or in combination with energy delivery. This chapter will explore some of these emerging aspects in the existing system and speculate on the future shape of the delivery and the role of compensators.

12.1 Power Quality Sensitivity

There is much evidence that power quality can be of significant importance particularly to industrial customers. The cost to US industry of voltage dips is estimated to be US$10 billion per year. The cost of a single severe voltage dip to one semiconductor manufacturer in Singapore is estimated to be Sigapore$1 million per event. There have been a number of surveys to identify the customer perceived cost. The advantage of quantifying the cost to customers is important to determine the level of investment that is appropriate to correct supply quality complaints or to price a premium quality tariff for selected customers. Another motivation is to understand and to help customers understand what is a realistic level of supply quality without major tariff changes. There have been many papers aiming to quantify the level of quality actually occurring and the costs to various classes of customers [1-6].

The areas of industry reported as having high sensitivity to quality are listed in Table 12.1. Examples of why some of these processes are particularly sensitive to voltage dips are

Table 12.1. Highly sensitive industries

Continuous process industries	Complex machines parts	High Technology Products	Safety and security related industries
Paper, fiber and textile, plastics, metals etc.	Large pump forging, automotive parts etc.	Data processing centers, banks, telecommunications, broadcasting etc.	Hazardous processes, chemical processing, hospitals and health care facilities, military installations, Power plant auxiliaries etc.

- Paper mills: Dip causes speed changes, at a normal speed to 20 m/sec even small differences can break the web and require a plant restart.
- Textile mills: Major flaws in the pattern and loss of production.
- Steel rolling mills: speed differences in a chain of rollers, wire, strip or bar can cause the product to buckle and end as a twisted wreck on the mill floor.
- Petrochemicals: Loss of process and process contamination requiring purging.
- Wafer fabrication: Loss of process resulting in scrap, resetting and startup.

The depth and duration of a dip can have different effects on different equipment. Examples of recorded events which have caused equipment malfunction are listed in Table 12.2 [1-3].

Table 12.2. Examples of dips causing malfunction

Equipment	Voltage remaining	Duration of dip (ms)
Photoelectric emergency stop	78%	21
Undervoltage relay	67%	18
Personal computer	60%	160
PLC	59%	400

12.1.1 Costs of Power Quality

There have been many surveys of different groups of customers in different countries on the cost of interruption. Such costs reported from Finland for the year 2001 are given in Table 12.3 [1].

The surveys indicated that for more than 50% of customers there were no economic losses while 1-2% reported extremely high losses i.e. 20,000-70,000 Euros per event. To avoid overestimation the loss levels were thresholded to 20,000 Euros. The resulting costs for a deep sag are shown in Table 12.4.

12. Future Directions and Opportunities

Table 12.3. Costs of Power Quality in Finland

Customer Category	Type of event	Cost per event (Euros)
Domestic	Short interruption	1
Agricultural	Short interruption	1
Industrial	Voltage sag of 50% for 200 ms	1200
Commercial	Short interruption	210
Public	Short interruption	280

Table 12.4. Costs of sag in Finland

Customer Category	Cost per sag (Euros)
Domestic	1
Agricultural	1
Industrial	1060
Commercial	170
Public	130

Overall when the frequency of interruption to each class of customers was considered, the industrial sector showed the overall greatest total losses while some of the rural customers showed high losses due to the very high frequency of sags in that rural area.

In Canada a survey [4] of the effect of the duration of interruption for industrial and commercial customers showed a significant initial cost rising with duration. Typical cost data are given in Table 12.5. Note that this cost of interruption is normalized in terms of the annual energy consumption, so the data indicates that one 8 hour interruption has a normalized cost equal to C$0.25 while the normalized cost for energy over the year may be C$0.10.

Table 12.5. Cost of interruption in Canadian$/kWhr

Customer type	Duration of interruption			
	1 min	20 min	4 hr	8 hr
Industrial	0.0003	0.0015	0.003	0.25
Commercial	0.0001	0.0015	0.003	0.15

The above survey did not have a strong response in the government institutions and office buildings GIO sector. A later survey [5] showed the following relationship shown in Table 12.6. The normalized cost $/kW refers to the cost of this interruption peak annual demand; thus a customer with a peak demand of 1 MW would have an average cost for a 4 hour interruption of $21.36M.

A US survey of large customers was made detailing the aspect of the business where the lost costs were incurred [6]. This is shown in Table 12.7. Studies of this type can be used to quantify the effect of outages on customers and in a service mode, quantify the appropriate level of investment for mitigation. For more market driven arrangements it provides a basis for negotiation on the power quality service component of supply contracts.

Table 12.6. Cost of interruption in government and office sector in Canadian$

Interruption duration	Average cost per customer per interruption	Average cost consumption – normalized $/kWhr	Average cost demand – normalized $/kW
2 s	147.7	0.000352	1.6586
20 min	469.9	0.001097	3.2655
1 hr	1067.9	0.002687	7.2297
4 hr	3764.1	0.009372	21.365

Table 12.7. Cost of power quality for large customers in US$

Scenario	Costs
4 hour outage without notice	74835
1 hour outage without notice	39459
1 hour outage with notice	22973
Voltage sag	7694
Momentary outage	11027

12.1.2 Mitigation of Power Quality Impacts from Sags

A voltage dip is initiated by faults on the supply lines or in customer premises. Its depth and duration determine the effect of a voltage dip. The fault impedance, line impedance and source impedance determine the level of a voltage dip. Changing source impedance to faults can reduce dip level. Over 70% of faults are single line to ground faults. The dips caused by these can be reduced by increase of the zero sequence impedance of the source. Inclusion of a reactor or resistor in the neutral leg of distribution transformers can increase this zero sequence impedance. Other fast acting controls can insert additional impedances for the duration of the fault. One side benefit of limiting fault current is that it increases the chances that the arc of the fault will self extinguish without a circuit breaker having to be opened. There is a limit to the impedance that can be included in the neutral leg because of safety concerns about step and touch potential as a very high impedance will cause voltages in the proximity of the fault to rise significantly.

The duration of the dip is determined by the response of the protection system. For a fuse we often have a protection rating given as $K = I^2 t$ which means that as the fault current rises, the time for the fuse to break falls. For circuit breakers a similar characteristic is employed often with a definite minimum time to operate enabling a better design of grading so that only the breaker closest to the fault opens even when breakers further up line see the same fault current. In terms of improving power quality the issue is to reduce

12. Future Directions and Opportunities

the operating time of the breakers as much as possible consistent with retaining the grading against upstream protection. With modern digital protection the uncertainties of operate times are reduced so the safety/uncertainty margins that have been traditionally used in grading can often be reduced.

Scottish Power was looking to improve its quality of supply to improve its ability to attract sensitive industrial customers to preferentially choose to set up in their area. The method it chose to improve the voltage sag figures was through careful tuning of the protection settings to reduce the duration of dips [7].

The success of the program can be seen in Figures 12.1 and 12.2 that show a dramatic reduction in the duration of faults particularly in the long duration region. The records showed no fault longer then 400 ms after the retuning of the protection grading. The retained voltage during the fault is shown in the legend.

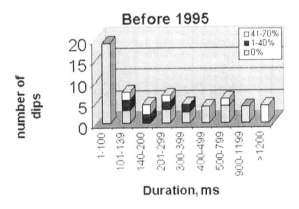

Figure 12.1. Voltage Dip pattern before retuning

12.2 Utility Based Versus Customer Based Correction

One principle in dividing responsibility for fixing a problem is to apply the issue of fault; if someone causes a problem then the responsibility for correction of the problem should lie with the originator. For many of the power quality problems the cause can be diffuse and hard to lay at the feet of any one customer. In that case another principle is to pass the responsibility to those in the best position to provide a solution. Let us examine the application of these principles to different power quality problems.

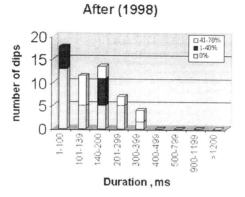

Figure 12.2. Voltage Dip pattern after retuning

12.2.1 Dips and Outages

One viewpoint of power quality is that "if I am buying a product and it does not perform to my needs I should complain to the manufacturer." In this vein all power quality disturbances are the fault of the utility and the utility should correct them. For domestic customers one of the strongest complaints is often the requirement to reset ten or more clocks in appliances, stereos, microwaves etc. following a 2 second outages. At another level is the work that is lost with the tripping of the home computer when a moderate voltage dip is experienced. Another pathway for complaints on power quality is the fax machines which print a record of voltage dips. Thus the quality of supply has not changed, but the customers are now more aware of any deficiencies in supply.

The network solution to these problems for residential customers may be to have multiple supplies to each domestic transformer with a battery energy storage system with inverter to handle the change over. The customer solution may be to purchase appliances with battery backup to maintain the auxiliary functions such as time and status during outages, to install a small UPS on the computer to permit a graceful shutdown. This customer solution would address the short outages which cover 90% of the complaints.

In an industrial area the needs for security of supply may have most of the customers purchasing back-up generators sized for the maximum possible demand. These customers would also need UPS for each of the critical loads such as computers and line process control. The network solution may be to provide alternative sources of supply as well as backup generation for a group of customers. There could also be a provision of protected supply for critical loads.

12. Future Directions and Opportunities

For the domestic customers there is little preparedness to pay extra for improved quality of supply which could avoid the longer term outages. The network solution for small outages may be much more expensive than local solutions. One of the difficulties is that issues of robustness of appliances to power quality are not commonly on the mind when purchasing. Information on whether your new washing machine is dip sensitive is not always readily available at the store when the attention is often on price. In Australia attention is drawn to energy efficiency by a star rating system. Each of the refrigerators has a clear label of one to five stars depending on the energy efficiency compared to others in the same category. A star system for supply robustness may be of high value to a customer in a rural area experiencing frequent voltage dips but of low value to a person living in the CBD of a city experiencing one significant dip event per year. One of the significant causes of dips in residential premises is the DOL (direct on line) starting of air conditioners when there is a shared transformer for several customers. The high currents that can be experienced during starting can cause a significant voltage dip through the low voltage lines and transformer. This dip will then affect the appliances in neighboring properties. Development of tight standards for peak starting current may require the use of variable speed drives which avoid the severe starting requirements. One positive side effect of the variable speed drive solution would be the improvement in the options for demand side management.

Following from the Californian supply crisis in 1999-2001 or from local experience of the cost of loss of supply more businesses may be prepared to put a cost to the quality and reliability of supply. This awareness of value will then create a market for improved performance which can be made more cost effectively in a custom power park with a shared form of quality solution than with each business providing its own solution.

12.2.2 Harmonic, Flicker and Voltage Spikes

When there are specific identifiable harmonic sources then the best solution is to require suppression at the source at the cost of the owner. When there is a broad rise in the level of harmonics without a well identified source the decision on a solution will depend on the impact. When there is little identifiable rise in interference with operation of customer equipment the utility may not see justification in installation of compensation equipment. In the presence of legislation requiring limited harmonic levels or where there are significant widespread identified impacts on customer equipment the responsibility falls to the utility to find a solution such as a filter or compensator.

When the impact is limited to a few specific items of equipment the cost-effective solution may be to reduce the sensitivity of that specific equipment through filters or active compensators. Major problems with flicker are often easily associated with particular sources and responsibility identified. Correction of flicker at each customer premises would not be expected to be cost-effective and a source or network solution would be required.

Voltage spikes often arise from faults on the supply network from lightning strikes, line switching or high voltage lines falling on low voltage lines. All these aspects would be best addressed by the network company.

12.3 Power Quality Contribution to the Network from Customer Owned Equipment

When electricity distribution companies own compensators, there is a clear role for the use of the compensator in reactive support, voltage dip and harmonic correction. Inverter connection for customer equipment is required for interfacing of uninterruptible Power Supplies, and distributed generation options such as microturbines and fuel cells. In most of these cases the main justification for the installation of the inverter was for low harmonic transfer of real power. The real power requirement can frequently be much less than the inverter rating which makes the excess capacity of the inverter potentially available for grid support and power quality enhancement. For example if an inverter of rating 50 kW is available but the customer real power transfer needs at one time are 30 kW then up to 40 kVAr is potentially available for grid support.

12.3.1 Issues

There are a number of issues that come into play associated with grid support issues using customer equipment.

1. There needs to be sufficient economic incentive for customers to agree to grid friendly controls which do not interfere with their own real power needs.
2. Voltage support controls are a decentralized form of grid friendly connection, but have the potential for interaction with other inverters in close proximity.
3. Voltage dip support can be valuable but should not interfere with the main feeder protection.
4. For transformer connected inverters, the transformer impedance should be carefully considered to determine the value of grid friendly connection to the network.

12.3.2 Addressing the Barriers to Customer Owned Grid Friendly Inverters

The main methods to tackle the adoption of a grid friendly inverter standard are to enable differential tariffs in regulated markets or a preferential connection agreement for more open markets. To be truly reflective of benefits the grid friendly benefit would need to be tailored to the locality and presence of needs for support or correction. When customers are aware of the incentives and are seeking compliant inverter systems then there will be benefits to manufacturers who offer equipment, which meets the grid friendly requirements. When these connection standards are agreed upon across regional and national boundaries through standards, there will be a healthier marker for compliant equipment.

12.4 Interconnection Standards

Connection of photovoltaic inverters is one area most advanced for general connection of inverters to the grid. Most of these standards aim at avoiding the worse potential of inverter impact on the grid through the limiting of harmonic injection and power factor limits. These standards also address the issue of inverter disconnection when the feeder supply is removed to avoid the possibility of islanding operation.

Grid friendly connection refers to inverter connections which aim at fundamental voltage support, transient suppression and harmonic correction. One method to encourage the adoption of grid friendly controls is through tariff structures. In regions where there is potential for a utility to avoid the costs of network reinforcement through operation of customer inverters, an agreement on what would be of benefit to a utility needs to be broadly defined to operate without significant manual intervention from customers. The voltage support should be built into the equipment and a special connection rate applied if the equipment meets the grid support standards. Issues to be defined would be the degree of correction of dips supplied compared with the nominal equipment rating. Decisions would need to be made to reward availability or to reward actual performance. One option is to give a tariff rebate if an inverter model, which provides reactive support up to the level of kVA rating beyond real power requirements, is available. The other approach would be to record reactive support correlation with system need. The first mentioned approach of having the equipment available may be preferable for many smaller sized units. The monitoring approach would be expected at the much larger commercial support contract.

12.5 Power Quality Performance Requirements and Validation

In this section we highlight the requirements of the different customers and the regulator requirements to meet the customer demands.

12.5.1 Commercial Customers

When purchasing a product such as premium power, the customer needs to know that he is getting what he has paid for. If the utility provides alternative supply connections and backup generators to create a superior quality of supply then marketing of that product required quantification of the resulting performance.

Even standard connection agreements may include penalties for performance at less than defined quality of supply. This type of agreement would then form a basis of when a utility should invest in particular improvements which address quality of supply. These agreements are becoming more common where there is a competitive supply market. In UK which has been offering choice of supplier for over 10 years, the quality of supply and the time of day costing of that supply become subject to detailed agreements.

In many markets commercial customers need to have a recording meter to relate to the power used and the time of day so the actual usage can be compared with what was agreed and costing applied. These same meters are increasingly including dip and outage monitoring. Typically the recording will show when and for how long the voltage has fallen below a threshold such as 85%.

12.5.2 Regulator Requirements

Smaller and residential customers are not presumed to have a strong individual negotiating position with respect to quality of supply. The collective interest is often protected by a regulator who defines the minimum standards for performance. These standards initially focused on speed of response to complaints or connection requests. Performance standards in the area of power quality are becoming stronger in some jurisdictions. The basic level of performance reporting are acceptable levels of

- SAIDI system average interruption duration index
- SAIFI system average interruption frequency index
- CAIDI customer average interruption

12. Future Directions and Opportunities

Specific minimum performance on these indicators can be required on average across the network or could be specified as a minimum for every feeder.

The biggest influence of the average minutes lost per customer per year can initially come from addressing planned outages and doing more live line maintenance. Eventually the response to unplanned outages needs to be addressed. In this area more repair crews help but network automation is finding a stronger role in automatically reconfiguring a network to isolate a faulted section and restore supply.

Dips and momentary outages are aspects that are being specified in a limited number of regulation areas. As the technology to report on quality performance develops, there will be greater adoption of performance specification by regulators. The first level of reporting would be on some level of system average dip and outage performance. The issues that arise are that good monitors are very expensive so very large numbers of measurements would not be feasible. There are monitors as part of metering of commercial customers but these need not represent the customer average performance of the network. Issues to be addressed then are to develop a sampling strategy which would adequately represent the performance of all customers to a defined accuracy.

12.5.3 An Example

One case where a service quality agreement was reported was in [8]. In 1995 Detroit Edison entered into a 10-year agreement with several auto manufacturers. Power quality levels were a portion of the agreement because of concerns that a captive customer may not receive fast response to outages from the utility. The total number of outages over a number of sites was defined and a payment for exceeding the defined outage rate set roughly at the cost to that section of the manufacturing process of the outage. The payment ranged from $2000 to $297000. The target outage rate was agreed upon to reduce by 5% per year. The performance of the supply initially did not meet the target and payments were made to the customers. The advantage of the defined cost of quality was that projects to improve quality now had specific drivers and could be solidly justified. The approaches were by additional maintenance on specific lines, replacement of problem equipment and installation of animal deterrents. Additionally a "best-feed" operating configuration was implemented where one of two services served the entire load while the other was used as hot standby to reduce interruptions. In one site a static transfer switch was used to reduce sags to 0.25 sec.

Overall this concentration of effort led to a 60% decline in penalty payments over the first 5 years. A voltage sag penalty payment was introduced in 1998 and to manage the risk a greater detail of knowledge of system performance was required.

12.6 Shape of Energy Delivery

For large commercial customers it is likely that they stick to core business and leave large scale energy supply to specialist companies. These supplies are currently provided from fossil fuel such as coal, gas and nuclear with some contributions from hydro wind and other renewables. As the pressure for environmental responsibility for greenhouse gasses rises and the trend against nuclear continues there will be increasing pressure for generation from renewable sources. These trends will not dramatically change the large generation and transmission process currently employed. The main pressure in these cases will be for improved power quality where this has a significant effect on business efficiency. This increase in power quality will only be achieved by careful consideration of network faults and the use of power electronics to avoid or correct the fault.

For residential and small business customers there is the potential for a rise in the use of fuel cells and local photovoltaics. Unless there is a dramatic change in the costs of energy storage, good reliability will be achieved only by use of a local distribution system to share the energy. Power quality standards on these grid connection inverters will be even more important to avoid multiplication of distortion effects. In this case the control of power quality is likely to be addressed both at the energy source and potentially at the neighborhood energy storage center.

Several energy players are seriously discussing moving to a hydrogen economy where wind and wave energy is converted to hydrogen and later used in fuel cells. Once again this will require use of inverter systems to provide the required ac waveform. DC reticulation over short distances such as within buildings may be used once the speed of power electronic controllers is able to handle the fault recovery. The usual difficulty with dc has been

- DC is difficult to transform between voltage levels for efficient transmission
- Arc faults on dc systems do not have natural current zero points where the arc is easily extinguished making protection much more difficult.

The first issue of voltage changing of dc is easily handled at present for low power with switch mode supplies. No major breakthroughs appear likely

12. Future Directions and Opportunities

for handling hundreds of MW of dc transformations cost effectively. The current dc links are voltage transformed on the ac side and then controlled rectification and inversion applied. The second issue can be addressed by having the sources all drop rapidly to zero to extinguish the arc and then reconnect. The combination of these effects can mean that building level dc may become feasible but a total changeover for all lines is unlikely.

Another choice is higher frequency ac. The usual objection is that line impedance is largely inductive and higher frequencies would just give larger voltage drops. Higher frequencies can be used in generation such as microturbines or in specialized loads but if we continue to use overhead lines there is little benefit in changing frequency apart from limited opportunity for a change in transformer size.

12.7 Role of Compensators in Future Energy Delivery

With the expected expansion of distributed generation the role of power quality enhancement could shift from dedicated hardware designed by the utility to address specific problems to one of general enhancement supplied by a wide base of connected customers. Here each of the distributed generation connection inverters can be encouraged to have grid enhancement features with most of the responses in reactive injection being in response to local difficulties. The use of real power to support network operational requirements would be a feature negotiated with customers. One mode of operation would be for the distribution utility to communicate to customers' equipment on a particular feeder of a need to reduce line overload. For an appropriate reward, some will respond by load reduction or deferral, others will have the capacity to respond by increased generation.

Provided the economic conditions favor the development of distributed generation technologies the distribution system could be changing from a conduit for channeling power from large central generation to large number of passive customers. The new role would be focused on controlling the reliability and quality on behalf of a mix of customers many of whom have controllable real and reactive capacity. If this can be addressed well there should be greater robustness but having smaller regions continuing to operate even when faults have resulted in system segmentation. The power quality enhancements will come from faster fault isolation as well as from active response limiting the damaging fault current. Voltage dips and harmonics would have many converters operating to mitigate these effects. While much of this response would be automatic and localized, an overall control scheme would permit a cost effective response to exceptional system demands.

12.8 Conclusions

There are real costs from poor power quality. The impacts differ significantly between customer classes and individuals. Market opportunities exist to retain customers or to sell premium grade power. To address these needs and to respond to expansion of distributed generation better inverter controls and integration with network operations are developing issues. The future shape of energy delivery will be more than the conventional pattern of large fossil fuel stations sending energy through graduated levels of lines but will become a more integrated system of multiple sources configured where it is possible to address power quality as well as energy delivery.

12.9 References

[1] P. Heine, P. Pohjanheimo, M. Lehtonen and E.Lakervi, "Estimating the annual frequency and cost of voltage sags for customers of five Finnish distribution companies," *CIRED*, paper 2.25, June 2001.

[2] C. Guidi, J. Espain and J. Garcia, "Relationship between quality of service demanded and tariff," *CIRED*, paper 2.27, June 2001.

[3] C. J. Melhorn, T. D. Davis and G. E. Beam, "Voltage sags: their impact on the utility and industrial customers," *IEEE Trans Industry Applications*, Vol. 34, No. 3, pp. 549-558, 1998.

[4] R. Billinton, G. Tollefson and G. Wacker, "Assessment of service quality worth," *Third Int Conf on Probabilistic Methods Applied to Electric Power Systems*, pp. 9-14, 1991.

[5] J. Gates, R. Billinton and G. Wacker, "Electrical service reliability worth evaluation for government, institutions and office buildings," *IEEE Trans Power Systems*, Vol. 14, No. 1, pp. 43-50, 1999.

[6] M. J. Sullivan, T. Vardell and M. Johnson, "Power interruption costs to industrial and commercial consumers of electricity," *IEEE Trans Industry Applications*, Vol. 33, No. 6, pp. 1448-1458, 1997.

[7] I Hunter, "Power quality issues: a distribution company perspective," *Power Engineering Journal*, pp. 75-80, April 2001.

[8] A. Dettloff and D Sabin "Power quality performance component of the special manufacturing contracts between power provider and customer" *Proc. Ninth International Conference on Harmonics and Quality of Power*, Vol. 2, pp. 416-424, 2000.

Index

Active filter
   series, 372
      hybrid, 375
   shunt, *see* DSTATCOM
Arc furnace, 91
Australian standards, 48

CBEMA curve, 39
Converters, *see* Inverters
Custom power (CP), 18
   status of application, 134
Custom power park, 21, 131

DC offset, 36, 45
Deadbeat control, 200, 328, 362, 398
Distributed generation, 22, 407
   angle stability, 410
   anti-islanding protection, 422
   dip during motor start, 417
   grid friendly inverters, 409
   harmonic effects, 419
   islanding, 422
      dedicated operation, 423
   issues, 410
   optimal locations, 411
      eigenanalysis, 411
      interaction index, 414
   power quality, 417, 437
   protection, 435
   ROCOF, 424
   voltage flicker, 420
   vector shift, 423
Distribution line compensation
   inverter based series, 429
   series, 427
   series real injection, 429
   shunt, 427
   tap changers, 426
   tap changer and series, 433
   tap changer and shunt, 432
   voltage sensitivity, 425
Distribution STATCOM (DSTATCOM), 19
   capacitor filter, 299
      rating, 301
   control, 302
   current control using phasors, 310
      back projection, 317
   dc capacitor control, 292, 308, 319
   load compensation
      non-stiff supply, 123, 296
      single-phase loads, 242
      stiff supply, 121, 291
      three-phase loads, 245
   reference current generation
      delta-connected loads, 265
      equal current, 280
      equal power, 282
      equal resistance, 278
      general algorithm, 268
      instantaneous symmetrical
         component, 259

p-q theory, 249
  star-connected loads, 260
  three-phase three wire, 264
  zero sequence compensation, 258
state space model, 303
structure
  for non-stiff source, 299
  for stiff source, 288
switching control, 302
voltage regulation, 126, 321
  output feedback control, 327
  state feedback control, 322
  dc capacitor voltage control, 325
Dynamic voltage restorer (DVR), 20, 127
  dc capacitor supported, 340
  fundamental frequency characteristics, 341
  harmonic distortion, 354
  rating, 350, 368
  rectifier supported, 336
  reference generation
    balanced supply, 346
    instantaneous symmetrical components, 355
    unbalanced and distorted supply, 348
  output feedback control, 360
  state feedback control, 364
  state space model, 361, 365
  structure
    with capacitor filter, 359
    with LC filter, 359
    three-level switching, 366
  voltage restoration, 370

Electric power quality, 27
  contribution from customer equipment, 450
  cost, 444
  dips and outages, 448
  impacts on end users, 4
  monitoring, 7
  performance requirements, 452
  sensitivity, 443
  standards, 6
  terms, 29
  voltage distortion, 449

Flexible ac transmission systems (FACTS), 8
Fuel cells, 23

Gas turbine, 23
Grid friendly inverters, 409, 451
Grid interconnection, 24
  standards, 408, 451

Harmonics, 36, 42, 72
  DIN, 84
  distribution factor, 86
  effect on distribution system, 44, 72
  Fourier series, 73
  Fourier transform, 75
  interharmonics, 72, 75, 82
  limits, 104
  sub-harmonics, 72
  THD 37, 84
High voltage dc (HVDC) transmission, 8
HVDC light, 9
Hysteresis control, 141, 187, 301

IEC standards, 6, 48, 85, 91, 105, 420
IEEE standards, 6, 85, 105
Induction generator, 438
Interline power flow controller (IPFC), 18
Interruptions, 33, 35
Inverters
  chain, 167
  CSI, 137
  H-bridge, 138
    dynamic equation, 140
  multilevel, 162
    5-level, 164, 175
    4-level, 163
    3-level, 162
  multi-step, 155
    48-step, 162
    12-step, 157
  resonant, 148
    state space model, 150
    ZVS, 148
  three-phase, 143
  VSI, 137

Linear quadratic regulator (LQR), 193, 307, 313, 323, 367, 395, 404

Index

Load balancing, classical, 93
    closed-loop, 98
    current, 102
    open-loop, 94

Mirror root locus, 205

Negative phase sequence (NPS), 47, 98
Notching, 37, 45

Open loop voltage control of inverters
    sinusoidal PWM, 169
        bipolar switching, 170
        carrier wave, 169
        frequency ratio, 170
        modulation index, 170
        modulating signal, 169
        over modulation, 172
        unipolar switching, 173
    multilevel, 175
    single-phase, 169
    three-phase, 174
    space vector modulation, 178
Overvoltage, 35

Passive harmonic filters, 106, 375
Pole shift controller, 202, 305
Poor power factor, 41
Power
    instantaneous reactive, 68, 250
    instantaneous real, 67, 250
Power factor, 86
Power frequency variations, 39
Power outages
    analysis, 55
    CAIDI, 57
    CAIFI, 56
    customer impact, 92
    MAIFI, 58
    SAIDI, 57
    SAIFI, 56
Power quality, *see* Electric power quality
Power semiconductor devices, rating
    FCT, 153
    IGBT, 153
    MOSFET, 153
    GTO, 153

Reciprocating Piston Engine, 22

Renewable energy, 23
RC snubber, 38, 119

Sampled error control, 185
Sequential LQR (SLQR), 203
Sliding mode control, 192
Solid state circuit breaker (SSCB), 19
    topology, 220, 222
    coordination, 223
Solid state current limiter (SSCL), 19, 117
    topology, 216
    operating principle, 217
    coordination, 223
Solid state transfer switch (SSTS), 19, 120
    topology, 225, 231
    make before break, 226
    transferring during fault, 228
Stability of switching control
    first order system, 183
    higher order system, 189
Static compensator (STATCOM), 14
Static synchronous series compensator (SSSC), 16
Static var compensator (SVC), 10, 49, 98
Sustained interruptions, 35

Thyristor controlled braking resistor (TCBR), 17
Thyristor controlled phase angle regulator (TCPAR), 18
Thyristor controlled reactor (TCR), 11
Thyristor controlled series compensator (TCSC), 12
    fundamental reactance, 14
Tracking convergence
    controller, 195
    reference, 198
Transients
    impulsive, 29
    oscillatory, 31

Unbalanced load, 46
    effect on distribution system, 47
    instantaneous symmetrical components, 64
        on-line extraction, 357
    symmetrical components, 60

on-line extraction, 76
Undervoltage, 35
Unified power flow controller (UPFC), 16
Unified power quality conditioner (UPQC), 20, 130
  left-shunt
    characteristics, 388
    configuration, 381
    state space model, 403
    structure, 401
  right-shunt
    characteristics, 381
    configuration, 380
    dc capacitor control, 401
    harmonic elimination, 398
    state space model, 393
    structure, 391

Voltage flicker, 38, 90
  standards, 91
Voltage imbalance, 35
Voltage sag/swell, 33, 86
  detection algorithms, 232
    instantaneous symmetrical components, 234
    symmetrical components, 232
    two-axis transformation, 233
  Detroit-Edison score, 88
  mitigation of impacts, 446
  reduction, 108
  rms variation, 89
  VSLEI, 88

ZnO arrester, 216, 219